D0340523

Water Wasteland

This book is printed on one hundred percent recycled paper.

Ralph Nader's Study Group
Report on Water Pollution

WATER
WASTELAND

by David Zwick
and Marcy Benstock

GROSSMAN PUBLISHERS   NEW YORK   1971

Copyright © 1971 by The Center for Study of Responsive Law
All rights reserved
Published in 1971 by Grossman Publishers
44 West 56th Street, New York, N.Y. 10019
Published simultaneously in Canada by
Fitzhenry and Whiteside, Ltd.
SBN 670–75169–3
Library of Congress Catalogue Card Number: 77–112516
Manufactured in the United States of America

All royalties from the sale of this book will be given to The
Center for Study of Responsive Law, the organization established
by Ralph Nader to conduct research into violations of the public
interest by business and governmental groups. Contributions to
further this work are tax deductible and may be sent to the
Center at 1156 Nineteenth Street, N.W., Washington, D.C., 20036

Task Force

DAVID R. ZWICK, Editor and Co-Project Director
  B.S., United States Coast Guard Academy,
  3rd-year law student and graduate student in Public
    Policy, Harvard University
MARCY BENSTOCK, Associate Editor
  B.A., Radcliffe College
  M.A., Economics, Graduate Faculty, New School for
    Social Research
JOHN C. ESPOSITO, Co-Project Director
  Consultant, Center for Study of Responsive Law
  Co-author, *Vanishing Air*
RALPH NADER, Special Consultant

MARTIN A. BECK
  Senior, Harvard College
JEREMY BLUHM
  Sophomore, Harvard College
CAROL A. COWGILL
  A.B., Duke University
  3rd-year law student, University of Chicago
PHILIP L. GRAITCER
  B.A., Duke University
  D.D.S., Temple University School of Dentistry
MICHAEL C. HARPER
  B.A., Harvard College
  1st-year law student, Harvard University
WILLIAM R. HERMAN
  B.A., Yale University
  3rd-year law student, Harvard University
CURTIS L. KEHR
  B.S.E., Princeton University
  Graduate student, Operations Research, Columbia
    University
JERROLD L. NEUGARTEN
  B.A., New College (Sarasota, Florida)
  2nd-year law student, Harvard University
KENJI ODA
  B.A., New College (Sarasota, Florida)

LEWIS M. POPPER
  B.A., Harvard College
  3rd-year law student, Harvard University
STUART L. RAWLINGS III
  B.A., Stanford University
  M.A., University of Pennsylvania
  2nd-year law student, George Washington University
BARBARA A. SHORE
  B.A., Radcliffe College
  3rd-year law student, Harvard University
ELLEN SILBERGELD
  B.A., Vassar College
  3rd-year doctoral student, Environmental Engineering, Johns Hopkins University
BARRY A. WILLNER
  A.B., Lafayette College
  2nd-year law student, Georgetown University Law Center

Special Projects: Linda Hotchkiss, Larry Silverman,
                      James Sullivan, Lisa Heywood,
                      Patricia Teare, Thomas Roach,
                      Shelby Bryan, Andrea Hricko
Production Directors: Connie Jo Smith, Susan Fagin
Research Assistants: David Cincotta, Doreen Moses

Contents

Introduction by Ralph Nader

For all the environmental concern of the past two years, the focus for action has rarely included the Federal government's pollution control agencies. Citizens and student conservation or ecology groups expect so little from these long-faltering, if not pathetic, bureaucracies that they have largely discounted their role. The speeches and discussions on campuses during Earth Day, April 22, 1970, almost ignored these agencies as irrelevant and irredeemable. The institutionalization of politics-ridden and industry-indentured traditions had turned the active citizen away from any hopeful expectations for a viable Federal impact on polluters. This skepticism is especially deep in the area of water pollution, where the sources are less visible and less susceptible to clear calls for regulatory action than the automobile exhausts and smokestacks that spew pollution into the air.

It is time for a detailed analysis of the role of the Federal Water Quality Administration (FWQA), now called the Water Quality Office of the Environmental Protection Agency (EPA) since it was moved nearly intact from the Department of Interior to EPA in December 1970. David Zwick and his associates have provided this first comprehensive evaluation of the agency. When the study began in the summer of 1969, the Task Force at first encountered the most intransigent responses from any Federal agency studied to date by our colleagues. Assistant Secretary of Interior Carl Klein, then running the agency, placed severe restrictions on interviews, required monitors to be present, and in other ways hampered efforts to have routine citizen access to government. This was the same Carl Klein who, in a speech given a few weeks earlier in April 1969, urged an "informed public, channeling its energies through citizen groups" to "persuade, prod, and guide government" with the following exhortation: "Local, State and Federal governments will respond to pressure from citizen groups, but the pressure must be well-directed and unremitting and, most important, it must have an element of anger and urgency. My last word to you is this: don't let up on government. Hammer it. Worry it. Keep after it."

Although Assistant Secretary Klein's restrictions discomfited FWQA Commissioner David Dominick, our appeals within the Department went unheard until sessions were initiated with White House Director of Communications Herbert Klein and Secretary of Interior Walter Hickel. Thereafter, FWQA showed no more than the normal secrecy that marks the prevailing bureaucratic climate in Washington, and agency staffers, particularly in the field, became genuinely helpful. But the initial experience both lowered further the morale of those in FWQA who believed they were being repressed from contact with citizen-inquirers—while special interests had full entree—and increased the determination of the Task Force to persevere.

The Federal role in water pollution control began on a temporary trial basis in 1948 and became permanent in 1956. Its effectiveness to date can be concisely assessed by the virtual absence of any evidence that the seven laws passed and more than three billion dollars spent by the Federal government has reduced the level of pollution in any of our country's major bodies of water, so that they are once again suitable for human use as fish habitat, recreation spot, or drinking water supply. The Task Force Report shows that surprising agreements exist with this ultimate test—failure of the program's impact on the nation's lakes, rivers, estuaries, and bays. The tracing of this failure in the Report is meticulous, and from this detail comes the alternative course for the future outlined in the chapter on conclusions and recommendations.

Beginning with the drafting of the water pollution legislation, the Federal effort grew into a complex charade. The built-in procedural delays exceeded the professional avarice of the most adamant corporate lawyers. The generic delegation of initiatory moves to the states insured the availability of a "divide-and-rule" tactic by industry vis-à-vis already subservient and underequipped state agencies. A 72-year-old Federal law banning the dumping of industrial pollution into navigable water ways went almost completely unenforced until its 70th birthday when the first of some 30 injunctive actions were brought against a fraction of the approximately 40,000 daily violators. Hundreds of millions of dollars of construction grant subsidies flowed from Washington to local government for waste treatment plants which industry promptly used to dump more waste through. This subsidy to local government turned into a subsidy to factories that increased water pollution—and at the taxpayers' expense. The Kafkaesque tapestry extends into the mockery of Federal enforcement conferences, the never-ending deadline extensions for the weakest of pollution controls, the secured trade secrecy over what lethalities industries dump in the public's waterways, the clear evidence of serious and worsening contamination of drinking water, the damage to other people's property and property rights by industrial, municipal, and agricultural pollution without even any compensation, the loss of livelihoods for thousands of commercial fishermen, and the emergence of water so laden with ignitable wastes that rivers such as the Buffalo and Cuyahoga are declared official fire hazards.

What has this economy wrought in its pell-mell rush to lay waste to water? From the once idyllic rivers of Maine to the Hudson and Delaware Rivers, to the Chesapeake and the Potomac, to the Savannah and the Gulf of Mexico and the Mississippi and Great Lakes, to the Columbia and other waterways of the Pacific Northwest, the tearing down of America into a water wasteland hurtles on.

Water is the most precious, limited natural resource we have in this country. The technology that ruins this water can also save it. But because water belongs to no one—except the people—special interests, including government polluters, use it as their private sewers. Toilet-training polluters will require the replacement of short-sighted, cowardly, or venal men and institutional abuses with a legal system which sees prevention as the cardinal ethic. That system must be enforced with sanctions that reach beyond the institution to the personal irresponsibility of those presently in charge.

There are many people in this country—old enough

to remember when water was a many-splendored thing or young enough to want to have such memories—who want answers. There are scientists, engineers, lawyers, and economists who could provide many of these answers, using available but unused pollution prevention technology. A humane technology is economic as well. The longer we ignore its application, the more horrendous the costs to human health and well-being. And sooner, rather than later, the costs will descend with crushing reality on those insensitive perpetrators of this pollution who no longer will be able to escape the consequences of their own exudations.

May this Report help to herald a new awareness that moves beyond perception of the problem to develop the realignment of forces that will make contamination of water a crime against humanity and a depredation that is more costly for its generators to continue than to stop. Someday water pollution will be seen in retrospect as a colossal waste of resources, as more efficient industrial processes and recycling come into use. Then those officials and citizens who for years researched and pleaded their lonely case, while incurring opprobrium, ostracism, demotion, or eviction at the hands of entrenched powers, will be relieved that common candor need no longer be considered uncommon courage.

RALPH NADER
April 1971
Washington, D.C.

Preface

For the past 21 months a Task Force of young people, mostly graduate students, with backgrounds in law, engineering, economics, medicine, and science has been studying the nation's water pollution problem and our efforts to combat it. Most of the Task Force members received subsistence support or less during the summer months from the Center for Study of Responsive Law, and contributed their considerable efforts without compensation the remainder of the time. This Report is the result.

The study began in the summer of 1969 when an initial group of 10 came to Washington, D.C., to investigate the Federal Water Quality Administration,* the government agency with primary responsibility for controlling contamination of the water. The summer was spent wading through file cabinets and official documents, interviewing hundreds of persons in and out of government at the Federal, state, and local levels, sending out questionnaires to industries and state pollution control agencies and evaluating the responses, visiting numerous Federal offices in the field and regional and state control agencies in various parts of the country, and meeting continuously to coordinate the study effort.

When the summer of 1969 ended, we had what we considered to be a good solid foot in the door of this massive subject—but much more was needed. A smaller group, joined, at various stages along the way, by several newcomers to the Task Force, continued to work through the 1969–70 academic year and the summer and fall of 1970. The study perspective was broadened to take in the interrelationship of industry, citizen efforts, and

---

* Its name in 1969 was the Federal Water Pollution Control Administration. The name was changed to Federal Water Quality Administration in 1970, and in 1971 it became the Water Quality Office of the Environmental Protection Agency.

local government with the Federal water pollution program.

In order to keep abreast of events that were, in at least a superficial way, moving very quickly all the while, we completed first those chapters dealing, in general, with areas changing most slowly and worked our way toward the topics that were changing more rapidly. Research and writing on Chapters 8, 9, 11, and 13 were essentially completed in late December 1970. Chapters 2, 6, and 12 were finished in January 1971. Chapters 5, 10, and 14 were completed in February and the remainder—Chapters 1, 3, 4, 7, 15, 16, 17, 18, 19, and 20—were put to bed in March 1971.

The Task Force would especially like to acknowledge the help we received from the employees of the Environmental Protection Agency's Water Quality Office. Though they were uncooperative at the beginning and on occasion constrained by what the Task Force regarded as needless restrictions on the release of certain information, agency staffers proved always willing to assist where they could and often genuinely enthusiastic about this effort to inform the public. The Task Force believes the exemplary attitude they showed in this regard was due in no small part to the tone of cooperation which Water Quality Office Commissioner David Dominick established early in our study. It is also worth noting, however, that on those occasions when Task Force members telephoned to the Water Quality Office but did not identify themselves to agency staffers except as interested members of the public, the assistance we received was typically rendered with a commendable display of sincere interest.

Ralph Nader's introduction tells the story of the restraints put on us and on the FWQA at the beginning of our study. The result of these restrictions was that many hours of agency staffers' valuable time (as well as our own) which could otherwise have been spent working on cleaning up the water was instead consumed in clearing documents for release (a process that often involved the writing of several letters) and arranging interviews (a process that typically involved the writing of several intra-agency memoranda). Task Force members were sometimes given as little as an hour's notice or less of impending interviews, and the result was that we occasionally kept agency staffers sitting in the "interview room" far from their offices (where they might at least have been getting some work done), waiting until their interviewer showed up. The most egregious waste of public resources in the whole scheme was the $20,000 or more annual salary being paid each of the two Department of Interior agents who, for over a week, were assigned the primary duty of sitting in on every conversation a Task Force member had with any water pollution agency employee.

Mr. Nader's intercession eventually cleared the way for the study to proceed without Assistant Secretary Klein's interference. But the experience contained a valuable lesson. The rationale typically given for denying members of the public access to the information that public tax dollars paid to have collected, and to the inner workings of government generally, is that agency routine will be upset and employees' time taken away from their jobs. Once the barriers of bureaucratic secrecy were lifted, the Task Force's investigation did not interfere in any way with the everyday functioning of the agency and consumed only negligible amounts of the time of any of the agency's employees. The only interference with the smooth working of the agency was caused by the effort to exclude us.

Our report is uncompromising, and often harsh, in the judgments it makes. But we believe it is fair. It is our hope that it will help clean up the water and add generally to sound public policies.

DAVID R. ZWICK
April 1971
Washington, D.C.

Part One

THE DANGER

1

Water Wasteland

The frog does not
Drink up
The pond in which
He lives.
                —*American Indian Proverb*

On June 22, 1969, the Cuyahoga River burst into flame. Thousands of gallons of oil from a source in Cleveland —a source never identified—ignited suddenly, causing extensive damage. Cleveland residents had long joked that the Cuyahoga was the only body of water in the world that should be declared a fire hazard. On a quiet Sunday afternoon in June the local joke became a reality.

The drama of the Cuyahoga and other burning rivers is only the most visible manifestation of the vengeance with which our waterways can repay years of exploitation and neglect. There are many less visible but more dangerous consequences of water pollution. We may be near the time when there will not be enough usable water left in the United States to fulfill our basic needs. While the adventurous American can still find patches of clean air, he can find almost no significant body of clean water. Practically no major American waterway in the United States is free of serious pollution.

Water pollution in the strictest sense—man's alteration of existing rivers, oceans, lakes, and streams—has long been with us. Because waterways provide natural modes of transport and dilution, man has always relied on the water around him to carry away and assimilate waste. As long as these wasteloads were reasonably small, the natural self-purification processes of the waterways—dilution, oxidation, biological degradation—could cope with the addition of foreign matter. This is no longer true. We have overburdened the natural assimilative capacity of the waters by depositing too much of the materials that our rivers, lakes, and estuaries could normally have handled in smaller quantities. Moreover, complex new technological processes have introduced exotic substances that even an otherwise clean waterway could not naturally cope with. These developments, largely the result of chaotic industrial, agricultural, and residential growth, have degraded most of the large and small watercourses of the nation. Many of our waters have, as a result, been rendered almost unfit for anything but navigation and continued dumping.

Our waters are being contaminated by: *

—*organic wastes* from agricultural, industrial, and domestic sources, which remove oxygen from the water;

—*biological nutrients,* such as phosphates in detergents and human excreta, which accelerate the growth of aquatic plant life, thus increasing the oxygen demand on waterways;

—*disease-bearing organisms* from domestic, agricultural, and industrial sources;

—*temperature increases* from cooling water used in power plants and industrial operations. Waste heat (thermal pollution) can injure plants and fish directly and also reduces the oxygen content of the water;

—*synthetic chemicals,* both organic and inorganic, found in pesticides, plastics, detergents, etc.; *mine runoff,* such as acids and minerals; and *radioactivity* from fallout, nuclear power generation, reprocessing nuclear fuel, etc. These pollutants may be toxic, carcinogenic (cancer-producing), mutagenic (causing genetic damage), or teratogenic (causing birth defects) to all forms of animal life, including ours.

While most people are sickened when they see water pollution—floating debris, putrid green algae, oil slicks, dead fish washing rhythmically against the shores of our lakes and rivers—for the most part these losses seem

---

* Some of these categories overlap. Nutrients may be organic, treated sewage may raise the temperature of receiving waters, etc.

far removed from their day-to-day lives. Few remember the day when it was prudent to swim in the lower parts of the Delaware or the Hudson River; we find our esthetic and recreational pleasures elsewhere. But polluted water cannot be avoided; we cannot cut ourselves off from unpleasant aspects of our environment. We are too much a part of it. As Dr. Paul B. Connerly, President of the American Public Health Association, said, "If the fish are dying, the people are not far behind." [1]

We are only beginning to appreciate the pervasive effects of water pollution. As with so many other areas of environmental despoliation, this is a result of a blind obedience to the precepts of "progress"—an unquestioning faith in economic growth, especially growth resulting from the application of science. Many new applications of science, as well as older ones that have been accelerated or made more "efficient," have been introduced without careful pre-assessment of the consequences. Ecologist Barry Commoner described what we are doing this way:

We have been massively intervening in the environment without being aware of many of the harmful consequences of our acts until they have been performed and the effects—which are difficult to understand and sometimes irreversible—are upon us. Like the sorcerer's apprentice, we are acting upon dangerously incomplete knowledge. We are, in effect, conducting a huge experiment *on ourselves*. A generation hence—too late to help—public health statistics may reveal what hazards are associated with these pollutants.[2] [Emphasis in original.]

In this chapter we shall examine some of the pollutants and some of the consequences.

Drinking Water

One might begin to appreciate the magnitude of the potential crisis in national health by placing himself on the shores of the Potomac or Mississippi River, Lake Erie or Lake Michigan, and considering this bit of intelligence: he and millions of other Americans draw their most important foodstuff—drinking water—from these poisoned lakes and streams. No doubt there is a vague awareness that toxic chemicals from industry, pesticides, herbicides, hormones in animal feed, human sewage, and thousands of other contaminants enter our waterways. Most Americans assume that "somebody" removes these substances before their water comes out of the tap. This is not the case.

C. C. Johnson, former Administrator of the Consumer Protection and Environmental Health Service (CPEHS) * of the Department of Health, Education and Welfare (HEW), estimates that 90–95% of our municipal water purification systems were originally designed to kill bacteria only; they cannot eliminate all toxic chemicals or all viral pollution. Chemicals and viruses can, and do, pass through municipal purification systems into drinking water. A nationwide survey of drinking water by HEW's Bureau of Water Hygiene (BWH—now in EPA's Water Quality Office) released in July 1970, the most comprehensive ever taken in this country, revealed that 30% of the samples taken at consumers' taps contained quantities of chemicals exceeding limits recommended by the Public Health Service.[3] (The PHS standards are themselves not entirely adequate; they neither cover many of the "recently discovered" pollutants nor take into account potential long-term hazards of some of the "older" pollutants that have only recently come to light.†) The Director of the Water Utilities Department of Dallas, Texas, Henry J. Graeser, was not exaggerating when he said: "We are surely moving towards the time when a major waterborne problem, either from ingestion of long-term materials or from a severe epidemic, is going to appear upon the scene and create a national hysteria and a crisis in water hygiene." [4]

Local water authorities have tried to combat the rise in the number and quantity of pollutants by increasing the amount of chlorine used in water "purification." The Cincinnati Water Works, for instance, increased the number of gallons of water pumped from the Ohio

* CPEHS was moved in December 1970 to the Environmental Protection Agency (EPA).
† In the current edition of the PHS standards there is no mention of mercury, for example. The U.S. Water Resources Council in a 1968 report said the standards "cover only traditionally-known contaminants. The standards are predicated on the use of relatively unpolluted water as a raw water source." (*The Nation's Water Resources*, Part 5, Chapter 4, p. 2.) The standards are currently undergoing revision, for the first time since 1962. Prior to 1962 they had been revised only twice since 1925.

River by 40% between 1955 and 1970. During that period its use of chlorine increased almost 200%. The U.S. Department of Health, Education and Welfare keeps no comprehensive statistics on chlorine use, but officials the Task Force talked to acknowledged that Cincinnati is not an isolated example. The increase in the use of chlorine has outstripped the rise in the demand for water. In some places chlorine is used to give drinking water a "sparkling quality." But despite the fact that a glass of water may look and smell as if it had been scooped from an olympic swimming pool, it may not be much safer than one scooped from the scummy Potomac. Safe water supplies can be assured not by outmoded remedial "purification" but by preventing dangerous substances from entering the nation's streams.

As the current director of the BWH, James H. McDermott, has noted, "Aside from occasional irritation associated with taste and odor problems, overconfidence or apathy seems to pervade the public's attitude with respect to drinking water." [5] The general public ignorance about the state of our drinking water stems in large part from the efforts of state public health officials and others in the "water industry." Many state health officials seem to have become security-conscious administrators who believe that maintaining public trust is at least as important as protecting the water itself. Consequently, local public health authorities all too frequently respond to queries concerning drinking water hazards with an ill-founded assurance. Questionnaires from the Task Force elicited glowing descriptions of the high quality of our drinking water. Fearing panic or a loss of public confidence, local authorities—in sharp contrast to Federal officials—tended to minimize the hazards. There are even attempts to cover up documented cases of waterborne disease. The Task Force was told of a dysentery outbreak in a Southern town that affected several thousand people. Even the Federal government could get no information on the outbreak until a promise of confidentiality was extracted from the U.S. Surgeon General—a promise that extended even to the name of the town.

Yet the greatest threat to our drinking water is not disease-carrying organisms but the massive quantities of

sophisticated chemicals being poured into our waterways. The enormous growth since the 1940's of industries producing synthetic organic chemical products such
as detergents, plastics, and synthetic fibers has added to
the already heavy pollution load from older industries—
steel, paper, food processing, textiles, electrical and
transportation equipment, petroleum, non-ferrous metals. Little is known about the estimated 12,000 potentially toxic chemicals now in industrial use.[6] Hiding
behind the veil of trade secrets, industry has been almost
uniformly reluctant to inform public authorities about
the chemical characteristics of the contaminants it is
spewing into streams, lakes, and oceans; local officials
charged with safeguarding drinking water supplies do
not routinely check for chemical residues; we get a peek
at the possible dangers of these chemicals only when
hazards have become manifest.

The situation is likely to get worse. An estimated 500
new chemicals are produced each year [7] and introduced
into manufacturing processes without public information concerning the extent of their dispersal throughout
the environment, or of the dangers of that dispersal.
This permissiveness toward new hazards is the logical
extension of our historically lax attitude toward older
sources of pollution. It may be costing us dearly.

Many of both the new and the older, "conventional"
chemicals finding their way into the nation's water have
alarmed scientists. Drs. Wilhelm C. Hueper and W. D.
Conway, experts on carcinogens, warned in 1964 of the
cancer threat posed by excessive exposure to organic
and inorganic chemicals in the water:

The most common and often prolonged, and, therefore, the
most dangerous contact with carcinogenic pollutants of
water occurs when water thus contaminated is used for drinking purposes and in the preparation of food. It is here important that most of the agents . . . (arsenicals, chromium,
radioactive substances, chlorinated hydrocarbon pesticides)
are retained in the body and may accumulate in certain
organs, such as the liver, skin (arsenic), bones (radioactive
matter), or fat tissue (chlorinated hydrocarbons), from
which they later on may gradually be released thereby causing a continuous or prolonged exposure of the tissues with
these carcinogens. . . . . [The] cancers not infrequently [be-

come] manifest months, years or decades after such contacts have ceased and often the causative agents may have totally disappeared from the tissues.[8]

Investigators in Holland have found death rates from cancer higher in cities that get their drinking water from polluted rivers.[9] According to Dr. Samuel S. Epstein, chief of the carcinogenesis and toxicology laboratories at the Children's Cancer Research Foundation in Boston, "there is growing recognition that most human cancers are probably caused by environmental carcinogens," and that many of them might be prevented by keeping carcinogenic chemicals out of our water, air, and food.* [10]

Some carcinogens (nitrosamines, for example) do double duty as teratogens (birth defect producers), since they are capable of crossing the placenta during pregnancy and causing deformities in unborn children. A water pollutant need not be carcinogenic to be teratogenic, however. No governmental, private, or academic agency keeps comprehensive records on the incidence of birth defects in the United States.[11] But according to Dr. Henry A. Schroeder, professor of physiology at the Trace Element Laboratory at the Dartmouth Medical School, "there is a good correlation of certain qualities of municipal water supplies [e.g., high concentrations of trace metals] and deaths from congenital abnormalities in the U.S." [12]

In addition to encouraging cancer and birth defects, many of the chemical pollutants in drinking water may cause genetic damage. Here too it is difficult to establish a precise cause-effect relationship, since the evidence (which might range from such visible mutations as dwarfism, mongolism, albinism, and polydactyly [six fingers] to more ill-defined abnormalities, such as increased susceptibility to leukemia, cancer, and other diseases, or a shortened life span) [13] may not appear for several generations. But as the Mutagenicity Panel of the HEW Secretary's Commission on Pesticides pointed out in 1969, "the risk to future generations, though difficult to assess in precise terms, is nevertheless very real"; once it appears, "genetic damage is irreversible by any

* Dr. Epstein told us he knew of no efficacious carcinogens for which non-carcinogenic replacements were not available.

process that we know of now." [14] Dr. James F. Crow, head of the Genetics Study Section of the National Institutes of Health when the group's first major conference on the chemical risks to future generations was held in 1966, warns that

a number of chemicals—some with widespread use—are known to induce genetic damage in some organisms. . . . Even though the compounds may not be demonstrably mutagenic to man at the concentrations used, the total number of deleterious mutations induced in the whole population over a prolonged period of time [e.g., by ingesting the chemicals in drinking water] could nevertheless be substantial. Such increase in mutation rate probably could not be detected in a short period of time by any direct observations on human beings.[15]

The potential carcinogens, teratogens, mutagens, and just plain poisons in drinking water that are currently causing the most concern among public health officials are the heavy metals—mercury, lead, cadmium, etc. While at least seven metals * are found normally in the human body in trace amounts (less than 0.01% of the body) and are believed to be essential to human life, Dr. Schroeder, the trace elements expert, reports that 28 additional elements, most of them metals,† have been found in the tissues of modern Americans, "some in concentrations large enough to cause concern as to possible innate toxicity." [16] In addition to possible connections to cancer, birth defects, and mutations, these metals are associated with a broad range of other human illnesses. Organic mercury, which tends to be retained in both animals and humans, can cause headache, fatigue, and irritability; more severe exposure can cause tremor, slurred speech, blurred vision, deafness, numbness of limbs, loss of coordination, diarrhea, personality disturbances, and kidney damage. About 10% of the mercury taken into the body goes to the brain, where damage is likely to be permanent since brain cells do not regenerate.[17] The effects may not show up for many years. Like many other forms of poisoning from water pollutants, mercury poisoning may be misdiagnosed

* Chromium, manganese, iron, cobalt, copper, zinc, molybdenum, and possibly strontium and nickel.
† E.g., arsenic, lead, cadmium, aluminum, titanium, silver.

(e.g., as encephalitis) when it does finally appear,[18] and thus fail to be recognized as a hazard to public health.

While some waterborne mercury comes from nature, a large part is added by man. It runs off of farmland, where it is used to treat seed grain. It is dumped in the waste water from plants that manufacture chlorine, electrical equipment, paint, pharmaceuticals, and pulp and paper. If it is deposited in waterways from which drinking water is drawn, it will probably reach the consumer. It is a persistent chemical and the effects of present mercury pollution could last from 10 to 100 years. Gordon C. Robeck of the Robert A. Taft Sanitary Engineering Center in Cincinnati told the Task Force in July 1970 that information on the removal of mercury from drinking water is sparse. (Although an apparently effective removal procedure has been developed in the laboratory, it is far too complicated and expensive for most municipal water supplies.) In some places mercury levels have already approached or exceeded the PHS drinking water limit of .005 parts per million parts of water (ppm). The Rio Grande River in Colorado, an area where mercury is mined, was found in the early summer of 1970 to contain mercury at .0048 ppm.[19] A sample of the Detroit River, into which Wyandotte Chemical's chlorine caustic soda plant at Wyandotte, Michigan, was discharging 10 to 20 pounds of mercury daily,[20] was found around the same time to contain .030 ppm of mercury, six times the PHS limit.[21]

A Federal biologist, Victor Lambou, summed up the case against mercury: "I don't think you can find any chemical that is potentially worse for man. It is toxic, it is cumulative, it is persistent." [22]

Mercury is only the most notorious of the metal toxins invading our water supplies. The PHS currently considers 26 trace metals potentially toxic; many of them may also be carcinogenic.[23] One of them, arsenic, causes injuries similar to those from mercury. Like mercury poisoning, slow arsenic poisoning is easily misdiagnosed (e.g., as hypertension) and its effects may not appear for years. Symptoms include weakness, headache, weight loss, muscle pains, skin inflammation, damage to the brain or central nervous system, and cancer of the skin or liver. The PHS, in its 1969 edition of *Drinking*

Water Standards, reports that "chronic poisoning from arsenic may be insidious and pernicious," and indicates that arsenic may induce cancer when ingested via drinking water.[24] If the findings of the BWH survey are representative, arsenic exists in excessive amounts in almost 100 municipal water systems in this country. It was recently found in "substantial quantities" in the lower Mississippi by Federal investigators who had been searching for mercury there.[25] Like mercury, arsenic is not removed by treatment of drinking water.

The main sources of arsenic in water supplies are pesticides, animal foods, tobacco,[26] the production of certain metals, and household detergents.* Cases of arsenic poisoning from drinking water have already been documented. Wells in some rural parts of the country, such as Lane County, Oregon, are thought to contain the highest concentrations. Two nearly fatal cases of chronic arsenic poisoning prompted a study in Lane County in 1962–63; 16% of the Lane County wells tested contained arsenic concentrations over the maximum acceptable PHS limit of .05 ppm. Five had more than 10 times the PHS limit. The main source appeared to be arsenic-bearing rocks, although the runoff from cherry`orchards sprayed with arsenic-containing insecticides had also seeped into groundwater.[27] Excessive levels of arsenic in the drinking water in Fallon, Nevada, prompted Federal concern in 1969, but state authorities argued that the drinking water was not dangerous. Dr. Ernest Gregory, chief of Nevada's Bureau of Environmental Health, told the Task Force that Nevada's arsenic levels were equaled and even exceeded in many other Western states. This was confirmed recently by the Bureau of Water Hygiene's 1970 survey, which discovered excessive levels in four Western areas, "mostly in California."

* The Food and Drug Administration (FDA) concluded in late 1970, after a University of Kansas team headed by Dr. Ernest Angino traced high concentrations of arsenic in the Kansas River to household detergents, that the quantities of arsenic found in common detergents do not pose an "imminent hazard" in the household. But the FDA overlooked the possibility that continued use of such detergents, together with other sources of arsenic contamination, might raise arsenic concentrations in many of our waters above the PHS drinking water limit. The detergents with the highest arsenic concentrations were Biz, Dash, Tide XK, and Borateem, all of which were found by the FDA to contain more than 30 ppm.

Because lead-based paints have long been recognized as a health hazard—especially to children, who may eat flaking paint—the use of lead pigment in paints has been banned since the 1940's. No comparable attention has been paid to the excessive quantities of lead—as much as six times the maximum PHS limit of .05 ppm—recently found by BWH in tap water in 28 U.S. communities. In New York City, lead concentrations ranged as high as .14 ppm—three times the PHS limit.* Like other heavy metals, lead when ingested can lead ultimately to acute brain disorders. Cadmium, another toxic metal found in municipal water supplies, has been linked to hypertension and arteriosclerosis.[28] "There is little doubt in my mind that cadmium pollution [of both air and water] is a major factor in human high blood pressure, of which 23 million Americans suffer," [29] Dr. Henry Schroeder told the Senate Subcommittee on Energy, Natural Resources and the Environment in 1970. Dr. Schroeder's summary of the potential effects of heavy metal pollution made other forms of contamination seem innocuous by comparison:

I must emphasize that environmental pollution by toxic metals is a much more serious and much more insidious problem than is pollution by organic substances such as pesticides, weed killers, sulfur dioxide, oxides of nitrogen, carbon monoxide and other gross contaminants of the air and water. Most organic substances are degradable by natural processes; no metal is degradable. . . . Those metallic and elemental pollutants we have with us now are here to stay for a long time.[30]

Agricultural chemicals that get into the water from farm runoff—such as nitrate fertilizers, pesticides, herbicides, and hormones—are also poisoning tap water. The rapid growth in the use of nitrogenous ammonia fertilizers in the past 10 years has probably been the major cause of health hazards from nitrates.[31] Nitrate levels in water sources have risen dramatically in some parts of the country with the use of the new fertilizers,[32] adding

---

* Lead is discharged directly as a water pollutant by several industries, but a more serious source of lead contamination in water is currently believed to be the fallout from air pollution (e.g., from automobile exhaust). Drinking water can also pick up lead, copper, cadmium, etc., from water pipes. The lead in New York's drinking water is believed to come primarily from water pipes.

to long-standing nitrate pollution from animal wastes, human sewage, and industrial wastes.

High nitrate levels in water and food can cause acute methemoglobinemia in infants ("blue babies"). The 2000 cases of "blue babies" reported in North America and Europe over the past 25 years (around 8% of which were fatal) are estimated by Dr. Elliott F. Winton, a disease specialist at BWH, to reflect only 10% of the actual total.[33] Experimental evidence also suggests that babies whose mothers drink water with high nitrite concentrations while pregnant may develop methemoglobinemia in the womb, and have trouble gaining weight and be subject to increased mortality for a short time after birth.[34]

Acute cases of "blue babies" may be only the tip of the iceberg. Scientists have found chronic, sub-clinical methemoglobinemia in children drinking water with nitrate concentrations only slightly above—or even below—the recommended PHS limit of 45 ppm. The chronic form of the disease has been shown to result in behavioral disturbances and mental deficiency.[35] A 1961 study found that nearly half the 800 children in one Russian nursery school who were drinking water containing only 20–40 ppm nitrates developed sub-clinical methemoglobinemia; [36] a 1970 study reported a correlation between nitrate levels under the 45 ppm PHS limit in downstate Illinois and the appearance of sub-clinical methemoglobinemia in newborn babies and their mothers.[37]

Finally, nitrates can break down into nitrites, which may interact in the stomach with secondary amines from drugs, food flavorings, or decaying meat to produce nitrosamines. Some nitrosamines are carcinogenic, teratogenic, and mutagenic.[38]

While drinking water is not the only source of the nitrates people ingest—baby food, for example, is another—nitrate pollution of the water clearly contributes to the total nitrate intake of human beings. A number of special investigations have uncovered excessive nitrate levels in drinking water in many states, including California, Illinois, Colorado, Nebraska, Missouri, Iowa, Minnesota, and Kansas. The *average* nitrate concentrations in the San Joaquin Valley in California, for example, are almost as high as the PHS unsafe limit.[39] A report of the City Council of Delano, California, where concentrations of nitrates are among the highest in the state, notes that "85 other large water systems have wells with over 45 parts per million [the PHS limit], and 26 have over 90 ppm [twice the limit]." [40] The first quarterly progress report on the Delano study in April 1970 estimates that over 200,000 residents of California drink water that contains nitrates in excess of the PHS limit at some time during the year.[41]

Recognizing the extent of the hazard, at least two communities in California have placed printed warnings on water bills mailed to consumers. New parents in one community are given additional special warnings against the ingestion of tap water by infants. Dr. Thurston E. Larsen of the Illinois State Water Survey told the Task Force that pediatricians and hospitals in parts of Illinois have been issued special warnings five or six times within the past three years.

Pesticides also threaten drinking water supplies. Like most new chemical contaminants, they are not removed by standard water purification processes. DDT, for example, may be carcinogenic and mutagenic, yet the pesticide has been allowed to contaminate the environment so thoroughly that all U.S. children are now likely to have DDT in their tissues at birth.[42] DDT usually enters water from agricultural runoff, but it can enter water systems in large quantities in many ways. An Olin Chemical Corporation DDT production plant in Alabama dumped DDT into a ditch that emptied into a nearby wildlife sanctuary. The sanctuary includes a reservoir holding drinking water for the city of Decatur, Georgia.[43]

The source of some herbicide contamination is bizarre in its irony—herbicides have been used to kill pond weeds in fresh-water reservoirs. The herbicides 2,4-D and 2,4,5-T have both been proved conducive to birth defects; [44] 2,4,5-T contains a contaminant (dioxin) roughly 1,000 to 10,000 times more potent in causing birth defects than thalidomide.[45] While thalidomide caused a worldwide panic, the reckless use of herbicides has not yet generated a similar concern.

During the summer of 1970, following warnings against use of 2,4-D and 2,4,5-T in Vietnam, it was discovered that the main reservoir in Nassau County, New York, was being treated with 2,4-D. A public outcry resulting from this disclosure led to an announcement that the practice would cease. But there is no indication that other water authorities have followed Nassau County's lead, or that they are carefully monitoring contaminants from other sources. Up to September 1970, Baltimore, Maryland, for example, was applying 2,4,5-T to a municipal golf course that drains into a drinking water reservoir.

Another severe threat to drinking water safety is posed by the hormones contained in agricultural runoff. Diethylstilbestrol (DES), a synthetic hormone added to feed in order to fatten animals, is only one example. DES and other hormones may be excreted in substantial quantities; the animal waste then drains into waterways, carrying the hormones with it. DES is considered so dangerous a carcinogen that it was banned in chicken feed by the Food and Drug Administration in 1960,[46] and in cattle feed in several European countries around the same time.[47] U.S. cattlemen have continued to use the hormone, however, and in September 1970 Sweden and West Germany outlawed imported U.S. beef because of DES residues. The U.S. Circuit Court of Appeals judge who upheld FDA's ban on DES in poultry feed in 1966 ruled that "there is no known threshold or safe level for DES . . . . experts in cancer research testified that based upon clinical experience and to the extent practicable, no quantity of DES, regardless of amount, should be added to the diet." [48] A detailed study of DES is planned by the U.S. Agricultural Research Service. The results may be years in coming. In the meantime, large quantities of this carcinogen may be entering the drinking water of millions of Americans.

Not all cases of drinking water contamination are matters of life and death or serious illness. A researcher at the Federal Water Quality Administration, Dr. Aaron Rosen, has said that the taste and odor problem "is perhaps the most widespread, most difficult problem there is in supplying good drinking water—not safer, but good." [49] Approximately one-fifth of those water officials responding to a recent questionnaire from FWQA reported that blue-green algae, the most common cause of foul taste and odor in drinking water, have created problems for their treatment plants.[50] There have been published accounts of vomiting and gastric upsets apparently caused by the odor of drinking water, and many complaints about soapsuds coming directly from drinking water taps.[51]

If the taste and odor of your water are not enough to make you sick, its appearance may. Some drinking water has an unusual color or a cloudy appearance and contains visible organisms. Thin red wriggling bloodworms, which generally indicate polluted water, have shown up in at least 17 widely scattered drinking water systems in the United States,[52] including New York City's and Washington, D.C.'s.[53] Although health officials say that the worms are harmless, the reaction of many consumers must have been expressed by C. C. Johnson of CPEHS: "Why should we have to pay for purified worms?" Nor is it safe to assume that all worms are "purified." Nematode worms, which survive ordinary chlorination,[54] and which may shield pathogens from water treatment,[55] have been reported in a large number of treatment plants. The fact that snails, sow bugs, copepods, water fleas, cyclops, and a 30-inch water snake have also found their way into drinking water systems is hardly calculated to inspire faith in water purification.

Despite the soothing words of local water officials, there are signs of an erosion of public confidence in the safety of drinking water. In 1969 consumers paid $80 million to avoid tap water, and sales of bottled water are increasing at the rate of 15–20% a year. Such major beverage and food companies as Coca-Cola, Borden, and Nestlé have already entered the bottled water field.[56] This may supply the needs of the affluent and alert, but millions of others are less fortunate.

Several years ago one author noted that local governments "are clashing with each other in jurisdictional disputes for the last unpolluted streams."[57] That battle is likely to escalate. New York City, which is closely adjoined by abundant supplies of fresh water, transports its drinking water from 125 miles away. We may soon have to ration drinking water routinely—even as we now do during dry spells in various parts of the country. Insofar as safe drinking water is concerned, we may be entering a permanent "dry spell." Economist Kenneth E. Boulding's description of the scarcity resulting from environmental pollution is particularly apt: "The closed economy of the future might . . . be called the 'space-man' economy, in which the earth has become a single spaceship, without unlimited reservoirs of anything, either for extraction or pollution. . . ." [58] This closed-system phenomenon may occur first with drinking water. Or it may already have occurred without our knowing it.

The Food Chain

When man tugs on one thread holding together his environment, he soon finds that everything else is attached —including himself. The food chain is an excellent example of man's dependence upon even the lowest forms of life. Aquatic invertebrates feed on small organisms called plankton. The invertebrates are in turn eaten by higher forms—fish and other animals—that are sources of food for human beings. This is, of course, a simplified description of the awesome, precarious natural process called the food chain, which is so delicate that the severance of any link could ruin it. But the links are even more fragile than this. When contaminants are passed up the food chain there is a tendency for the poisons to concentrate in geometric orders of magnitude. That is, relatively innocuous quantities of non-biodegradable, fat-soluble organic pollutants concentrated in plankton-sized beings can combine in lethal quantities in higher forms of life.

A classic instance of concentration up the food chain in Clear Lake, California, was described by Rachel Carson in *Silent Spring*. DDD—a relative of DDT—was applied to the water to control flying gnats in 1949, 1954, and 1957, in concentrations no higher than .02 parts per million (ppm). Shortly after the last application, not a trace of DDD could be detected in the water. But crop after crop of plankton continued to accumulate residues of 5 ppm long after the treatment had stopped. Fish eating the plankton concentrated the residues in their fat to levels ranging from 40 to 300 ppm, and fish-eating fish stored DDD concentrations up to 2500 ppm. While the residues did not kill the plankton or fish, grebes (one species of bird feeding on the fish) with DDD concentrations of 1600 ppm were gradually killed off. By 1960 no grebes were left on Clear Lake.[59] The contaminated fish might just as easily have been eaten by human beings.

These dangers are multiplying as lower forms of life increase their resistance to poisons. Dr. Denzel B. Ferguson, an expert on pesticides and chairman of the Zoology Department at Mississippi State University, has reported that mosquitoes and fish in the Mississippi Delta now carry agricultural contaminants in concentrations up to 120 times the amount that once would have killed them. "The time may come," he warned, "when we will be more concerned over the fish that survive a pesticide kill than over those that are killed." [60]

The food chain stretches thousands of miles; distance from the source of pollution is no guarantee of safety. Federal marine biologists told *The New York Times* in October 1970 that the livers of Alaskan fur seals were found to contain up to 116 times the amount of mercury considered safe for human consumption. Dr. George Y. Harry, Federal marine biologist in Seattle, said he considered it "astounding" that large residues had been found in aquatic mammals. "It's possible," he explained,

that the mercury stemmed from industrial wastes [along the Oregon and Washington coasts, about 2000 miles away] and moved up to the food chain from plankton-sized creatures, to salmon and pollock, to the fur seals.[61]

One need not be a seal liver connoisseur to be concerned about mercury in foods. In 33 states it has been found in fish in concentrations much higher than in drinking water.[62] An FWQA official has estimated that "as much as 10% of U.S. inland waters may be spoiled or endangered for fishing for years and decades to come because of mercury pollution." The Food and Drug Administration announced midway through a swordfish-testing program in February 1971 that 87% of its samples had been found unfit for human consumption as the result of mercury contamination.* While the FDA also reported with relief that only 3.6% of its tuna samples had been found to contain more than the .5 ppm mercury limit,[63] Dr. Alex B. Morrison, deputy director general of Canada's Food and Drug Directorate, suggested in February 1971 that the .5 ppm standard itself might not provide enough of a margin of safety.[64] Nineteen other species of fish are currently being tested by FDA for mercury contamination.

The dangers at the human end of the food chain were vividly demonstrated at Minamata Bay in Japan more than a decade ago. Minamata Bay has the dubious distinction of lending its name to acute mercury poisoning —"Minamata disease." During the 1950's a factory producing vinyl chloride dumped huge quantities of mercury salts into the bay. One hundred eleven people were poisoned between 1953 and 1960, either from eating the contaminated fish or from eating animals that fed on the fish. Forty-three persons died and many more suffered blindness, loss of the use of limbs, and brain malfunctions. Nineteen congenitally brain-damaged children were born to mothers who themselves had been little affected.[65]

The FDA takes great pains to reassure Americans that the levels of mercury being found in fish in the U.S. are far lower than the concentrations that killed and maimed the residents of Minamata Bay (mercury concentrations 30 to 80 times the FDA limit), and that the average American eats only one-fifth as much fish as the average Japanese.[66] Yet the fact remains that mercury poisoning is cumulative, and that certain sectors of the U.S. population eat disproportionately large amounts of fish. Dr. Bruce McDuffie, the chemistry professor at the State University of New York who was the first to discover mercury contamination of tuna and swordfish in December 1970, reported in January 1971 that dieters who had been eating large quantities of the fish had five times as much mercury in their systems as

---

* In May 1971 the FDA warned the public not to eat swordfish because 89–95% of samples tested contained mercury "substantially" above the limit.

a group that had eaten little or no tuna or swordfish.[67] Dr. Barry Commoner has pointed out that 45% of all tuna is consumed by 9% of tuna eaters.

Mercury is only the latest example of a general affliction in our environmental "policy"—a policy built on assumptions with little basis in fact. For years it was assumed that industry would permit very little of an item so expensive as mercury to reach its wastewater, and that the very small amounts reaching the rivers, lakes, and oceans would, because of the weight of the metal, fall inert to the bottom. Neither of these assumptions was correct. Of the 6 million pounds used annually by industry, up to one-half is lost in the environment, much of that in wastewater. And instead of lying harmlessly at the bottom, the metal interacts with bacteria to form the highly toxic methyl mercury that dissolves in the water and then enters the food chain.

The fallacious assumptions that have permitted so much mercury to enter our ecosystem parallel our approach with other pollutants, an approach that permits the use of our environment for private dumping until the contaminants have been proven dangerous by the public. The public, moreover, gets little help from government in meeting this burden of proof. The first revelations concerning mercury did not come from government sources but from Norvald Fimreite, a student at the University of Western Ontario.

Remedial action to control mercury pollution in the U.S. has been random and sporadic (see Chapter 12). But even the most vigilant after-the-fact approach would not have been enough. Dr. John Wood, who led the University of Illinois research team that worked extensively on the problem of the conversion of mercury to methyl mercury, told a television audience on the CBS program *Is Mercury a Menace?* that

it's not good enough to say that we should stop putting mercury in the environment. It's too late. There's so much material that's been deposited. What we have to do is come up with good methods for removing it from the environment. If we don't come up with those good methods, then you can anticipate for example in the St. Clair system, that it would be thousands of years before people would be able to eat fish from that particular area. If it's possible to re-move 95% of the mercury from that system, it would still be hundreds of years before people can eat fish out of that system.

As for current remedies, Environmental Protection Ad-ministrator William Ruckelshaus put it simply on the same program: ". . . we don't know at this stage . . . just exactly what we do to get it out of the environ-ment."

The problems of chlorinated hydrocarbon pesticides like DDT in the food chain are similar in many ways to those of mercury. Pesticide residues are now found in nearly every animal and fish in the world—even in Antarctica, where there has never been a known direct application of pesticides.[68] The chemicals have been carried in the wind, on particles of dust, in water runoff, and in living organisms to every corner of the globe.[69] Ecologist Charles F. Wurster, Jr., has said that DDT and its residues are "probably more widely distributed than any other man-made chemical" and can probably be found in every human being in the world.[70] High con-centrations of chlorinated hydrocarbon residues have been found in edible fish. A national pesticide study con-ducted during 1967 and 1968 by the U.S. Bureau of Sport Fisheries and Wildlife found DDT in 584 out of 590 samples of fish taken from rivers and lakes across the nation.[71] Concentrations as high as 45 ppm per whole fish, nine times the FDA limit, were found. Dieldrin, a pesticide more toxic and more persistent than DDT, was found in 75% of the samples.

The poisoning of fresh-water fish has been fairly com-mon. In 1969 the FDA seized 28,150 pounds of con-taminated Lake Michigan coho salmon. As Senator Gaylord Nelson (D.-Wisc.) noted in introducing legis-lation to control pesticides in 1969, "This . . . proves the tremendously dangerous persistence of these pesti-cides. To . . . reach the salmon, the DDT and dieldrin probably traveled hundreds of miles . . . and was con-sumed through the normal food chain of up to a half dozen organisms." [72] In December 1970 it became ap-parent that the danger extended to salt-water fish, when the FDA impounded 1260 pounds of DDT-tainted kingfish. The measured residues were 19 ppm, almost four times the FDA maximum.

"Peaceful" Nuclear War

With the nuclear age has come one of the most dangerous and persistent forms of environmental contamination. Uranium mining and processing, the fallout from nuclear testing, the production of nuclear weapons material, medical research, industrial use of radioactive material, nuclear power generation, and the reprocessing of spent reactor fuel have already added radioactive wastes to natural background levels of radioactivity in the water. According to Federal water quality officials, most surface waters now contain detectable concentrations of strontium 90 and cesium 137, man-made radionuclides,* which fallout has deposited in lakes and streams across the country. And the "peaceful" use of nuclear power is expected to increase dramatically. The 17 nuclear power plants now in operation provide less than 2% of U.S. electric power. But electric utility companies, seeking to supply a growing demand for electricity—consumption has doubled every decade since 1940—are pushing hard to expand generating capacity through nuclear power. Ninety-two nuclear power plants are in construction or on order,[73] and the Atomic Energy Commission expects 950 to be operating by the year 2000. By that time, if AEC projections are fulfilled, nuclear plants will be generating 60–70% of all U.S. electric power.[74] With the growth in power production by nuclear-fission reactors will come a proportional increase in radioactive waste.[75]

The critical difference between radioactive wastes and other environmental contaminants of chemical origin is that there is no method of treatment which can counteract the innate biological harmfulness of radioactive substances.[76] Each radionuclide decays at its own fixed rate, no matter what is done to it.[77] Some become biologically harmless within a matter of minutes. But others are extraordinarily long-lived. Strontium 90 and cesium 137,

---

* "Radionuclide" is the general term for an element in unstable form which emits radiation as it decays. The term is often used interchangeably with "radioisotope."

for example, have half-lives * of 28 and 30 years, re-spectively—which means it takes six to 10 centuries for them to decay to safe levels.[78]

The low-level radioactive wastes that nuclear power plants, AEC installations, and fuel-reprocessing plants discharge directly into waterways are the most obvious and immediate cause of concern to environmentalists and water pollution experts. Yet this liquid effluent from leaking equipment, floor drains, and the cleansing of equipment and reactor coolants contains only a tiny fraction of the radioactivity generated at nuclear in-stallations. Though their volume is less, the "high-level" (intensely radioactive) wastes from the reprocessing of spent reactor fuel—which are stored in tanks and buried underground—account for more than 99% of the radio-activity in waste materials now on hand.[79] Isolating the growing pile of buried radioactive garbage from under-ground water supplies, the surface waters they feed into, and the biological environment could well become the most serious radiation problem in the future.[80]

Neither the tanks used to store high-level wastes nor the AEC sites where most of them were buried were completely adequate for isolating the radioactive mate-rial from the environment on a long-term basis, accord-ing to a 1966 report by one of the AEC's advisory committees, the National Academy of Sciences' Commit-tee on Geological Aspects of Radioactive Waste Disposal. The Committee was dissolved and replaced by a new one in 1968, and the AEC refused to release the report until Senator Frank Church (D.-Idaho) called attention to it in March 1970.[81]

The huge concrete-encased steel tanks (some over a million gallons in capacity) [82] where the boiling-hot wastes from reactor fuel reprocessing are allowed to cool and are stored have a life expectancy of no more than 10 to 20 years.[83] Twelve of the 149 older "single shell" steel tanks stored in concrete vaults near the surface of the ground at the AEC's Hanford installation (Richland, Washington) have sprung leaks in the last 26 years.

---

* A half-life is the time required for a radioactive substance to lose half its radioactivity; a period of two half-lives would reduce the quan-tity of radioactivity to one-quarter the original amount. (U.S. Atomic Energy Commission Understanding the Atom Series, *Radioactive Wastes*, p. 10, and *Atomic Power Safety*, p. 4.)

"Every time one of those tanks leaks, radioactivity goes into the ground," one AEC official explained to the Task Force. At the AEC's Savannah River plant near Aiken, South Carolina, five cases of tank failure have been recorded so far. In one incident in 1959 the intensely radioactive waste spilled over the huge five-foot-high heavy steel "saucer" that the tank sits in and actually reached the water table some 30 to 50 feet below. While the AEC has started converting high-level liquid wastes to solid form to reduce their mobility, 1976 is the earliest the AEC plans to finish the project at two of its four major installations—Hanford and the National Reactor Testing Station (NRTS) in Senator Church's home state of Idaho. Savannah River and Oak Ridge (Tennessee) will take even longer. In the meantime most intensely radioactive waste is still being stored in the more dangerous liquid form in underground tanks.[84]

But even solidification won't contain highly radioactive waste for the hundreds or thousands of years the material remains biologically hazardous.[85] Only "safe natural containers [i.e., geologic formations] totally and permanently separated from the zone of fresh water must be used," [86] warned the NAS Committee in 1966. The Committee found none of the four major sites at which the AEC is still storing approximately 80 million gallons [87] of intensely radioactive waste (mainly from nuclear weapons production) suitable for long-term storage, with the possible exception of Oak Ridge.[88]

The AEC has considered shipping some of these wastes to a safer spot—the salt mine in Lyons, Kansas, where the growing volume of high-level waste from all the nation's nuclear power plants is expected to be buried after 1976.[89] Salt is highly impervious to groundwater flow, and storage in the mine is considered the safest plan for the long-term containment of radioactive waste.[90] But hauling the waste to Kansas would cost an estimated $1.5 to 2 billion for Hanford and Savannah River alone.[91] The NAS Committee had found it necessary to dissent in its 1966 report

from the working philosophy of some operators, though certainly not that of the AEC, that safety and economy are factors of equal weight in radioactive-waste disposal. . . .

The Committee remains convinced that economics is a criterion secondary to that of safety.[92]

Ignoring the Committee's recommendation, the industrial corporations * that operate the four AEC installations are still looking for cheaper methods of long-term storage.[93]

The proposal at Savannah River—"Project Bedrock" —is to pump highly radioactive liquid waste directly into the bedrock below the Tuscaloosa aquifers,† [94] which, as the NAS Committee noted in 1966, "are certainly going to be drawn on increasingly for human consumption in approaching decades." [95] Doubting that a permanently leakproof chamber could be constructed in the rock and worried that the highly radioactive liquid might leak into the aquifer, the Committee recommended that the AEC stop even investigating bedrock storage.[96] That was in 1966. Now the Hanford installation is considering deep-cavern storage too.[97]

NRTS in Idaho is also considering storing some of its solidified highly radioactive wastes on-site.[98] NRTS lies over the Snake Plain aquifer, "one of the largest of the country's remaining reserves of pure fresh water." [99] Some 1.3 billion gallons of groundwater move slowly south below the AEC installation each day,[100] flowing ultimately into much of the Pacific Northwest.[101] Contamination of this priceless resource would be nothing new for NRTS. FWQA investigators in October 1968 found that the groundwater below NRTS had been degraded by low-level radioactive wastes "discharged to seepage pits, lagoons and directly to the underlying aquifer through disposal wells." [102] Their recommendation that the latter practice be stopped was carried out in 1970. But it will be at least 10 years before another of the radioactive hazards they discovered is moved off-site: the local burial ground (every major AEC installation has one, as do five states where private com-

---

* An Atlantic Richfield subsidiary is the main operating contractor at Hanford; Idaho Nuclear Corp. (a jointly owned subsidiary of Aerojet General and Allied Chemical), General Electric, Phillips Petroleum, and Westinghouse Electric at NRTS; Union Carbide at Oak Ridge; and DuPont at Savannah River.
† An aquifer is a water-bearing stratum of permeable rock, sand, or gravel.

panies have entered the field) * for radioactive solid waste. The trenches where radioactive residues and contaminated laboratory equipment, clothing, and other radioactive junk are buried at NRTS have been flooded periodically by melting snow; [104] dikes are now being built to minimize the danger of radioactive material being washed down into the groundwater.

The possibility of contamination of drinking water and food chains by the radioactive waste from nuclear reactors poses a serious hazard to human health. Exposure to radioactivity and, more generally, to ionizing radiation has been linked to sterility, birth defects, cancer, genetic damage, a shortened life span, and abnormal growth and development. The effect of a particular radionuclide depends on its half-life, how much of it is present, the energy and characteristics of its radiation, and its chemical affinities for special organisms, organs, or biochemical processes (the radionuclide strontium 90 is a bone-seeker; iodine 131 accumulates in the thyroid gland). Once a radionuclide enters human tissue, it bombards surrounding cells with ionizing radiation. While the precise mechanism is not understood, ionizing radiation has induced cancer and leukemia in man.

One of the most serious potential hazards from chronic exposure to low levels of man-made radiation is genetic damage. The genetic effects of radiation have two components, according to Drs. John Gofman and Arthur Tamplin of the Bio-Medical Division of the AEC-funded Lawrence Radiation Laboratory in California:

(1) Lethal effects that lead to death before maturity or that lead to sterility, and (2) effects that contribute to the general pattern of illness and mortality in adult life. In the population the present pattern of illness and mortality results from a complicated (and essentially unknown) interplay between hereditary factors and the environment. The mechanism by which radiation would be expected to influence these patterns is by altering the genes and chromosomes that determine the hereditary factors transferred to the child.[105]

---

* Nuclear Fuel Services operates one commercial burial site for low-level radioactive waste in West Valley, New York. Nuclear Engineering Company operates sites in Richland, Washington; Beatty, Nevada; Sheffield, Illinois; and Morehead, Kentucky.[103]

"How serious are the genetic effects of radiation?" ask Gofman and Tamplin: "No one knows!" [106]

The nuclear power industry and the AEC, which licenses nuclear power plants and nuclear fuel reprocessing plants and administers radiation standards, maintain that the amount of radioactive waste allowed in lakes, streams, estuaries, and groundwater is too small to pose any threat to life and health. AEC regulations place limits on the maximum permissible concentration (MPC) allowed in the water for each of the radionuclides discharged.[107] The MPC limits are derived from another, more basic, standard, the maximum permissible dose (MPD) of radiation that the AEC and various standards-setting organizations calculate individuals can safely be exposed to from man-made, non-medical sources (i.e., excluding cosmic rays, X rays, etc.).* The AEC emphasizes that actual discharges of radioactive wastes are usually far below its MPC limits,† and that even these amounts are diluted once the water carries the radioactivity away from the nuclear power or fuel reprocessing plant. In any case, say AEC and Federal water quality officials, they are careful in licensing and monitoring radioactive waste dischargers to make sure that people exposed to the radioactivity will not get more than the maximum permissible dose from water, food, and air combined. In short, the AEC says, don't worry.

---

* According to officials in the EPA Radiation Office, the relationship between the AEC's maximum permissible concentration (MPC) and maximum permissible dose (MPD) standards is as follows. If a standard-size man drank 2.2 liters of water each day (the average intake) for 50 years, and the water contained the maximum permissible concentration of radioactivity, then the amount of radioactivity which would accumulate in the man's body or in the critical organ would equal the maximum permissible 50-year radiation dose (MPD) which AEC regulations allow an individual to be exposed to from all non-medical man-made sources (e.g., the fallout in the air he breathes and the cheese sandwich he eats for lunch, the nuclear power plant wastes in his drinking water, the luminous substance in his wristwatch dial, his color TV set, etc.).

† The radioactive discharge from nuclear power plants averages 1–10% of MPC over the course of a year, according to one Federal water quality official. The only non-government nuclear fuel reprocessing company in operation so far, Nuclear Fuel Services, Inc. (West Valley, New York), released an average of 19% of MPC into Cattaraugus Creek in 1969. The measurements were taken at some distance from the original point of discharge, however, after the effluent had been diluted by several lagoons and creeks.

Critics of the nuclear power industry are not nearly so confident. The standards-setting organizations that advise the AEC repeatedly warn that *all* exposure to radiation is probably harmful, that all unnecessary exposure should be avoided, and that a calculation of risk versus benefit is always involved in setting the maximum permissible dose. Critics believe that the AEC, as the agency in charge of promoting the growth of the nuclear power industry, has been unduly impressed with the benefits of atomic energy and insufficiently wary of the risks. For example, Drs. John Gofman and Arthur Tamplin recommended in October 1969 that the AEC lower its radiation dose limits by at least a factor of 10.[108] Since then, lower limits have been endorsed by Nobel Prize winners Linus Pauling and Joshua Lederberg and other scientists,[109] including some who doubt Gofman and Tamplin's conclusion: that 17,000 additional cancer cases would result in the U.S. each year if everyone received the AEC's currently allowable radiation dose.[110] Gofman and Tamplin have since estimated that

it is possible that exposure to the present allowable levels [of radiation] could result in a 5% to 50% increase in the death rate [as a result of genetic damage], producing some 150,000 to 1,500,000 additional deaths each year in a future population of 300,000,000 people. Moreover, the evidence suggests that there would be (over and above the fatal diseases) a 5% to 50% increase in such crippling diseases as diabetes, rheumatoid arthritis, and schizophrenia.[111]

Apart from the fact that the AEC standards for maximum permissible concentrations of radioactivity in the water are derived from an MPD that may be too high to begin with, Dr. Tamplin has charged that the MPC standards do not adequately reflect concentration up the food chain. Long-lived radionuclides such as cesium 137, with a half-life of 30 years, remain radioactive while they concentrate in plants and animals; cesium 137 may be concentrated 10,000 times in the aquatic food chain. By Tamplin's calculations, "a man eating a pound of fish a week, grown in water at the [maximum permissible concentration of cesium 137], would receive

[an annual] dosage . . . 30 times the AEC guidelines."
A 75-pound child would get a dosage 60 times the AEC
guideline.[112]

The AEC stresses that radioactive discharges are normally a very small fraction of the maximum allowed.
But it is hardly reassuring that a nuclear power plant
may discharge only one-tenth the AEC's maximum permissible concentration of radioactive wastes, and that
they are further diluted by a factor of 10 some distance
from the plant, if the food chain can reconcentrate the
radioactivity by 10,000 times. The state of Minnesota
proposed limits on radioactive waste discharges 50 times
more stringent than AEC standards in 1969.

Unlike the serious present danger posed by chemical
pollutants in the water, the risks of radiation damage
from "peaceful" atomic power are believed to be small
so far—though no one can be sure. The critics' main
concern is the future, when the use of nuclear energy is
expected to expand exponentially. "Since the utilities are
operating more than 10 times below the standards, but
fight lowering them," says Dr. Barry Commoner, "then
the only rational explanation is that they think *they
might have to release more radiation in the future*." [113]

The Other Costs

Oh! the old swimmin'-hole! In the happy days of yore,
When I ust to lean above it on the old sickamore,
Oh! it showed me a face in its warm sunny tide
That gazed back at me so gay and glorified,
It made me love myself as I leaped to caress
My shadder smilin' up at me with sich tenderness.
                                        —James Whitcomb Riley

Indiana's Brandywine Creek, which inspired that
poem, was declared unfit for swimming in the summer
of 1969. Many other waters are more or less permanently off limits because of bacterial contamination. The
lower Hudson River is so laden with bacteria and other
filth that merely touching it poses a health threat. The
coliform (intestinal bacteria) counts in the Hudson vary
widely, but counts as high as 170,000/100 mililiters have
been taken by FWQA beneath George Washington
Bridge. This is about 170 times the safe limit.

Some places where unsafe coliform levels have been found have nevertheless been kept open for swimming— in Chicago and Connecticut, for instance. Eric Mood, an assistant professor of public health at Yale University, charged in July 1970 that Connecticut authorities were keeping polluted beaches open to avoid riots in sweltering urban areas. Mood claimed that water samples taken by the state health department showed that only one-third of the state's shoreline was completely safe for swimming. Most of the beaches that should not have been recommended for bathing, he said, were in or near dense urban areas. Mood was very generous in assessing the ethical question facing health authorities:

As a professional health official [he] knows he should close the beach for swimming. But as a humanitarian he knows that the socially deprived residents of his city have no other convenient outdoor bathing area to which they may seek relief from the oppressing summer heat of the urban slums in which they live.[114]

Water pollution threatens more than swimming; it imperils continued sport and commercial fishing in many parts of the United States. In 1969—the most recent year for which data are available—there was a record number of fish kills. FWQA's annual inventory of fish kills shows that over 41 million fish were killed in 1969 —more than in 1966, 1967, and 1968 combined. The upsurge was due to what is probably the largest recorded kill ever, at Lake Thonotosassa, Plant City, Florida. Twenty-six million fish died as the legacy of years of dumping by four food-processing companies. The companies—Salada Foods, Sugar Rose Canning, Paradise Fruit, and Florida Sip—paid no fines or damages and did not restock the lake. Instead, they began pumping their wastes through the city's treatment system.

With events like these occurring on a smaller scale nationwide, pollution threatens the continued existence of many fish species. A state conservation official has said of Lake Michigan: "As long as DDT and dieldrin are found in the lake at present levels it is questionable if there will be any reproduction of most species of fish. . . . There will be a demise of some species [in Lake Michigan]." [115]

Already, large numbers of commercial fishermen have been driven out of business. A 1966 survey showed that almost 2 million acres of shellfishing areas had been closed to harvesting.[116] A 1968 study showed that in the Chesapeake Bay area alone, pollution was responsible for an annual loss of nearly $3 million for the shellfish and finfish industries.[117] DDT contamination has been blamed for the demise of the sardine industry in Monterey, California.[118] In Bishops Harbor, Florida, people complain bitterly that a series of spills from the local Borden Chemical Company plant is driving the "little guy" out. "People used to come here to fish and I did a real good business," said the 60-year-old owner of a fish camp. Now the place smells of rotting fish killed by the chemical spills; the waters are dead and weed-choked, and the owner has put a "For Sale" sign up. "This place means 40 years of work to me," he said. "Now I can't sell it to anybody." [119]

The impact of thermal pollution from the waters used for cooling in some manufacturing and nuclear power plants is potentially devastating for fish, wildlife, and humans. Nuclear plants can use as much as half a million gallons of water per minute to dissipate the enormous quantities of heat produced. Cooling waters drawn from rivers, lakes, and estuaries have been poured back at temperatures as high as 115°.

Elevated water temperatures greatly increase the survival of unwanted organisms, while decreasing the life span and number of more desirable animals and plants. The higher the heat, the more bacteria survive, including bacteria that cause disease in fish and men. Viruses thrive in fish subjected to increased heat. Also, the toxicity of many pesticides to fish is greater at higher temperatures, as a result of heightened metabolism.

In addition, an environment of higher average temperature may change the type of vegetation, as subtropical or tropical plants replace temperate- and cold-water natives. The effect of this change is incalculable. Likewise with fish; the results are similar; certain species cannot survive thermal change, particularly while in egg or fingerling states. The critical time is that of spawning; if conditions are not correct, the oyster, for example, reabsorbs its eggs and produces no offspring. The result for the shellfish industry is disastrous.

Water pollution has also been directly and indirectly responsible for the death of many thousands of birds. Oil spills can slaughter birds by suffocation, exposure, and starvation. When a bird hits oil, its feathers—which normally insulate the bird—coagulate, exposing the skin to cold water. The bird may die from exposure, or become water-logged and drown; it may also be poisoned by volatile petroleum distillates that seep through the skin, or it may simply become mired and choke to death; it may manage to struggle to shore, but, unable to forage for food, slowly starve.

Birds of prey, like the bald eagle, peregrine falcon, and many others, are endangered by poisons in the fish they eat. Ecologist Dr. Charles F. Wurster, Jr., estimates that "we could lose 50 to 100 species of birds in the next 25 years" to pollutants such as DDT.[120] A study of brown pelicans in the Channel Islands off Santa Barbara, California, found that DDT residues killed all the eggs in 284 of the 296 pelican nests found.[121] DDT has been blamed for some birds' inability to lay eggs with shells hard enough to protect their developing offspring. Some American bald eagles have laid eggs without shells.[122]

The Coming Struggle for Water

More than aggregate figures, meaningless in their magnitude, the passing of the peregrine falcon, the extinction of numerous species of fish, and the plight of the Bishops Harbor fishermen are eloquent signs of the beginning of the end of our environment. Some losses can be translated into economic terms. Economist Edwin Johnson of FWQA has estimated that water pollution costs the nation $12.8 billion annually. A single child born severely retarded because of chemical contamination of the water can cost society $250,000 in remedial training and custodial care.[123] While the cost of months in the hospital or an early death from cancer can also be computed on narrow actuarial grounds, no economist can put a price tag on genetic mutations that may damage generations to come. And the children deprived of the "old swimmin' hole" cannot buy it back.

The decline in water quality is likely to accelerate, unless we step in to reclaim what is ours. Unless water use patterns are drastically altered, "demand" for water will soon outrun usable supplies. The maximum amount of fresh water that will ever be available for withdrawal from natural sources (i.e., not counting additional use gained from immediate treatment, internal recycling, and reuse), given present technology, is an estimated 600 billion gallons a day (bgd).[124] In 1960 we withdrew some 275 bgd to meet our needs—domestic, agricultural, manufacturing, electric power generation, and mining.[125] By 1980 it is expected that we will be using about twice that amount—560 bgd or more.[126]

The result of this profligate use—and waste—of our waters will be shortages of this indispensible commodity. During the next few years there will be a mounting struggle to retain the little "clean" water still available. But very soon there will be little left worth fighting over. What is needed is a national commitment to assure that all people have ample supplies of safe, pleasing water for all legitimate uses. As the following pages show, however, we have not made even the most modest beginnings toward achieving that goal.

Part Two

POLLUTERS AND PROTECTORS

2

People, Profits, and Pollution: Municipal and Industrial Pollution

No one has the right to use America's rivers and America's waterways, that belong to all the people, as a sewer. The banks of a river may belong to one man or one industry or one State, but the waters which flow between the banks should belong to all the people.
—President Lyndon Johnson, upon signing the Water Quality Act of 1965

Depending on whose figures you believe, the average American uses somewhere between 45 and 100 gallons of water a day—3 gallons each time the toilet is flushed, 30 to 40 gallons for a bath, 20 gallons in the washing machine, 10 gallons to wash dishes by hand, and varying amounts on gardens, housecleaning, and drinking.[1] Suburbanites soak up far more water per capita than city residents do, especially in summer, when the symbols of conspicuous consumption—green lawns, swimming pools, dishwashers, multiple baths and showers—receive heaviest use. Ghetto dwellers demand the least; when the plumbing doesn't work, it's hard to use water. But whatever the volume used, nearly all the water that enters a household clean goes out again as sewage.

Our capacity to pollute has outstripped our ability and will to provide municipal waste treatment. In 1900 only one million persons were served by sewage treatment; the total human wasteload, both treated and untreated, streaming from municipal sewers into lakes and streams across the country had a pollution equivalent to the raw untreated sewage from 24 million people.[2] We have been building more municipal treatment plants since that time, but not nearly enough. By 1968 little more than half the U.S. population—just over 130 million people [3]—was served by any form of municipal sewage treatment; the overall human wasteload pouring directly into our nation's waters from sewered cities and towns alone was at least three times greater than it was in 1900.[4]

The Federal government considers more than half the municipal treatment now in use inadequate on several counts.[5] Nearly 30% of the "treated" waste receives only "primary" treatment.[6] A primary treatment plant simply screens and settles out the largest solid chunks before sending the sewage straight to the river, still heavily laden with an assortment of oxygen-consuming organic pollutants, inorganic industrial wastes piped into city sewers, viruses, and disease-carrying fecal bacteria. (Some cities chlorinate their primary effluent to kill the bacteria, but typically only during the summer water recreation months.) Though just over 65% of existing plants also provide secondary treatment [7]—a biological process that removes 60–85% of the oxygen-demanding inorganic waste—even secondary treatment is not enough in some places to keep municipal pollution from increasing. And most treatment plants are too small to handle the wastes of the population they serve. The cities solve that problem either by skimping on treatment (running the sewage through too fast to get as much pollution out as the treatment process normally would) or by treating some of the sewage and running the rest into the river raw (called "bypassing" the plant). Over 1000 new communities outgrow their treatment systems every year.[8]

Municipal pollution gets worse when it rains. Many of the nation's newer sewer systems have storm sewers that are completely separate from the regular sewer system. Separate storm sewers carry what is known as urban runoff—the oils, rubber particles, pesticides, heavy metals, organic pollutants, and just plain trash from the filthy streets and sidewalks—directly to the river without treatment, even though its concentrations of polluting substances are sometimes as high as or higher than domestic sewage.

Contrast the separate storm sewer scheme, considered enlightened, with the older plan: combined sewers. Every time it rains in New York City, Rochester, Detroit, Cleveland, Chicago, and Boston (as well as many smaller urban centers of similar vintage), street runoff pours into the regular sewer system and all the wastes converge on the sewage treatment plant together. The treatment system overflows and what runs into the river untreated is not only urban runoff but enormous quantities of raw human sewage and industrial wastes.

Americans invested an estimated $1.2–$1.3 billion in construction of municipal sewage treatment plants and interceptor sewers in 1970 [9]—less than they shelled out for hair spray, hair tonic, hair dye, lipstick, eye makeup, and deodorant that year. The sum was well under half the amount ($2.8 billion) that the Federal government conservatively estimates should be spent each year between 1970 and 1974 just to provide secondary treatment for most of the urban population (only 75% of the total U.S. population) by 1975.[10] And that $2.8 billion estimate doesn't include any money at all for controlling stormwater overflows (which would cost overall an estimated $15–$49 billion [11]); for anything beyond secondary treatment (which doesn't adequately remove phosphates, for example, by far the most serious single municipal pollutant); or for collection sewers (an estimated $1.2 billion per year [12]); or the costs of buying land for treatment plants, or operating and maintaining them (an estimated $1.4 billion total for five years in 1968 dollars [13]). Unless spending priorities are drastically changed, municipal pollution will go on increasing much, much faster than our efforts to control it.

Industrial Pollution

Pollution is good for business. All the sanctimonious rhetoric of corporate responsibility cannot veil this crucial fact. Water is as essential to industrial operations as any of the other factors of production—labor, raw materials, etc. But there is an important difference: companies must pay for their labor and raw materials, but they pay little or nothing for the water they use in such huge quantities—the water they pollute with cyanides, oils, suspended solids, and other industrial wastes.

American industry used 10 times more water in 1970 than municipal users; and industry helps itself to a bigger share of the available water with each passing year.

WATER USE—MUNICIPAL, INDUSTRIAL, AGRICULTURAL [1]
(In Billions of Gallons Per Day)

	1950	1960	1970[b]	1980[a]	2000[a]
Municipal	14	21	25	29	42
Industrial mining, power, manufacturing	77	140	250	363	662
Agricultural (irrigation)	82.6	113.6	143	167	184

[1] Figures based on information supplied by Colonel Atkinson, American Water Works Association. These figures were compiled in 1965 by the Department of Agriculture and, according to Colonel Atkinson, represent the latest full-scale survey of water needs, existing and projected.
[a] Projections by American Water Works Association.
[b] Projections by the Task Force based on AWWA data.

Most of this water eventually returns to the nearest lake or stream, laden with contaminants. Thus whole stretches of rivers are appropriated as free industrial dumping grounds, though this special "use" of our water doesn't show up in the water use tables. The public is, in effect, subsidizing industrial water use—and water pollution—by failing to penalize corporate polluters for the damage they cause, and business firms have understandably considered this to be an ideal situation. They don't especially like it when their own free water comes down the stream so polluted that it has to be cleaned up even for industrial purposes. But once the firms have used the water, they have no incentive to bear the costs of cleaning it up again; it is cheaper to let the public pay. The costs are exorbitant, though people have borne them passively so far—higher taxes and sewer charges for citizens whose cities pick up industry's share of the waste treatment tab; higher water bills for restoring polluted water to drinkable quality; nauseating odor; direct injury to property; and incalculable losses in health, recreation, and natural beauty. Not until industries' *own* incomes begin to sink in inverse proportion to the contaminants they put out will they find it in their interest to stop polluting.

Municipal polluters face most of the same economic incentives, of course; from the standpoint of the polluter —industry and city alike—money spent on water pollution control is money down the drain. There the similarity ends, however. Industry's excretions are more toxic, more numerous in kind and quantity, and more difficult to treat than domestic sewage.

In 1968 America's industrial sector "manufactured," before treatment, enough organic waste (i.e., of animal or vegetable origin) to have consumed more than 29 billion pounds of oxygen in decomposition (this "biochemical oxygen demand," or BOD, is the standard unit for measuring the polluting capacity of organic wastes) —over three times the total organic waste output from all domestic sources—plus additional unmeasured billions of pounds of non-oxygen-consuming pollutants. Major industrial malefactors include:

Industry sends some of these wastes into the rivers via municipal treatment systems, and the result is that industrial pollution is now the biggest part of the municipal pollution problem. Except in those heavy-polluting industries that consume the greatest amounts of water— primary metals (mainly steel), pulp and paper, chemicals, petroleum refining, and electric power—companies frequently tie into already overloaded municipal treatment systems. FWQA figures indicate that between half and three-quarters of all the oxygen-demanding wastes in the nation's municipal systems now come from industry.[14] These wastes include oils, metals, pesticides, poisons, dyes, chemical catalysts, and other esoteric toxicants that conventional municipal wastewater treatment cannot handle; municipal treatment systems have been known to break down altogether under the strain of industrial wastes.

Here too industrial polluters get a subsidy, and are encouraged to inundate the sewers (and ultimately the rivers) with an ever increasing load of waste. If data from a comprehensive 1970 survey hold true for the entire U.S., industries are assessed nothing for sewage treatment in 14% of the nation's communities; in all but

Industry	Wastewater * 1963[1]	Standard Biochemical Oxygen Demand ** 1968	Settleable and Suspended Solids ** 1963[1]
Wastewater treated (if at all) on premises:			
Chemical & allied products	3,700	14,200	1,900
Paper & allied products	1,900	7,800	3,000
Primary metals	4,300	550	4,700
Petroleum & coal products	1,300	550	460
Wastewater treated mainly in municipal treatment plants:			
Food processing	690	4,600	6,600
Textile mill products	140	1,100	(2)
Machinery	150	180	50
Transportation equipment	240	160	(2)
Rubber & plastics	160	60	50
All other manufacturing	450	470	930
All manufacturing	13,100	29,670	18,000
For Comparison: Sewered Population of U.S.	5,300	8,500	8,800

* In billions of gallons
** In millions of pounds
(1) 1968 figures not available.
(2) Not available.
Source: *The Cost of Clean Water 1968*, Federal Water Pollution Control Administration. Summary Report, p. 17; and EPA Water Quality Office.

13% of the remaining communities, factories are assessed no extra charge (above a nominal flat fee or rate based on volume of water used) no matter how great the volume of wastes or how high the cost of treating them.[15]

Burdensome as it is, the glut of liquid waste that industries pour into municipal systems is only a fraction of their total waste output—less than a tenth of their wastewater and less than a quarter of their BOD.[16] The rest of industry's discharge water gets dumped, still crammed full of factory filth, back into the lakes and streams, where it takes an increasing toll with every jump in GNP. Industrial pollution first surpassed domestic pollution sometime between 1900 and 1920. By 1960 the amount of industrial waste plunging into our nation's overburdened watercourses had risen to over double the contribution from residential sources.[17] During the 1960's industrial production increased three times as fast as the population,[18] and by 1968, when the latest official look was taken, industry was responsible for somewhere between four and five times more water pollution than domestic sources.

BOD PRODUCTION AND DISCHARGE, 1968
MUNICIPAL AND INDUSTRIAL SOURCES

	BOD Produced		*BOD Discharged to Waterways*	
	In Million Pounds/ Day	*As % of Industrial- Municipal Total*	*In Million Pounds/ Day*	*As % of Industrial- Municipal Total*
By Human Population ("municipal")	23.2	19%	8.2	17.6%
By Industry (total)	98.9	81%	38.3	82.4%
Discharged through municipal treatment plants	18.9		7.6	
Treated by industry or discharged without treatment	80.0		30.7	

Source: Cost Effectiveness and Clean Water, Annual Report to Congress by EPA, March 1971, p. 12.

To leave it at that would be grossly misleading, however, since industrial and domestic wastes actually share few common pollution properties. Quantitative comparisons between industrial and domestic discharges thus inevitably understate the pollution damage being caused by industry's lethal leftovers. Cadmium, for example, does not decompose in water as domestic sewage does; it goes into solution, where it remains indefinitely, and simply becomes progressively more dilute as it flows on. Statistics cannot show the fact that five pounds of this deadly poison is more dangerous than 500 pounds of domestic solid material. Nor do aggregate weight statistics reveal that industrial pollutants may be corrosive, carcinogenic, teratogenic, or mutagenic. Thus, as impressive as the aggregate figures may be, it is difficult to appreciate the hazards posed by industrial water pollution without taking a closer look at some of the polluting industries and their special brands of environmental violence.

The industries described below are only a sampling of the many distinguishable industrial groups that pollute. Some of them pollute a lot and others a great deal less. But each in its own way is hastening the destruction of the water we depend on to sustain life.

Steel

The steel industry is the king of the manufacturing water consumers, using upward of four trillion gallons of the precious fluid a year. Steelmakers need water to cool and condense hot metal (i.e., to carry away "waste" heat) and to "scrub" gases before using them for fuel; they also blast finished sheets of steel with a high-velocity water spray to knock off the waste scale clinging to the steel surface. The average steel plant produces one million ingot tons of steel each year. For each ingot ton a plant puts out, it also generates 125 pounds of suspended solids, 2.7 pounds of lubrication oils, 3.5 pounds of free acids (like sulfuric acid), 12.3 pounds of combined acids (like metal sulfates), eight ounces of emulsions, and between one and two ounces of such poisons as phenol, fluoride, ammonia, and cyanide. Each ingot ton produced raises the temperature of water used for cooling by 10°. Most operating steel mills donate at least half the total amount of these lethal by-products to the public.[19]

Synthetic Textile Finishing

The Baltimore plant where London Fog raincoats are made discharges the chemicals it uses for sizing and waterproofing into a small stream called Jones Falls. Downstream from the plant no protozoan or algal life survives. The Londontown Manufacturing Company and its cohorts in the rayon, acetate, nylon, acrylics, and polyester finishing industry are not nearly so thirsty as their cousins in primary metals. Altogether they use only 17.2 billion gallons of water a year. They make up for their relative temperance, however, by mixing in a respectable assortment of harmful wastes before sending their water back to the streams. There are the usual organics (177 million pounds of BOD each year), suspended solids (178 million pounds), and dissolved solids (292 million pounds). The salt bath applied to rayon after dyeing can produce an effluent of 12,000 parts per million salt—a third as salty as seawater. But the finishing industry's real showcase pollutants are in the dyes themselves. Beta naphthylamine is one of a number of known or suspected carcinogens in dyes; monochlorobenzene, one of the dyeing chemicals used on polyester, produces toxic fumes in the carrying water. Most of these textile artists give their wastes little more treatment than Londontown does.[20]

Poultry Processing

After a batch of chickens had been slaughtered, bled, plucked, eviscerated, and plunged in vats of water to cool, employees of the small family-owned processing plants on Maryland's eastern shore used to hose down the plant and dump all the wastes into small rivers like the Nanticoke near Salisbury, Maryland. The blood, guts, and chicken offal and the rotting carcasses of chickens that had died before their time would gradually decompose, giving off the familiar stench of chicken gone bad, and depriving the water of precious oxygen. But disembodied chicken feet, feathers, and heads, which do not degrade completely, would float downriver or wash ashore. When the State Health Department finally got after the chicken processors in 1967, many

of the Maryland plants installed the most widely favored
pollution control device in the industry: screens. In some
cases these are just lengths of chicken wire strung across
a ditch leading from the plant to the river. In heavy
rains the swollen water carries a torrent of poultry
wastes over the makeshift screens. When the wastewater
ebbs, feathers, offal, and trapped chicken pieces are left
to molder in the ditch. Carcasses, feathers, and blood
are still being found a mile away from the poultry plants
on Maryland's eastern shore. While the chicken pluckers
don't compare to the big polluters on a national basis,
the feathery fallout from a single processing plant can
make a stream as unfit for human use as a gigantic steel
mill.

Organic Chemicals

"Pollution results from mechanical failure and human
errors," writes Ben Edwards of Union Carbide Chemical
Corporation. "If you spill something at home, it creates
a problem, but not quite as large a problem as a chemical spill creates for a plant." [21] In November 1970 a
pipline broke at Union Carbide's plant in South Charleston, West Virginia. Hundreds of thousands of gallons of
acrylonitrile, a highly toxic chemical, poured into the
Kanawha River. When the company found the break
nine hours later, they notified pollution control authorities, who immediately shut down the drinking water
system in the downstream community of Nitro, West
Virginia.

Union Carbide's public relations department prefers
to overlook the occasional "mishap" and focus on pollution control efforts like the "Crusade for Cleanliness"
currently in full swing at its plant in Institute, West Virginia. One of the "highlights of the Institute plant's environmental program" is its cooling water system, which
the company describes this way:

Most of the water taken from the Kanawha River is used
for cooling equipment such as heat exchangers. This amounts
to more than 300,000,000 gallons a day. That's enough water
for a man to take a bath a day for the next 33,000 years.
No existing industrial waste water treatment plant could
treat that much water every day but *there is very little con-*

centrated waste in this stream. It flows directly back to the Kanawha River after serving its cooling purpose.

But plant operators and engineers are very much aware of the chances for contaminating cooling water so it is monitored by an instrument called a total carbon analyzer. . . . No organic chemicals can reach the river without being detected on the total carbon analyzer. [Emphasis added.] [22]

The waste may not be concentrated, but 33,000 years of bathing in Union Carbide's cooling water could leave quite a nasty ring around the tub. In an October 1970 deposition taken for the Federal Power Commission, F. Douglas Bess, Union Carbide's Assistant Manager for Environmental Pollution Control, Chemicals and Plastics, conceded that just a little stray dirt from the plant was slipping into the Kanawha after all, including: BOD (25,000 pounds/day), acids (33,000 pounds/day), alkalinity (44,000 pounds/day), materials in solution (570,000 pounds/day), suspended solids (30,000 pounds/day), oxidizable nitrogen (400 pounds/day), and chlorides (180,000 pounds/day).

Union Carbide, Dow Chemical, DuPont, Monsanto, American Cyanamid, Allied Chemical, Celanese, and their competitors in the synthetic organic chemicals business together use over 3.7 trillion gallons of water a year.[23] Organic chemicals represent the fastest-growing segment of the chemical industry. They make possible modern miracle products ranging from synthetic fibers to aerosol oven cleaners; such ecological hazards as pesticides, herbicides, detergents, fertilizers, food additives, lead additives for gas, shampoo containers, Plexiglas, Saran Wrap, and hundreds of other plastics; and, ironically, chlorine and other chemicals for water pollution control.

This highly innovative industry develops new ways to wage chemical warfare on water users so fast that pollution control authorities can barely keep track, let alone keep up. No one can detect or treat many of the complex new compounds or know what their effects are in water; the rivers serve as giant test tubes for the corporate chemists. Oils, acetaldehydes, and acetones befuddle control officials while hydrogen sulfide, phenols, cyanides, and various poisonous biocides frighten them. While pollution control regulations for chemical companies are relatively strict because of the obvious danger of poisoning drinking water, the rules are still too frequently honored in the breach. And accidents do happen. That's what worries the citizens of Nitro, West Virginia.

Pulp and Paper

When the Owens-Illinois pulp mill was built on Jumping Gully Creek in Valdosta, Georgia, in 1953, the company vice president, a former engineer named William Webster, designed its water treatment system himself. His massive lagoons, at first laughingly called "Webster's nature ponds" by local wiseacres, have consistently removed more than 90% of the BOD from the effluent the mill discharges into the Withlacoochee River. Owens-Illinois, makers of Lily paper cups, is one of the very few pulp and paper companies that have consistently installed the best pollution control systems available— and pioneered in developing new ones—usually in advance of state requirements.[24]

More typical of the pulp and paper industry is International Paper, makers of Flushabye disposable diapers. International Paper is the largest paper company in the world and the largest private landholder in the U.S.[25] In 88 years of discharging dirt, wood fiber, wood chips, and other solids from its Kraft mill at Ticonderoga, New York, International (with the help of some other pulp mills, now defunct) built up a 300-acre sludge mat on the bottom of Lake Champlain, 12 feet thick in places.[26] Slimy sludge mats like these not only smother the rich and varied aquatic bottom life, but occasionally get restless. For a week in June 1967 any Androscoggin River fish that happened by the International Paper plant in Jay, Maine, were instantly felled by what a state report called "masses of organic matter rising from the bottom" of the river.[27] In 1970, after repeated badgering by state and Federal pollution control officials and by the State of Vermont across Lake Champlain, International Paper opened a sparkling new mill at Ticonderoga. But it shrugged off responsibility for the 300 acres of bottom muck, and goes on polluting elsewhere. The International Paper mill at Mobile, Alabama, takes in half as much water as the Scott Paper mill nearby, yet discharges 10 times as much BOD to the Mobile River.[28] The company's pulp mills on the Hudson River (at Corinth, New York), the Niagara River (North Tonawanda, New York), and the Mississippi River at Natchez give their wastes no treatment whatsoever.[29]

In 1969, International Paper, Owens-Illinois, and their competitors in the pulp and paper industry produced 576 pounds of paper for every man, woman, and child in the United States [30]; with less than 6% of the world's population, Americans consume nearly 45% of the world's total output of paper.[31] Over 2.1 trillion gallons of water are used to reduce wood to fibers and rearrange them into sheets of paper. Hydraulic jets shoot the bark off logs; water carries the fibers from one stage of the process to the next, soaks, cooks, washes, and dilutes them, and cools the pumps and machines. Without water, there could be no pulp and paper; as a raw material it is second in importance only to the fiber itself.

Of the 2.1 trillion gallons of water used in this industry in 1967, 1.9 trillion were dumped back more or less laden with sludge-forming solids, reddish-brown sulfite liquor and other cooking and bleaching chemicals, oxygen-hungry wood sugars, murky brown and black residues and other discoloring matter, mercury (used as a fungicide or slimicide for wood chips and in the manufacture of some paper-bleaching chemicals), paper-coating chemicals, and thick yellow-brown foam, which rides several feet high on rivers where the pulp is made from resinous Southern woods.[32] Nearly two-thirds of the pulp mill effluent of the 24 largest U.S. paper companies gets less than secondary treatment.[33]

How far industry still has to go to clean up its pollution has been obscured by a flood of environmental advertising. "Today's bonanza" is what *Advertising Age* calls the public's new concern about the environment, and Madison Avenue has embraced the issue as the latest gimmick for attracting consumers' flagging attention. The ads boast of million-dollar corporate expenditures purportedly for pollution control. The figures look impressive—as long as the company abides by the iron rules of environmental advertising: (1) never, even upon request, break down the round numbers to reveal how many "pollution control" expenditures went for projects that turned a handsome profit and only incidentally happened to reduce pollution; (2) never give net cost figures after special tax write-offs; (3) never compare the amount spent with the amount that *could* or *should* have been spent. The fact is that the money invested so far is hardly detectable in the overall corporate budget. When industrial pollution control expenditures are matched up against industry's profit-making capital expenditures, the comparison reveals a contemptuous lack of concern for the environment and the public.

1969 WATER POLLUTION CONTROL EXPENDITURES AS A
PERCENTAGE OF TOTAL CAPITAL EXPENDITURES BY INDUSTRY*

Industry Group	Ratio in Percent
All Manufacturing	2.2
Primary Iron and Steel	8.0
Paper and Allied Products	6.5
Fabricated Metal Products	3.0
Instruments and Photo Equipment	2.1
Petroleum and Coal Products	1.8
Chemical and Allied Products	1.8
Stone, Clay and Glass Products	1.6
Rubber Products	1.2
Primary Nonferrous Metals	1.0
Textile Mill Products	.8
Food and Beverages	.8
Machinery, except Electrical	.6
Electrical Machinery and Equipment	.4
Aircraft	.2
Autos, Trucks and Parts	.1

* Figures are for a sample of 206 companies which responded to a 1970 Conference Board survey.
Source: Additional Statistical Material to Accompany Article on "Industry's Current Pollution Control Costs" by Leonard Lund in the *Conference Board Record,* April 1971.

The hollowness of industrial big-spending boasts becomes more evident when total water and air pollution control capital expenditures for each industry are compared with gross revenues; the most generous industry group spends a grand total of 0.69% of its revenues on controlling pollution.

Company executives spent little more on pollution control equipment in 1969 than they cheerfully forked over for business and personal aircraft to fly around in. Manufacturing corporations spent over 20 times as much

AS A PERCENT OF REVENUE, 1969

Industry Group	I 1969 Gross Revenue * (billions of dollars)	II 1969 Air & Water Pollution Control Capital Investments ** (billions of dollars)	Column II as % of Column I
All Manufacturing	694.6	1.281	0.19
Paper & allied products	20.6	.143	.69
Primary iron & steel	27.7	.179	.65
Petroleum refining	58.8	.260	.44
Stone, clay & glass products	17.4	.063	.36
Chemicals & allied products	55.5	.140	.25
Transportation equipment, excluding motor vehicles & aerospace	6.1	.015	.25
Primary nonferrous metals	21.4	.041	.19
Instruments & photographic equipment	16.4	.025	.15
Fabricated metals products	35.2	.044	.13
Machinery, except electrical	57.7	.051	.09
Autos, trucks & parts	59.9	.055	.09
Aerospace	26.4	.022	.08
Food & beverages	93.0	.058	.06
Textile mill products	21.8	.010	.05
Rubber	16.9	.009	.05
Electrical machinery & equipment	67.0	.032	.05

* *Quarterly Financial Report for Manufacturing Corporations*, First Quarter 1970, Federal Trade Commission—Securities and Exchange Commission.

** *McGraw-Hill Survey of Business Plans for Plant and Equipment* (Annual) 1970.

on advertising in 1969 as they did on water pollution cleanup. Given their free choice, industrialists will always find such "productive" purchases more attractive than filters, settling tanks, pumps, and pipes which give them neither pleasure nor profit and which the public never sees.

No one knows this better than the manufacturers of pollution control equipment. Buoyed several years ago by projections of a $2 billion annual market for their products by 1970, they have become sadder and wiser about industrial intentions; yearly sales by 1969 (to both industries and cities) had not yet passed the $800 million mark. And much of that was for equipment to purify *incoming* water to be used in manufacturing, not to clean the wastewater going back out.[34]

Pollution control manufacturers and consultants get to see industrial resistance to installing available control measures every day on the job. One equipment manufacturer gave the following example in response to a Task Force questionnaire:

[Our company] conducted tests and demonstrations at [steel company X]. This plant discharges eight to ten million gallons of water a day into a stream. This water is highly contaminated with iron and mill scale. Our demonstration was successful. After the demonstration was completed we were told that there was no money available for the project, so to this date nothing has been done and the polluted water continues to flow into the stream.*

He could only conclude:

We have found it very hard to place our equipment out in the field, mostly because industries refuse to spend any money for this cause. The attitude toward cleaning up the waste water is very negative, and they feel that they are paying for something in which there is no profit available to them. Therefore, unless they are forced into doing something about it, my opinion is that they are going to continue to stall either by pulling strings or denying that there is any purification available.

Ironically, the companies trying to corner the expected pollution control market are some of the nation's

---

* The names of both the pollution control company and its client were withheld at the manufacturer's request.

major industrial polluters and they know best of all—
from personal experience—what it will take before that
market ever becomes a reality. As Leo J. Weaver of
Monsanto Chemical Corporation's pollution control spin-
off, Enviro-Chem Systems, Inc., frankly conceded, "We
are living in a fool's paradise if we think industry will
do anything until it is forced to." [35]

3

Plus ça Change, Plus c'est la Même Chose: The Federal Water Pollution Control Program

Primary Federal responsibility for combating environmental violence resides with the Environmental Protection Agency (EPA). Formed in a December 1970 executive reorganization, EPA brings together under one umbrella many of the environmental programs previously spread throughout the executive branch (e.g., air pollution and solid waste disposal from the Department of HEW; pesticide regulation and research from the Departments of Agriculture, HEW, and Interior; radiation control from the Atomic Energy Commission; and water pollution from Interior). EPA is an independent agency, not part of an Executive Department; it reports directly to the President. In charge of EPA is Administrator William Ruckelshaus, formerly an Assistant Attorney General in the Justice Department. He serves at the pleasure of the President.

EPA's authority for implementing Federal water pollution legislation is exercised on a day-to-day basis by one of its five major subdivisions, the Water Quality Office. (Two other EPA subdivisions, the Pesticides Office and the Radiation Office, share water pollution control responsibilities with the Water Quality Office in their special areas.) The Water Quality Office's 2300 employees (around half of them professionals) and $1.1 billion * annual budget make it the largest and wealthiest

---

* $1 billion of the $1.1 billion is for sewage treatment plant construction grants to municipalities.

sub-unit in EPA (which has around 5600 employees and $1.4 billion appropriations in fiscal 1971). The bulk of the water pollution unit's professional expertise is in sanitary, civil, and chemical engineering.

Federal laws give EPA and its Water Quality Office responsibilities in research, pollution control planning, technical and financial aid to state and local pollution control agencies, subsidies for sewage treatment, and enforcement. Executive orders and agreements also mandate the Water Quality Office to help other Federal agencies abate the pollution their activities cause and to give technical guidance to agencies outside EPA whose authority touches on water pollution control—the Army Corps of Engineers, for example. EPA's Water Quality Office has its headquarters in Arlington, Virginia, a Washington suburb; it has 10 regional offices (each with a staff ranging from 99 to 328) to carry out field work and coordinate with state and local control agencies; many of them have smaller stations under their control for pollution monitoring in specific areas and other technical tasks. There are also eight major water pollution research laboratories located throughout the country.

The Water Quality Office has had many other homes and gone by many another name since the first Federal water pollution laws were passed in 1948. It has been successively a small branch of the U.S. Public Health Service, the Public Health Service's Division of Water Supply and Pollution Control (DWSPC), and a separate agency—the Federal Water Pollution Control Administration (FWPCA)—first in the Department of HEW and then later in the Department of Interior. In April 1970 the Interior pollution control agency exchanged its old sobriquet for one more in tune with the times: the Federal Water Quality Administration. And in the latest December 1970 executive branch shakeup, FWQA's box on Interior's organization chart was dropped onto the organization chart of the new Environmental Protection Agency and renamed the Water Quality Office.*

---

* Also under the wing of the new Water Quality Office is the former Bureau of Water Hygiene from HEW, which recommends drinking water standards.

The only significant difference between EPA's Water
Quality Office (alias FWQA, FWPCA, and DWSPC)
and its most recent predecessors is that it has yet another
title and a new Administration official in overall charge.*
(In an attempt to cut through some of the confusion inherent in this forest of initials, we will normally call
the water program by one of its two most recent names
—the Water Quality Office or the Federal Water Quality Administration; the other names will be used only
when they seem essential to avoid possible misunder-
standing.) In making its move to EPA the pollution
control program picked up no additional authority. EPA
simply inherited the legislative authority of the organizations it adopted, along with their personnel and their
budgets.

In the case of the Water Quality Office, EPA also inherited a great deal of atrophy left over from the water
program's earlier days. The Water Quality Office had
just begun to recover from a near-terminal case of
bureaucratic politics when it made the move. The water
pollution program has never had the kind of sustained,
capable leadership that might have pulled it together,
straightened out communications, and gotten the various
branches—enforcement, operations, research and development, and administration—to work together as a
unit to clean up the nation's water. And the old ways of
doing things have survived every attempt at reorganization so far—people working at cross-purposes, floun-
dering in confusion, guarding their bureaucratic fief-
doms, wasting enormous amounts of time trying to find
out the simplest things or clear up contradictions, and
jockeying for position in what always seems to be the
impending new scheme of things.

The Water Quality Office's fragmented style got firmly
established during its Public Health Service days in the
early 1960's, when the program's immediate supervisor,
Division Chief Dr. Gordon McCallum, left each of the

---

* The Administrator of EPA has the option of organizing the Protection Agency along functional lines. For example, he could put all
enforcement personnel in air, water, solid waste, etc., together in a
single EPA Enforcement unit. Even in that case, the constituent parts
of EPA would retain a good deal of their old character. And for the
time being, EPA plans to keep the Water Quality Office entirely intact,
just as it was when it was a separate agency in Interior.

pollution unit's jealous senior career officials to function with virtual autonomy. The head of each branch—enforcement, research, municipal construction subsidies, etc.—was, in effect, accountable to no one but himself for cleanup results achieved or not achieved. McCallum, the titular head of the pollution unit, served mainly as a referee when disagreements arose.

In 1965, when the water program was about to be set up as an independent agency in HEW, Secretary Anthony Celebrezze and his Assistant Secretary for Administration, Rufus Miles, decided to find out what was wrong with the program. Experienced management analysts were recruited to do a study. The study group's conclusions—now five years old—could have been written today:

The [program] lacks effective central direction, coordination, and control; organizational and administrative discipline; the means (including staff), the procedures and the executive-level disposition to develop and exercise national-level supervision over the implementation of short- and long-range policies, plans, and programs for the conduct of its multitudinous and far-flung Federal water pollution control affairs.

If politicians had seen this bleak evaluation, they would undoubtedly have called for "bold and effective leadership." But politicians also say—privately—that you can't win them all. So the new agency got James Quigley as its first Commissioner.

Quigley stepped into the Commissionership in 1966 from a job as the HEW Assistant Secretary charged with oversight of the water pollution program, a position he had held since 1961. He had been given that earlier assignment at HEW because he was—quite literally—a loser. A Democratic Congressman from Pennsylvania, he lost his seat to his Republican opponent in 1960. He had, however, campaigned hard during the election for President Kennedy, and a place had to be found for this loyal worker. So he was made an Assistant Secretary in charge of the water program, several bureaucratic notches down in the Public Health Service. Even back then, these circumstances struck many as inauspicious for imparting a sense of mission for water cleanup. All this is not to say that Quigley knew nothing about water. He had once been a junior Naval officer.

The year Quigley moved on from Assistant HEW Secretary to become Commissioner of the newly formed FWQA would have been convulsive for the pollution control program no matter who had been in charge. When it moved in 1966 from the Public Health Service, the agency lost more than a quarter of its middle- and top-level management, PHS uniformed officers who chose not to give up their service benefits by converting to civilian status. Just five months later the water program was moved again, this time to Interior, and more than ever it needed a firm guiding hand to keep it on course. But Quigley was hardly ever there. He spent much of his time on Capitol Hill, often departing in the morning for a discussion with one of his old Congressional cronies and not returning until the next day—perhaps to remain for no more than a few hours before heading back to the Hill again. He made frequent speaking trips back to his home district in Harrisburg, Pennsylvania, in a sad effort to cultivate a potential political comeback that was never to be. The few hours he put in at the agency were spent almost entirely in his own office, often behind a closed door with a "Do Not Disturb" sign which reportedly signaled that he was taking a nap.

One of the few times Commissioner Quigley was known by headquarters staffers to address himself to pollution control has become a minor legend at the agency. When one of the technical people brought in a flow chart for the Commissioner to inspect, Quigley took it from the engineer, threw a little red pillow down on the rug, and lay down to look at it. As he turned the chart around, he asked, "Where does it begin?" The engineer lay down on the floor with him to point out respectfully that Quigley was holding the chart upside down.

Quigley's desk was always piled with unsigned papers, and he reportedly waited so long to answer questions that most issues were dead by the time he resolved them. During his two-year reign (1966–68), FWQA's outposts around the country—the regional offices and their field stations—either set up independent operations or languished for lack of word from the center. One Regional Director, whose box on the formal organization chart is joined to the Commissioner's by a straight unbroken line, claims that at one point he heard nothing from Quigley for eight months. The Commissioner's personal and bureaucratic feuds with two headquarters associates with whom he should have been working closely— Quigley's administrative superior, Frank DiLuzio, the Assistant Secretary of Interior for Water Pollution Control, and FWQA's Assistant Commissioner for Enforcement, Murray Stein—further fractured control of the beleaguered agency. Quigley and Stein, whose offices were only a few doors apart, reportedly did not communicate during the entire time that Quigley was Commissioner, except by formal note or letter.

When Quigley resigned in January 1968 to take a job as Vice President of U.S. Plywood-Champion Company, Secretary of Interior Udall went to Texas to get his next Commissioner, Joe Moore. Moore is the one Commissioner ever appointed to FWQA who was qualified for the job when he got it. Head of the Texas Water Quality Board, Moore was knowledgeable about pollution and a tough but pleasant administrator who had a way of producing both work and affection from his subordinates. "The most serious need when I started at the agency," Moore told the Task Force, "was simply to get people to relate to one another. The entire organization was demoralized, the staff was shell-shocked. We had to get these administrative problems straightened out."

But Moore decided to take care of another of FWQA's problems first: its frayed relations with state agencies and with its regional offices and labs. Moore was away from headquarters even more than Quigley, and the feuding and lack of direction in Arlington got worse. Moore had barely begun to make inroads on the agency's administrative problems when the Nixon Administration replaced him with the current Commissioner, David Dominick. Joe Moore had been the third head of the program to spin through FWQA's revolving door in just over three years; by the time Dominick took over in March 1969, the agency was dizzier than ever.

Until his appointment Dominick had been the legislative assistant to Republican Senator Clifford Hansen of Wyoming. The new Commissioner had virtually no administrative experience and was largely ignorant of water pollution problems. Dominick's official résumé, the stuff out of which press releases are made, indicates that while working for Senator Hansen he "handled all staff matters relating to the Department of the Interior and natural resources and water issues." Translated, this means that he prepared Hansen's position statement opposing FWQA's requirement that states adopt a "non-degradation" water quality policy, under which all waters must be protected from deterioration even though their quality may exceed currently applicable standards. (Dominick explained to the Task Force that he had just been doing his job as Hansen's assistant and that he personally favored the non-degradation requirement.)

Dominick had other qualifications, however, that overshadowed these. He is the second cousin of Senator Peter Dominick (R.-Col.). His former boss, Senator Hansen, is a member of the Senate Interior Committee and an old pal of Richard Nixon. One who had already toiled in the vineyards of Republican politics, Dominick earned additional credits when Senator Hansen placed him on assignment to the Nixon campaign. The Commissioner is also a friend of Jim Watt, formerly the secretary of the natural resources committee of the U.S. Chamber of Commerce, now Deputy Assistant Secretary of Interior for Water and Power Development.

Dominick's first year or so as Commissioner again left FWQA without effective direction, and beset by friction between himself and his administrative superior in the Interior Department, Assistant Secretary for Water Quality and Research Carl Klein. As Dominick admitted freely in a June 1970 meeting with his headquarters staff, he spent March 1969 to March 1970 "in what might be called an extended 'training period' ": six months "traveling in the field" and six months more "educating myself . . . here at Headquarters." For his first 3½ months as Commissioner, the FWQA staff knew Dominick only as the author of three official notices: one forbidding the placing of candy wrappers in ash trays, one limiting the lunch break to 45 minutes, and one encouraging employees to purchase savings bonds, noting that they weren't "a bad investment."

Once Dominick issued directives that were somewhat more controversial, he came into conflict with Assistant Secretary Klein.

Klein had arrived at Interior a few weeks before Dominick through an entirely different set of political friends, and had played no part in choosing his subordinate, Dominick. A 52-year-old lawyer, Klein had served three terms as a state Representative from Illinois' 27th Congressional District in the heart of Chicago. As the aggressive chairman of a legislative commission on water pollution, Klein had earned a reputation as a clean-water advocate in Illinois. A loyal party worker, he got Republican Senator Everett Dirksen to sponsor him for Assistant Secretary. Klein is a large, gruff, undiplomatic man who made it as a Republican in Mayor Daley's Chicago; he towered over Dominick, who is restrained and discreet and blends well with the Nixon Administration. Thirty-two years old at the time of his appointment in 1969, Dominick is young enough to be Klein's son. But he proved to be more adroit than Klein in the kid-glove maneuverings of bureaucratic politics.

In the year and a half Klein was Assistant Secretary he managed, in his heavy-handed way, to make some needed changes at FWQA—notably, transferring responsibility for water quality standards to FWQA's enforcement office (see Chapter 14), and weeding out several entrenched officials who were holding up reforms. But the bull-like way Klein took charge of the agency undercut FWQA Commissioner Dominick's authority at every turn. When Klein wanted something done, he phoned agency staffers directly, bypassing Dominick. Klein did so, he explained to the Task Force, not to show who was boss but to get information more quickly. Dominick was a "very smart young man," he added. Whatever Klein's intentions, he put FWQA staffers in an impossible position.

Dominick, meanwhile, waged a persistent if diplomatic battle to salvage a measure of control. If Klein could give orders without informing Dominick, Dominick could issue a tactful memo "asking" FWQA personnel who received a "request for staff work" from the Assistant Secretary to please keep Dominick's deputy, John Barnhill, apprised of the nature of the task. One

FWQA official who received telephone instructions from Klein several times during a single Task Force interview complained that he didn't know what to do when he got orders from Klein. "Am I supposed to tell Dominick or not?" he asked with exasperation, concerned that either course would make the Commissioner suspicious of him.

The paralyzing effect that their divided allegiance had on FWQA staffers' work was aggravated by Klein's outright reversal of some of Dominick's orders. When the Task Force started its investigation of FWQA in July 1969, Dominick promised at the initial briefing he and his top officials gave that the entire agency would cooperate fully with us. A few days later, when Task Force members started showing up for their first interviews, they were greeted with the embarrassed explanation that all their scheduled interviews had been canceled in accordance with word "from above." Dominick had reportedly been told that his briefing had been a mistake. Agency personnel later told the Task Force that their first order to "give them as much assistance as you can" had been superseded by a "don't tell them anything," which was then amended in a series of complicated and often conflicting directives that gradually extended to FWQA personnel the First Amendment freedoms but left them thoroughly confused.

Shortly thereafter Klein fired John Barnhill, a longtime civil servant who was Dominick's second-in-command at the agency, without even bothering to consult the Commissioner. Dominick was so disturbed that, according to close associates, he considered resigning.

Barnhill's replacement, Bryan LaPlante, was a former deputy staff director of the Republican Senate Policy Committee. He had been put forward for the water pollution job by Republican Senator Gordon Allott of Colorado. When he arrived at FWQA he brought with him an officeful of party memorabilia, including photographs of Senator Dirksen and other Republican leaders in various poses and clusterings. Although he also brought badly needed administrative skill to FWQA, LaPlante had no background in environmental matters; like virtually every other high-level FWQA staff member hired under the Nixon Administration, his political ties were his most prominent credentials. One Senate staffer with a particularly apocalyptic view of the situation told the Task Force that FWQA was being turned into a "patronage agency."

Political hiring is nothing new in the pollution control program. But the Nixon Administration has gone about the job with an unprecedented vengeance. The knowledge that loyalty to the party in power or to powerful individuals within that party is the main criterion for determining who shall fill, and who shall be allowed to keep, the positions of leadership has increasingly demoralized FWQA staffers. The first hint that FWQA might come to be a rest stop for job-hunting Republicans arose, of course, with the appointments of Dominick and Klein themselves. The appointment of Robert L. McCormick as Assistant Secretary Klein's principal deputy was announced shortly thereafter. McCormick's background includes staff experience with the two Hoover Commissions on Executive Reorganization. He had done no previous work in the water area. But he had been research director for the Republican National Committee. McCormick was followed by Bryan La-Plante.

The latest addition to FWQA's political entourage is 26-year-old Peggy Harlow, who in July 1970 was put in charge of the Office of Correspondence Control and Congressional Liaison (now the Office of Congressional Relations), a 10-member group which lobbies for the water program before Congress. Peggy Harlow spent five years working her way through school on Senator Dirksen's staff, went on to become press secretary to Colorado's Republican Senator Dominick, and before coming to FWQA spent a year and a half as staff assistant in the Capitol office of Vice President Agnew. Her father, Bryce Harlow, was a Cabinet-rank Counselor to President Nixon from 1968 to 1970. He has now returned to his former job as director of governmental relations and chief lobbyist at the Washington office of Procter & Gamble, one of the nation's leading phosphate polluters. One FWQA wit observed that Peggy Harlow's appointment "smells like soap."

One Senate staffer told the Task Force that what bothered him most was that "they are even sprinkling party hacks out in the regions." A controversial plan for rotating FWQA's Regional Directors that went into effect in January 1970 was, according to one FWQA staffer, conceived partly to open up Regional Director slots for new Republican blood. The plan was to rotate directors from one region to another every three years; at least some of the directors were expected to resign rather than relocate. To avoid accusations that the Administration was trying to create more political jobs, the first new Regional Director appointed was to be a "career type" from the agency, with a background in the pollution field. After that, however, Regional Directorships were to be reserved for Republican faithful.

When the Chicago directorship fell open after the first reshuffling, the Republicans apparently felt less caution was needed. Valdas Adamkus, a 43-year-old free-lance engineer whose experience with water pollution consisted of losing a race for trustee of the Chicago Sanitary District in 1968, visited FWQA headquarters in January 1970 to discuss the possibility of taking over the Chicago job. Adamkus is past vice president of the Republican 13th Ward organization on Chicago's southwest side, Assistant Secretary Klein's old political base. He is a former president of the Lithuanian Republican League of Illinois, and in 1960 was executive secretary of the National Lithuanian-American Committee to Elect Richard Nixon President.

Besieged by the press after Adamkus' visit, Commissioner Dominick's assistant Bryan LaPlante confirmed that the Chicagoan was "being trained now for one of the regional director's jobs." [1] The *Chicago Tribune* lashed out at the Administration for using the directorship as a political plum, however and the post was given to Francis Mayo, a capable career official from headquarters. Adamkus' continued availability alarmed the *Atlanta Constitution,* which expressed its concern in an editorial that the Republican would be given FWQA's soon-to-be-vacated Atlanta Regional Directorship as a consolation prize. That speculation ended when Regional Director John Tholman, who had not been enthusiastic about his impending transfer, was permitted to stay on in Atlanta. Adamkus is still being "trained" at the Cincinnati Regional Office, where he has been since March

1970 a GS-15 engineer. (His Civil Service application was filed, according to LaPlante, by Klein himself.) The future of the patronage plan remains to be seen in 1971, when more Regional Directors are expected to announce their retirements.

If the last two years have done nothing to allay the disquiet FWQA staffers feel over the politicization of the pollution program, at least the debilitating conflict between Klein and Commissioner Dominick has ended. In September 1970, Klein abruptly announced his resignation after Administration officials let it be known that he would not be going along when the program was transferred to EPA in December 1970. Dominick has moved to EPA as Commissioner of the Water Quality Office and, proved to be a "fast learner," as some Congressional staffers put it who opposed his appointment because of his inexperience. In speaking with Dominick, the Task Force was impressed with his newly acquired but extensive grasp of water pollution, and with many of his ideas for setting the pollution control program on its feet. We came away convinced that he clearly perceives that energies which thus far have been wasted in internal disputes and uncoordinated efforts need to be redirected into a focused attack on the nation's water pollution problems. But after two years in office—longer than any past Commissioner—he has not yet produced the forceful leadership that will be needed to turn his perception into practical cleanup performance. In the meantime, the water program's chronic confusion has been compounded by its latest move. "Musical chairs, where do we go now, new suspicions, new personalities," as one headquarters staffer moaned several weeks after the shift to EPA.

Beneath Commissioner Dominick in the water pollution hierarchy are hundreds of dedicated and competent workers, genuinely concerned for the environment and frustrated by their agency's past inertia. Buoyed by the interest the public is finally taking in water pollution, but at the same time more knowledgeable than the public about the obdurate resistance that cleanup measures will continue to face, Water Quality Office staffers are waiting for their marching orders. But Commissioner Dominick still seems to be biding his time, cautiously preparing to make his move. His subordinates complained to the Task Force that the Commissioner is receptive to their ideas, but gives no indication of what will become of them; he listens, but nothing happens. As the chapters that follow illustrate, the time for making something happen is long overdue.

4

Pollution for Sale: Detergents

Roughly 6 billion pounds of soap and synthetic detergents were sold in the U.S. in 1969, grossing $1.7 billion for their manufacturers [1]—the big three (Procter & Gamble [Tide], Colgate-Palmolive [Axion], and Lever Brothers [All], with more than two-thirds of the market between them) and hundreds of smaller concerns. Nearly all the laundry and dishwasher liquids, flakes, and powders end up as municipal sewage.

The sheer volume of these products going down the drain means that the chemicals in them can have an enormous impact on water quality. Pure soap is considered safe; made of natural fats and oils, it is readily biodegradable (i.e., capable of being broken down biologically into harmless end products such as carbon dioxide and water). But the synthetic detergents that came on the U.S. market in the 1930's, and which by 1969 accounted for 84% of the industry's sales, are a different story. Each of their major ingredients—the surface active agent ("surfactant") that does the actual cleaning, and the phosphate and other "builders" that enhance the surfactant's cleaning power—has at one time or another caused obvious pollution problems. Still largely unknown are the effects of smaller quantities of other detergent additives—all the whiteners, brighteners, anti-corrosion chemicals, enzymes, and other ingredients added mainly for competitive purposes. Arsenic, a natural contaminant of phosphorus, has been found in phosphate detergents in concentrations as high as 10–70 parts per million (see Appendix F). Dr. Ernest Angino * early in 1970 measured arsenic concentrations

* Associate Director, State Geological Survey of Kansas, and Department of Civil Engineering, University of Kansas.

of 8 parts per billion in the Kansas River at Topeka—close to the recommended Public Health Service drinking water limit of 10 parts per billion.[2] Detergents, Dr. Angino concluded, were probably the major arsenic contributor.[3] At least one whitening agent used in detergents, amino-stilbenes, belongs to a chemical group that includes carcinogens. All these varied chemical additives in detergents, with their potentially grave impact on human health and the environment, are prime candidates for public control. But the soap and detergent industry has successfully resisted regulation for over a decade.

Rivers foamed and drinking water came out of the tap with a head on it in many places from the late 1940's to the mid-1960's because the sudsing surfactant used in synthetic detergents at the time, the petrochemical alkyl benzene sulfonate (ABS), was not biodegradable. The foam was not only unsightly—rising to heights of 20 feet on the Rock River in Illinois, for example [4]—it also clogged sewage treatment plants.

In the face of this evidence that its washday miracle was getting out of hand, the detergent industry was as glib as its television soap-sellers. In an unwitting rehearsal for the phosphate controversy to come, manufacturers insisted that the housewife demanded rich, billowing suds in her laundry as an index of cleaning performance. Trade group spokesmen protested at Congressional hearings that the foam was actually an environmental service, since it was a harmless but highly visible signal that water might be polluted by other substances.[5] Industry representatives tried to show that detergents were not wholly responsible for the foam on rivers, and pointed out that foam was only a small part of the pollution problem.[6] Standard Oil of California and other petrochemical producers with large investments in the ABS surfactant even came up with ingenious (if impractical) methods of sewage treatment that they said would remove the foaming ingredient from municipal wastewater.[7] Still, ABS and foaming rivers were doing little to enhance the sales image of detergents. So as early as 1951, according to Charles Bueltman of the Soap & Detergent Association, trade group for the industry, the detergent companies started looking for "an answer to the 'detergent foam problem.' " [8]

The search for a biodegradable substitute for ABS was still proceeding at the industry's measured pace over a decade later when state and Federal legislators, under mounting pressure, began writing bills to control detergents. In January 1963 Congressman Henry Reuss (D.-Wisc.) introduced a bill in the U.S. House of Representatives that would have required detergents to meet standards of decomposability set by the Secretary of HEW by June 30, 1965. By the time a companion bill was introduced in the upper chamber in March 1963, *Chemical and Engineering News* had given the legislation a 50–50 chance of passing.[9] Under the threat of Federal regulation, the Soap & Detergent Association hastily announced in April 1963 that the industry would "voluntarily" switch to biodegradable detergents by the end of 1965—provided their efforts were not "obstructed" by legislation.[10]

"All the big guns came in," one Senate staffer recalls, "and said nobody's ever jumped on us before, let us do our own self-policing." Still, a modified version of the detergent bill was included in the water quality bill the Senate passed in October 1963. But the lobbyists beat it back in the House. Procter & Gamble's Bryce Harlow was particularly persuasive, recalls Maurice Tobin, then staff assistant to the Rivers and Harbors Subcommittee of the House Public Works Committee. Harlow, who later served as an aide to President Nixon (1968–70) but has since returned to Procter & Gamble, was already a well-known figure on Capitol Hill when he went up to discuss the 1963 bill, having worked in Congress after World War II and served on President Eisenhower's White House staff. When the House reported the Senate water quality bill in September 1964, the detergent provision had been stricken out, thereby "avoiding another Federal enforcement empire," in Tobin's view.

The detergent-makers accelerated their ABS replacement program and came in with a biodegradable substitute, linear alkyl sulfonate (LAS), "ahead of schedule" in June 1965. Foaming tap water remained a problem in rural and suburban areas like Suffolk County, New

York, where cesspools and septic tanks could not handle the LAS much better than the ABS it replaced, and sudsy groundwater seeped into drinking water wells. But the foam receded on lakes and streams where full-scale treatment plants were located. The detergent industry was warmly applauded by water pollution officials for its "voluntary" efforts.

Unfortunately, detergents still contained ingredients with far more serious implications for water quality than the foam-producing surfactants: phosphate "builders." Builders enhance the cleaning power of detergents. The phosphate compounds added to detergents after World War II are excellent builders: they soften hard water by "tying up" minerals such as calcium, iron, and magnesium that would otherwise interfere with the cleaning action of the surfactant; they help remove oil and dirt from fabrics and prevent their redeposition on the material, and increase cleaning efficiency by making the wash water alkaline. Most heavy-duty detergents on the market contain 35–50% phosphate by weight (usually in the form of sodium tripolyphosphate, STPP); pre-soaks like Colgate-Palmolive's Axion and Procter & Gamble's Biz have even more (63% and 74% phosphate, respectively *). Altogether, over 2 billion pounds of phosphate compounds are added to laundry products each year.[11] Since sewage treatment plants have been doing a notoriously poor job of removing phosphates from sewage (Washington, D.C.'s Blue Plains secondary treatment plant, for example, removes about 20% of the phosphate) and a third of the U.S. population has no sewage treatment at all,[12] a large portion of detergent phosphates end up in groundwater, lakes, and streams. From 50 to 70% of the phosphorus † in municipal sewage is contributed by detergents. (The rest comes mostly from human excrement.) [13] Detergents are responsible for from 30–70% of the total phosphorus load on lakes and streams; an estimated half of the phosphorus deposited in the Potomac River and 35–50% of the phos-

* Percentages based on test results released by FWQA on September 6, 1970, are listed as Appendix G. Percentages may have changed somewhat since then.
† Phosphorus is an element; phosphates are chemical compounds containing phosphorus.

phorus added to Lake Erie each year come from detergents.[14]

Phosphorus stimulates the growth of algae and aquatic weeds in fresh-water lakes and rivers. Sudden and massive algal growths (called blooms) appear in many American waterways in early spring and summer. They are ugly growths, but the problem is not solely esthetic. These blooms age a body of water as they die and decay. Oxygen and other resources are exhausted in the oxidation of large amounts of dead organic matter. Slime and scum appear, flows are clogged, and the water is unable to support fish or other normal life forms. This natural aging process, by which lakes and rivers turn to swamps and then dry land over the course of centuries, is called eutrophication. When phosphorus and other nutrients for algal growth pour into our lakes and rivers from municipal and industrial wastewater and from urban and agricultural runoff, the natural aging process is speeded up, often many hundreds or many thousands of times. Lakes and rivers "die" an early death from over-enrichment, overblooms, and what might be called overkill in the algal life cycle.

Massive growths of algae can occur only if the plants get all the 15 to 20 nutrients they need in certain critical amounts. In the Great Lakes, the Potomac basin, and many other American waterways, phosphorus is probably the critical nutrient limiting algal growth—that is, the element in shortest supply relative to the algae's nutritional requirements. Even where some other element—e.g., nitrogen or carbon—may be the limiting nutrient, algal growth can theoretically still be curbed by choking off the phosphorus supply below subsistence level. And phosphorus in the form of phosphates is the one nutrient whose input to the water can be most easily controlled.[15] Man usually contributes a much larger share of the phosphorus in the water than he does of other nutrients. And he can reduce that supply substantially with known sewage treatment methods or by removing phosphates at their source.[16]

Lake Erie is the prime example of a eutrophied lake whose resources and capabilities have been so overworked by the continuous growth and death of algal blooms that its oxygen has been exhausted. Rotting

masses of dead plant life float on the surface and sink
to the bottom, where they putrefy and poison bottom-
dwelling life. Erie has seen some 150 centuries of its
natural life span disappear in less than 50 years under
the strain of industrial, agricultural, and municipal pollution.[17] In July 1965 a technical report prepared for a
Federal enforcement conference on Lake Erie noted
that the over-enrichment of the lake "caused by man-
made contributions of nutrient materials [phosphates
and nitrates] is proceeding at an alarming rate." Although the situation was critical, "eutrophication or
over-fertilization of Lake Erie . . . can be retarded
and perhaps even reversed," the technical committee
concluded, but only "by reducing one or more nutrients
below the level required for extensive growth. Soluble
phosphate is the one nutrient most amenable to reduction or exclusion from Lake Erie and its tributaries." [18]

The technical committee's 1965 recommendation for
slowing the death of Lake Erie was a conventional, if
ambitious, one: secondary treatment for all municipal
sewage running into the lake, with maximum phosphate
removal. By 1967, however, virtually no progress had
been made—not an unusual occurrence where recommendations for sewage treatment are concerned—and
both the phosphates and the algae were, as usual, increasing. At this point the technical committee took a
harder look at the source of over half the phosphates
being dumped in the lake and recommended that the
detergent industry and the Federal government seek to
replace the phosphates in detergents.[19]

A few months later, in July 1967, Secretary of Interior Stewart Udall and his Assistant Secretary for
Water Pollution Control, Frank DiLuzio, met with industry representatives to urge them to step up research
on phosphate substitutes.[20] The industry obligingly promised to do so, and that is approximately where matters
stand today. After all, finding a satisfactory replacement
for the non-biodegradable surfactant ABS took 14 years,
as Soap & Detergent Association vice president Charles
Bueltman reminded Congressman Reuss's Conservation
and Natural Resources Subcommittee in December 1969
—and "to find a non-phosphate substitute for today's

detergent phosphate," Bueltman warned, "is a much more difficult task." [21] Congressman Reuss had his own view of the matter:

If all the detergent industry shows with respect to phosphates . . . is the same sense . . . of urgency and responsibility that it showed in the case of foaming detergents, every lake and stream in America will be dead before they ever get around to using up their last supply of phosphates and getting on with the job.[22]

Despite this experience with the pace at which the detergent industry moves to safeguard water quality in the absence of the threat of coercion, the Nixon Administration, EPA, and Congress have all somehow been convinced that no Federal action should be taken to force a reduction in the phosphate content of detergents. They have even had trouble informing the public fully and effectively that phosphates contribute to pollution.

The stage was set for industry control of the Federal government's phosphate detergent program in 1967, when the major accomplishment of Interior Secretary Udall's and Assistant Secretary DiLuzio's July meeting with detergent industry representatives turned out to be the formation—in August of that year—of a Joint Industry/Government Task Force on Eutrophication. DiLuzio, who had arranged the July meeting, did not intend the Joint Task Force to impede action to cut down on phosphate pollution. "Nutrient pollution is complex," he conceded in his July 1967 talk to the industry representatives, and "we may never develop the absolute answer." But, he continued,

we cannot postpone action until every fact is in. The problem is severe and will not wait for perfect knowledge. If we wait for perfect knowledge, our lakes may be dead, estuaries completely clogged, and fish and wildlife habitat destroyed.
The problem demands action today, using the knowledge we have. And, I want to emphasize, I think *we do have sufficient knowledge to begin*. We know that phosphates are a serious pollutant, excessively fertilizing aquatic plants. We know the sources of these phosphates, in large part, and that they must be controlled.

*   *   *

> We . . . have sufficient knowledge *now* of the . . .
> eutrophication problem to know that action must begin im-
> mediately to limit phosphates added to our waters. [Em-
> phasis in original.] [23]

The detergent industry had other ideas. The Joint
Task Force, composed of four members from the Interior Department and nine from the industry and
chaired by Charles Bueltman of the Soap & Detergent
Association, immediately narrowed its scope to two lines
of attack that the industry thoroughly endorsed: phosphate removal at sewage treatment, and perfecting a test
for evaluating the effect of any given substance on algal
growth (the "Provisional Algal Assay Procedure," or
PAAP test). All the PAAP test research program was
intended to do was standardize and choose among testing procedures already well established when the Joint
Task Force started pushing the project.[24] While this time-
consuming search dragged on, the soft soap the Federal
officials on the Joint Task Force were subjected to at
their periodic meetings with the detergent-makers
washed away whatever urgency they might have felt
when DiLuzio gave his stirring address. "We believe
that the reduction or elimination of phosphorus from
detergents is desirable in concept but inappropriate for
implementation *at this time,*" FWQA Assistant Commissioner for Research and Development Dr. David Stephan
assured Soap & Detergent Association president E. Scott
Pattison in November 1969 (two years after DiLuzio's
vigorous call to action). Stephan further assured the
industry that FWQA would not consider phosphate replacement as even "a viable candidate method for pollution control along with waste treatment" until the so-
called PAAP tests became available.[25]

As early as May 1968 the editor of *Industrial Water
Engineering* saw exactly what the detergent-makers had
accomplished. The Federal water pollution agency
"needs industry help," he wrote,

> and it is to industry's advantage to give it. Consider, for
> example, how the Soap and Detergent Industry helped turn
> an ill-conceived plan for phosphate substitution into a pro-
> gressive program for studying eutrophication. More of this
> type of cooperative action is needed.[26]

And more was in store. The National Industrial Pollution Control Council (NIPCC) and its Detergent Sub-Council, chaired by Procter & Gamble's president and chief executive, Howard J. Morgens, have been giving President Nixon and his Council on Environmental Quality the detergent industry's views since NIPCC was set up as a private-industry advisory council in the U.S. Commerce Department in January 1970. And if the Administration needed any further advice on the detergent phosphate problem from 1968 to 1970, it had Bryce Harlow close at hand. The Procter & Gamble lobbyist was Nixon's aide at the time. Procter & Gamble is not only the nation's largest detergent-maker, with some 55% of the heavy-detergent market,[27] but is also the company with the highest phosphate content in its products.

The main course of action the detergent industry advocates for phosphate control—aside from studying eutrophication—is phosphorus removal at sewage treatment plants. "The real solution to any problem involving phosphates and eutrophication," says Pattison, "is adequate municipal treatment which will remove nutrients from all sources going into lakes and streams." [28] In the detergent-makers' view, whatever chemicals they put into their products are the responsibility of consumers; if the chemicals are found to be harmful after the fact, then they should be removed from sewage at public expense.

While the folly of waiting to solve environmental problems until the harm has already been done would seem to be almost self-evident, it is further underscored by the difficulty in solving the phosphate problem with sewage treatment alone. Fully 34% of our domestic sewage still gets no treatment at all before being discharged into supplies of surface or groundwater; another 22% receives only primary treatment.[29] The pace of treatment plant construction has been agonizingly slow, and there are small communities and scattered rural dwellings where sewage may never be treated, since the cost of installing facilities would be prohibitive.[30] Only the removal of phosphates from detergents at the source—in the manufacturing process—can guarantee their control in isolated rural areas and stave off further deterioration in water quality until treatment plants get built. As Congressman Reuss pointed out to Assistant Secretary of Interior Carl Klein at his 1969 hearings,

MR. REUSS: [Is it not a fact that] by and large the phosphate which shows up at sewage disposal plants comes from two main sources—household detergents and human waste?

MR. KLEIN: Yes, sir.

MR. REUSS: And household detergents are made by three major manufacturers?

MR. KLEIN: That is correct.

MR. REUSS: And human wastes are made by a couple hundred million manufacturers; is that correct?

MR. KLEIN: Yes, sir.

MR. REUSS: Well, doesn't it occur to you that it is easier to do something about three than about a couple hundred million? [31]

Even if all the necessary sewage plants would magically appear, there is no guarantee that they would be the best solution to the problem—or even an adequate one. For one thing, the more of a given pollutant there is to treat, the higher the costs of waste treatment usually are; phosphates are no exception. A 1969 report to the International Joint Commission on the Pollution of Lake Erie, Lake Ontario and the International Section of the St. Lawrence River estimated the cost of removing phosphates at sewage treatment plants would be reduced by a half to two-thirds if phosphates were taken out of detergents, saving $22 million *annually* in the Lake Erie–Lake Ontario–St. Lawrence River basin alone.[32]

Even if consumers want to pay these extra treatment costs, they may find that their heavy investment is not paying off. For example, the removal of arsenic by sewage treatment is not only very expensive but uncertain as well.[33] Compared to arsenic, phosphates are relatively easy to treat. Yet the highest phosphate removal rate that known waste treatment methods have consistently achieved in practice (and that engineers are willing to guarantee when they write their contracts) is 80–90%. The 99% removal rates some of these methods are said to have achieved under special conditions are "fairyland" figures, says Dr. George Berg of the Rochester (New York) Committee for Scientific Information (RCSI).* RCSI calculates that "the best sewage treatment you can buy"—the tertiary treatment which Rochester is installing—"won't do the job" of bringing phosphorus concentrations in the city's treatment plant effluent below the level likely to trigger algal growth. Only by eliminating phosphorus from detergents as well could phosphorus levels near the city's effluent pipe in Lake Ontario be brought below the critical level.

Disregarding the evidence of water quality experts that the phosphates in their products can make the difference between life and death for a lake or stream, the detergent-makers have stubbornly resisted not only the complete elimination of phosphates from detergents ("until a suitable substitute is found," they now hasten to add) but several stopgap measures for reducing the amount of phosphate pollution in the interim. One proposal, for example, is to market different detergent formulations in hard- and soft-water areas. The main purpose of phosphate builders in detergents is to soften hard water.† According to one Federal water quality scientist working on detergents, enough phosphate is put into the average heavy-duty (i.e., laundry) detergent to get relatively dirty clothes clean in areas where the water is hard. Yet, according to a 1962 report by the U.S. Geological Survey, only 27% of the population of the 100 largest U.S. cities is served by hard or very hard water.[34] In places like Baltimore, New York City, Atlanta, Memphis, and Portland, Oregon, where the water is soft, no phosphate at all is needed; even unadulterated soap is an excellent cleaner. In moderately hard-water areas like Detroit, Pittsburgh, St. Louis, and Washington, D.C., detergents need only half their present

---

* One sophisticated waste treatment plant, the Hyperion plant in Los Angeles, is said to remove 94–99% of phosphates from wastes it treats. (R. B. Bargman *et al.,* "Phosphate Removal by Activated Sludge Aeration," *Journal of the Sanitary Engineering Division, Proceedings of the American Society of Civil Engineers,* 1970, vol. 96, SA-1, p. 45.)
† Dr. John Singer, president of the Hampshire Chemical Division of W. R. Grace and Co., testified at the Reuss Subcommittee hearings in December 1969 that "Research conducted both in this country and in Europe indicates that [water softening] is the only significant characteristic attributable to detergent builders such as the polyphosphates and NTA."

phosphate content to clean as well as they do now. There are precedents for marketing different formulations in different areas; according to the House Committee on Government Operations, Economics Laboratory, Inc. varies the phosphate content of its automatic dishwashing detergents, Electrasol and Finish, according to the hardness of the water, and the Duz that Procter & Gamble sells in soft-water areas is a soap, while in hard-water areas it is a synthetic detergent.[35] Even halving the phosphate content in detergents would immediately reduce the total man-made phosphorus load on the average waterway by one-quarter. But the detergent industry has shown no interest in taking this beneficial step. "There is no geographically defined soft water or hard water area," asserted Procter & Gamble's William Krumrei at hearings before Senator Edmund Muskie's (D.-Me.) Air and Water Pollution Subcommittee of the Senate Public Works Committee in May 1970.[36] "There are!" was Cornell University ecology professor Dr. Gene Likens' incredulous reply.[37]

Industry spokesmen like Charles Bueltman never fail to justify their intransigence by claiming that "the levels . . . used are those which manufacturers have found necessary to satisfy the American housewife's demand for acceptable cleaning performance." [38] Aside from the fact that whatever demand exists has been created mainly by the industry's extravagant advertising (Procter & Gamble spent more on network television commercials in 1969 than any other advertiser, and Colgate-Palmolive ranked third [39]), the detergent industry has battled every effort to give consumers information that might lead them to modify their demand for a sparkling wash for the sake of cleaner waters. First the detergent-makers balked at the idea of even labeling detergents with their phosphate content. As Congressman Reuss pointed out to the Soap & Detergent Association's Charles Bueltman at the December 1969 hearings,

MR. REUSS: . . . a housewife who does want to do something about stopping the destruction of our lakes by using a detergent which has a smaller rather than a larger phosphate content is helpless to do that patriotic act; is she not?

"It is our complete conviction," Bueltman replied, "that the average housewife seeing a higher percent content will automatically equate this to better cleaning" and buy the high-phosphate product.[40] *

It seems unlikely that detergents would ever have been labeled with their phosphorus content if the industry had not been pushed into doing so. Congressman Reuss himself published a list of phosphate levels in various brand-name detergents prepared for him by a consulting firm, and the list was carried widely in the press. The Congressman's tabulation in turn stung the Federal water pollution agency into action, and it published a "more accurate" list of its own in May 1970, along with a press release explaining briefly why phosphorus is a "pollution problem." [41] This was a bold step for FWQA, which had been noticeably silent on the phosphate pollution issue. A second list followed in September 1970, along with an excellent four-page statement on detergents, biodegradability, and eutrophication. "It is the FWQA's position," the statement read,

that there be a reduction of phosphate levels in detergent formulations immediately, and the complete replacement of phosphorus compounds in detergents with less environmentally harmful substitutes as soon as possible . . . a substantial part of the increasing nutrient load in our lakes and rivers can be traced directly to the phosphate-based detergents used in home and industry. Since household use accounts for most of the consumption of detergent phosphate, it is clear that the household detergent, which has revolutionized washday for the modern housewife, is contributing substantially to the problem of nutrient enrichment.[42]

These were fighting words. Neil McElroy, Procter & Gamble chairman of the board and former Secretary of Defense while Richard Nixon was Vice President, took

* The "average housewife," or whoever is buying detergents these days, is in fact expressing a strong preference for low-phosphate laundry products. When Alexander's, a 10-store chain in the San Fernando (Calif.) Valley, posted phosphate contents and tagged soaps and detergents low in phosphates with "ecology preferred" stickers, total sales of well-known high-phosphate detergents declined 16.8% from mid-January to mid-April 1971, while the combined sales of relatively unknown brands of low-phosphate detergents rose 446%. (*The New York Times,* April 12, 1971.)

a trip to the White House to complain about the indication of phosphate contents and the wording of the statement. That was FWQA's last release on the subject. A Task Force member's request to FWQA's public information office for a copy of the September 1970 release on March 19, 1971, was turned down with the explanation that there were errors in the list and it was no longer being distributed.

Meanwhile, state and local ordinances were being passed requiring the disclosure of phosphate contents. This, in the opinion of one Federal official, was the crucial element in the industry's belated decision to label detergents nationally. New York State's labeling law was to go into effect on January 1, 1971, Chicago's on February 1. In November 1970, Soap & Detergent Association president E. Scott Pattison announced that household laundry and dishwasher detergents, "including all of the leading brand-name products," would "soon" be labeled according to the percent of phosphorus in the formula and the number of grams per recommended use level. "This action has been taken," Pattison explained, "to correct the largely erroneous information about phosphate content of products which appears in the many lists distributed by various organizations and publications to consumers." [43] * By writing its own labels, it should be noted, the industry also got a chance to "correct" the form in which the information had been presented on most lists, the percentage of phosphate as the compound sodium tripolyphosphate (STPP). The percentage of elemental phosphorus in any given product is roughly four times lower than the percentage STPP; Tide labeled "12.6% phosphorus" sounds a lot more innocuous than Tide with 50% phosphate.

There were only five other noticeable defects in the Soap & Detergent Association's "voluntary" labeling plan. No deadline was given for putting it into effect. There was no assurance that all detergent-makers—particularly the smaller ones—would participate. There was to be no indication on the package that phosphates contribute to water pollution—a tremendously important

---

* "We are also providing this information as a service to housewives to whom this subject may be a matter of interest," Pattison added.

feature.* Nor, for obvious reasons, do the detergent industry's voluntary labels indicate that consumers can use less of their phosphate products where the water is soft or the laundry lightly soiled. And, finally, no ingredients other than phosphates had to be listed.

On January 25, 1971, the Federal Trade Commission, responding to a directive from Congressman Reuss's Conservation and Natural Resources Subcommittee and petitions by William H. Rodgers, Jr., associate law professor at the University of Washington in Seattle, and several environmental groups in Washington, D.C., proposed a regulation requiring detergent manufacturers to "list prominently" all ingredients on every container of detergent, as a percentage of total weight and in grams per recommended use level, in descending order of predominance.[44] But that wasn't all. The second part of the proposed regulation, were it adopted, could revolutionize consumer choice where products that contaminate the environment are concerned. As Rodgers pointed out in his December 1970 petition to the FTC,

the consumption of phosphate-based detergents is promoted massively through commercial advertising. . . . What this advertising fails to do is to disclose the material fact that normal consumption contributes to the destruction of marine resources upon which all life depends. The advertising also fails to acknowledge that the consumption urged is in some parts of the country [i.e., where high-phosphate detergents have been banned] contrary to law.[45]

By restoring some balance to the soap-makers' advertising, the proposed FTC rule would, for the first time, give consumers a genuine opportunity to decide for themselves whether the blessings of phosphate detergents outweigh the harm they cause to the environment. The FTC rule would require the "clear and conspicuous" inclusion of the following warning in all advertising

* A consumer survey conducted for Sears, Roebuck Company in 1970 found that while 65% of the homemakers interviewed considered water pollution a serious problem, and 58% expressed a "serious" or "somewhat serious" concern about the effect of laundry detergent on water pollution, only 6% were able to identify phosphate as a contributor to the problem. Twelve percent thought detergent enzymes were pollutants. And 71% just didn't know which ingredients were to blame. (Hearings before U.S. Senate Subcommittee on Air and Water Pollution, "Water Pollution—1970," Part 3, pp. 1197–99.)

(including television commercials), all promotional literature, and on the main display panel of every box of detergent containing phosphorus:

Warning: each recommended use level of this product contains ___ grams of phosphorus, which contributes to water pollution. Do not use in excess. In soft water areas, use of phosphates is not necessary.

The television warning, moreover, would have to be "broadcast simultaneously on the audio and video portions, without background distraction."

The detergent industry, needless to say, is firmly opposed. "The Association, on behalf of its member companies, takes strong exception to the proposed FTC . . . rule," said a Soap & Detergent Association statement timed for simultaneous release with the FTC's announcement on January 25, 1971.

The proposed rule, if put into effect, would raise unnecessary doubts on the part of housewives, and in fact would be misleading.

Any need for phosphate information for the consumer has already been met by the industry. . . .

. . . it is not true that phosphorus contributes to water pollution. Phosphorus . . . contributes to eutrophication, not water pollution. . . .[46]

The detergent-makers were so upset by the proposed rule that they went (unsuccessfully) to court to try to prevent public hearings from even being held.[47] The Nixon Administration asked the FTC in April 1971 to hold off on labeling while EPA continued testing phosphate substitutes.[48] Nor is the proposed regulation worrying detergentmakers alone. If it went into effect, it would give every advertiser good reason to think twice before putting contaminants in his products which could enter the environment in any significant amount. "There have been a lot of calls to Congressmen on this one," *The New York Times* quotes an FTC official as observing.[49]

Industry opposition to outright bans on phosphates has been even stronger. Congressman Reuss put the detergent-makers on notice when he introduced a bill in June 1969 which would have required that detergents be free of phosphorus by June 1971, and several similar bills have been thrown in the hopper since then. Canada, which shares the Great Lakes with the United States, has prohibited the manufacture of laundry detergents with more than 20% phosphates since August 1970.[50] But the real scare came when the first U.S. municipal ordinance limiting and eventually banning phosphates in detergents was unanimously approved by the Chicago City Council in October 1970. The law, which went into effect on February 1, 1971, allows no more than 8.7% phosphorus in laundry detergents up until June 30, 1972. As of then the sale of detergent containing any phosphorus whatsoever will be illegal in Chicago.[51] The city of Akron, Ohio, under a state order to remove all phosphates entering the Cuyahoga River by 1972, unanimously passed a similar ordinance on December 8, 1970.[52] In January 1971 the detergent industry chose Akron [53] for the first in a series of suits to block enforcement of phosphate bans—perhaps, one official suggested, because Procter & Gamble, based in Cincinnati, is particularly powerful in Ohio politics. In an out-of-court settlement in June the Soap & Detergent Association, the big three detergent makers, and Economics Laboratory Inc., bought more time by agreeing to an 8.7% limit by August (six months after the partial ban had been due to go into effect)—and no total ban until the industry had a chance to air its views in a public hearing in Akron.

The Nixon Administration has taken the position, stated by Dr. Gordon MacDonald of the President's Council on Environmental Quality in January 1971, that phosphates should not be forced out of detergents "until we get an adequate substitute" that will not be harmful to humans or the environment.[54] This suits the detergent-makers just fine, since what constitutes an "adequate substitute" and/or a potential hazard has been left in large measure to the detergent industry to decide.

Until 1969 FWQA did not even try to provide itself with a basis for second-guessing the industry's assertions that no adequate substitute for phosphate was available. Somehow, when FWQA's phosphorus research and development contracts were handed out, all the money went into studies of eutrophication or sewage treatment projects—several of them carried out by the detergent industry.[55] No one thought to check into phosphate replacements. The detergent industry, lacking either a deadline or competition from independent investigators, pursued its own search for alternatives with all the urgency the Federal officials should have expected. Procter & Gamble, by its own account, devoted an estimated $3.2 million in 1969 to research "directed at the development of non-phosphate detergent builders" [56]—less than one-seventh the amount ($24.4 million) the company spent on television commercials alone to promote just eight of its phosphate detergents that year.[57] Colgate-Palmolive, whose technical director, Dr. Richard Wearn, testified in December 1969 that "we are working hard to get a [phosphate] replacement," [58] was at the same time busily introducing yet another new detergent to the market: Punch, with 44% phosphate.[59] When FWQA Commissioner David Dominick attempted to assure Congressman Reuss at his 1969 hearings that FWQA agreed "substitutes for phosphorus must be sought, and sought right now," Reuss exploded:

MR. REUSS: But you are leaving it to the industry, which uses phosphorus, and the makers of phosphates, to seek them. And obviously, our lakes will be dead before they do it.

They have a built-in conflict of interest; and as long as they can report to you once a year, "Sorry, boss, we haven't found any substitute for phosphate yet," . . . there is going to be no progress.

How can this be? You leave it to the industry; they enjoy things as they are. That's why they haven't found a substitute.[60]

Actually, the detergent industry does have a limited interest in finding a good phosphate substitute on its own. As the Soap & Detergent Association's Charles Bueltman said at the 1969 hearings:

If an ingredient can be found tomorrow that will perform all of the functions, all things being equal, the industry would probably move to that; *because it might make a better product—it cleans better, it might cost less,* all kinds of factors.* [Emphasis added.] [61]

* "So you would have had research on [phosphate replacements] whether there had been pollution or not," observed Congressman Floyd Hicks (D.-Wash.) to Lever Brothers' vice president for research and development Frank Healy. "That is correct," Healy replied.

The only trouble is, phosphates are cheap and tremendously effective ingredients. Finding a safe substitute to top them in cleaning power looks, at this point, like a long shot. And so does a voluntary massive conversion by the industry to any phosphate replacement the manufacturers consider slightly inferior. "Housewives are smart and have a lot of experience with laundry problems," says a Lever Brothers pamphlet; [62] these ladies know how to spot a wash that doesn't meet the competition. A large detergent-maker who went ahead voluntarily with a plan to take the phosphate out of all his products would—as the industry sees it—be charting his own demise. Congressman Paul McCloskey (R.-Cal.) confirmed the point at the 1969 hearings:

MR. MCCLOSKEY: . . . Do I understand your testimony thus far correctly—that phosphates as an effective agent in a detergent are such that essentially all three of the major companies in competition with one another have to maintain a certain level of phosphate content in order to compete? Is that correct?

MR. BUELTMAN [Soap & Detergent Association]: Yes, sir.

MR. MCCLOSKEY: So built into your industry is a desire in the present state of things to keep the phosphate level high—high enough for you to compete with your two competitors. . . .

DR. HEALY [Lever Brothers]: It is not the only factor that leads to the performance of a product, but it is one factor and we are all trying to make our product better than our competitors.[63]

The one phosphate detergent replacement considered generally harmless has been rejected out of hand by industry spokesmen: old-fashioned soap. Unlike many detergent surfactants, soap needs no phosphate builder to help it clean or to prevent the redeposition of soil. Soap does tend to form a deposit when it is used in hard water. The deposit can clog washing machines or form a dulling film on clothes. But the problem can be alleviated by softening the wash water with washing soda (sodium carbonate) or other non-phosphate water softeners. Water pollution officials have expressed qualms about what might happen to the water if the nation started dumping over 2 billion pounds of washing soda —or any other single chemical compound—down the drains.* But that is not the basis on which detergent industry spokesmen argue against a return to soap.

"U.S. supplies of fats and oils are inadequate to furnish needed raw materials for such quantities of soap" as would be needed to replace detergents, asserts E. Scott Pattison of the Soap & Detergent Association. To replace the current 5 billion pounds of detergent used annually, he argues, would require about 2.28 billion pounds of tallow—the white rendered fat from slaughtered cattle and sheep, used in soap, margarine, lubricants, and candles. Current U.S. tallow production is only 5 billion pounds a year. While to the untutored eye the difference between 2.28 and 5 looks more like a surplus than a shortage, it turns out the U.S. ships some 1.8 billion pounds of surplus tallow abroad each year to be used as food, and feeds another billion pounds to animals (Pattison's figures). Cutting into this surplus, warns Pattison, "would have a devastating impact on world feed and food supplies, particularly in underdeveloped countries." [64]

Fortunately for the starving people of the world, Pattison omits one crucial fact. Tallow is only one of the many natural fats and oils from which soap can be made. It is also possible to synthesize soap. And if it ever became necessary to dip into the edible tallow supply because phosphate detergents had been replaced with soap, we could start exporting some of the fish that would return to resuscitated lakes and rivers.[65]

Cleaning with soap instead of phosphate detergents would also, according to the Soap & Detergent Association's Charles Bueltman, "be equivalent to setting back health, cleanliness, and sanitation standards many

---

* EPA's Water Quality Office has only recently begun to investigate the effects on water quality of the ingredients in non-phosphate laundry products. Scientists tend to agree that there is relative safety in numbers in washing compounds—i.e., the more variety, the better for water quality. J. R. Vallentyne, Scientific Leader of the Eutrophication Section of the Freshwater Institute in Winnipeg, Canada, wrote to Dr. Samuel Epstein in September 1970: "My own personal feeling is that in reducing the phosphate contents of detergent products, the manufacturers should make serious attempts to diversify rather than to use any one single substitute. In other words, diversification with regard to environmental effects should be every bit as much a part of industrial thinking as it is in the creation of a stable economic base of operations." (Hearings before Senate Subcommittee on Air and Water Pollution, "Water Pollution—1970," part 5, p. 2043.)

years." [66] By "many years" he presumably means back to 1950, when soap was still outselling detergent 2 to 1.[67] "In automatic dishwashers . . . and in food processing equipment," a Lever Brothers pamphlet warns, "soap cannot provide the level of cleaning and sanitation required." [68] Required for what? The detergent-makers rarely specify. But even if it were true, as Soap & Detergent Association public affairs director Robert Singer claims, that "automatic dishwashing machines would be useless" without phosphate detergents,[69] or even that sanitation standards in dairies or beverage and food processing plants might suffer, it does not follow that soap is inadequate for clothes. And laundry products make up the overwhelming proportion of total detergent sales—90% in Chicago, for example.[70]

The industry simply cannot conceive of a return to soap, because detergents are "progress." "Soap had its chance 20 years ago," says Singer scornfully, "but detergents knocked it out of the box." [71]

In the detergent-makers' view, only one real contender for the role of partial phosphate replacement has come on the scene in the last half decade: nitrilotriacetic acid (NTA). A patent for NTA was taken out by the Hampshire Chemical Division of W. R. Grace and Company as early as 1962. By 1966 Procter & Gamble had begun gradually to replace a fraction of the phosphate in certain detergents * with NTA. It was used "as an adjunct to phosphates, in order to produce a product performance improvement," testified Procter & Gamble associate product development director John Bruck at the 1969 Reuss Subcommittee hearings. "It is a coincidence more than anything else that NTA is a potential compound to help solve this eutrophication problem," Hampshire Chemical president Dr. John Singer confirmed.[72]

The new ingredient was added to detergents as such chemicals always are: with no pre-market clearance by any Federal agency concerned with health or the environment. Procter & Gamble, the Monsanto Chemical

---

* In January 1970 NTA was being used in conjunction with phosphates in two Procter & Gamble products: Gain (10.3% NTA, 41.5% phosphate) and Cheer (11.8% NTA, 37.9% phosphate), according to *Phosphates in Detergents and the Eutrophication of America's Waters,* House Report No. 91–1004 by Committee on Gov't Operations, 1970, p. 35. For a list of detergents containing NTA see Appendix H.

Company (which later began producing NTA), and a
few other large, profitable, and respected firms did test
NTA on their own. "Studies have shown," testified Soap
& Detergent Association vice president Charles Bueltman in 1969, "that NTA, at the current low levels of
use, degrades biologically in waste treatment processes
and surface waters, and that it raises no aquatic toxicity
problems." [73] It was also found safe for humans, and unlikely to stimulate algal blooms.

By early 1970 the pressure was mounting for a rapid
and extensive phosphate replacement program. It was
only at this point that the Federal government started
evaluating NTA. But once the government belatedly
recognized the importance of checking out the new ingredient before it was introduced to the environment on
a massive scale, the project was given exceptionally high
priority by both the Nixon Administration and the Senate
Air and Water Pollution Subcommittee, chaired by
Senator Muskie, the leading contender for Nixon's job.
Working under "a political deadline" of December 1970
for deciding whether and how quickly NTA should be
added to detergents, FWQA laboratories started looking
for environmental effects, and various HEW labs began
experiments on potential health effects. "This testing
program involves policy matters of concern to the highest levels of government," wrote FWQA's Acting Director of Water Quality Research William Cawley, the coordinator of the program, to all the FWQA labs in May
1970:

Public requests for information should be treated accordingly.
. . . this NTA testing program rates top priority among
your on-going research programs . . . report to me . . .
which of your on-going research projects will be dropped
or deferred to accommodate the NTA work. *There is no
on-going project which has a higher priority than this work.*
[Emphasis his.] [74]

While the executive agencies raced to meet their December deadline, Dr. Samuel Epstein, chief of Boston's
laboratories of environmental toxicology and carcinogenesis at the Children's Cancer Research Foundation
and one of a panel of scientists advising the Muskie Sub-

committee, worked day and night on his own investigation. The findings that emerged from these various efforts were disturbing. NTA was not always biodegradable, it turned out, in overloaded treatment plants and, especially, in septic tanks. There was the likelihood that the NTA might chelate (i.e., pick up) toxic metals such as mercury, lead, or cadmium from washing machines, plumbing systems, raw sewage, and sediment; the NTA chelates then might liberate the metals into lakes and rivers or get into drinking water and circulate the toxic metals through the body.[75] And the National Institute of Environmental Health Sciences (NIEHS) of HEW's National Institutes of Health found that cadmium and methyl mercury administered to rats and mice simultaneously with NTA contributed to a tenfold increase in fetal abnormalities and fatalities in the test animals.[76]

In the meantime NTA kept going into detergents in increasing quantities, and raw-materials producers kept expanding capacity. By the summer of 1970 Procter & Gamble had committed some $6.8 million for manufacturing facilities for NTA. The ingredient was being added to roughly a third of the company's detergents (by volume), and $167 million worth of NTA had been contracted for by Procter & Gamble alone. The two leading NTA manufacturers, Hampshire Chemical and Monsanto, were selling all they could produce—150 million pounds a year between them—and were increasing their capacity.[77] The Ethyl Corporation was designing a large plant for NTA production, and other firms were gearing up for the huge demand projected for 1972.

Thus when EPA Administrator William Ruckelshaus and Surgeon General Jesse Steinfeld decided in December 1970 that the use of NTA should be suspended "pending further tests and review of recently completed animal studies," the industry was none too happy. The detergent-makers' "voluntary agreement" to hold off on NTA was announced by the two Federal officials on December 18, just one day before Dr. Epstein's widely pre-publicized report on NTA was released by the Senate Public Works Committee (of which the Muskie Subcommittee is a part). The soap-makers "greeted our decision glumly," Federal officials reported to *The New York Times*.[78] Still, the agreement was a compromise.

The detergent-makers were permitted to use up existing stocks of the raw material and go on selling whatever NTA detergents they had "in the pipeline." And the ban is only temporary. Both industry and government are still studying the chemical compound; as one detergent industry spokesman told the Task Force in March 1971, "NTA may yet come out with a clean bill of health."

What will be done in the months to come about phosphates in particular, and about the general problem of harmful chemicals released to the environment in consumer products, is still very much an open question. FWQA resigned from the Joint Industry/Government Task Force on Eutrophication—a step that Congressman Reuss's Subcommittee had strongly recommended —when the water pollution agency moved from Interior to EPA in December 1970. The long-awaited PAAP tests for evaluating the "eutrophicability" of a given substance are now almost ready for their debut. And since May 1970 EPA has had a $344,000, 18-month contract with the Gillette Company Research Institute to formulate and test a number of safe and commercially acceptable non-phosphate laundry detergents.[79] A research and development official in EPA's Water Quality Office assured the Task Force in March 1971 that Gillette was being asked to appraise all the new types of "nonpolluting" formulations that have recently come on the market, and that they were "not aiming for the ultimate in cleaning performance," but simply "a reasonable standard of cleaning ability."

On the minus side, the detergent makers have successfully delayed enforcement of some state and local phosphate bans by challenging them in the courts. The proposed FTC rule for labeling phosphate detergents with warnings, and particularly the requirement that the warning be featured in the soap-makers' commercials, is given little chance of surviving bitter industry opposition. The lessons of NTA are not reassuring. The safety of NTA "for both humans and the environment has received more consideration than any other detergent material we have ever produced," [80] stated the Monsanto Company in November 1970. If large and reputable firms with extensive research and testing facilities like Monsanto and Procter & Gamble can introduce chemical compounds like NTA into the environment on the basis of what outside scientists have later found to be unconvincing data, the prospects for sound decisions on these matters by other private firms accountable to no one but their stockholders do not appear promising. Of the hundreds of smaller detergent firms, few spend much researching the toxicity or environmental safety of their products, and even the largest concerns, if Monsanto is right, look even less carefully at their other compounds than anyone did at NTA.

The implications of the NTA story are clear. We must have laws that will permit us to catch harmful chemicals in our commercial products before they catch us. It is far too late to do that with phosphates now. Our next best alternative—and one we must take—is putting them out of circulation. We can ill afford to permit industry to deposit its harmful contaminants in the lakes and streams via the purchasing public. Choked and poisoned waters are too dear a price for whiteners, brighteners, jingles, and contests.

5

The Forgotten Polluters: Non-point Sources of Pollution

Many of our industries—agriculture, mining, and real estate construction, for example—do not generally send their wastes down to the nearest stream in a pipeline, as cities and manufacturers customarily do, depositing them at a few fixed and ascertainable locations. Instead, if they view their pollutants as wastes at all, these industries simply leave their unwanted matter to rest where it will, eventually to be blown or washed off acres of farmland, construction sites, or mines into America's rivers and lakes.

"Non-point source" polluters like these have always confounded the sanitary engineers who long ago staked out the water pollution field as their bailiwick. It is difficult to know whom to blame when you cannot follow a pipe back from the water's edge. And most non-point source pollution problems cannot be solved by laying a sewer or building a better waste treatment plant. The response of the sanitary engineering profession to this challenge was predictable, if unfortunate. It has simply channeled its efforts into the jobs it does best, building sewage treatment facilities. And lest other professions begin to intrude on their area, control authorities have always defined the field of "pollution control" narrowly enough so that the critical need for fresh ideas never became apparent. Until very recently pollution control officials and environmentalists alike have comfortably assumed that pollution problems will be solved as soon as recorded sources of industrial and municipal pollution are cleaned up.

The danger of this assumption was illustrated very

clearly by the results of a study that researchers from Rutgers University completed in 1969. They had been given state and Federal grants to estimate the economic costs of abatement in several small Eastern river basins. They were fairly sure that the state's discharge permits for municipal and industrial polluters, coupled with data from the state's program for monitoring these sources, would provide most of the information they would need. Still, the Rutgers team decided to begin by running a logical—but relatively unprecedented—experiment. Both the Federal government and local pollution control agencies had also been collecting, for some time, data on the water quality (as measured by dissolved oxygen content) in the rivers the group planned to study. If the amount of upstream discharge that the team's calculations showed would be necessary to produce the deterioration measured downstream was compared against the amount of discharge that state records attributed to identified industrial and municipal sources of pollution, the Rutgers researchers could see for themselves just how complete the state's ledger on polluters really was. They naturally expected a small discrepancy. But they were hardly prepared for what they discovered instead.

"This is a frightening thing," General William Whipple, Jr., director of the Water Resources Research Institute at Rutgers, told the Task Force. "To my utter stupefaction, I found that the pollution we didn't know anything about was more than twice as great as the amount we knew about." According to the group's calculations, only 39% of the pollution in New Jersey's Upper Passaic Basin could be traced to discharge sources known to local, state, or Federal authorities. And in the Millstone River and the Raritan River only 25% of the pollution could be accounted for.

Where did all this unidentified pollution come from? Some of it was what is known as "background pollution." It comes from natural nutrients in the ground and would presumably be there if the land had not been disturbed. And some of the mystery discharge, no doubt, belonged to polluters which the control agencies *did* have on record, but whose waste output had been underestimated. But after taking those factors into consideration, the bulk of the 61% and 75% unexplained contamination in these small New Jersey rivers could only have come primarily from man-made pollution of the non-point source variety.

Farmers, Ranchers, Loggers, and Chemists

One can only guess what the breakdown of those diffuse pollutants would be in the most (i.e., not very) rural parts of New Jersey. But chances are that those wastes which are usually lumped together under the broad category of agricultural pollution are doing greater damage than any of the other non-point wastes. It is true that when a given amount of pollutant enters the streams as agricultural runoff, it is less of a threat to the environment than it would be if the same contaminant were dumped into the water in the highly concentrated form characteristic of industrial and municipal effluents. This is why, throughout the urban East, the grossest pollution is to be found downstream from the major population and industrial centers. But at least in the West it is becoming increasingly clear that pollution control authorities have been listing the problems in inverse order of importance. There, agricultural pollution takes the greatest toll—as serious as much of the pollution in the East but in a part of the country where water resources are much more scarce. The disturbing fact is that in terms of sheer volume of waste output, America's ranchers, loggers, and farmers—with a mighty boost from the manufacturers of agricultural chemicals—are far and away *the worst polluters in the entire nation.* Taken together, they are now (and probably have always been) responsible for more water pollution than either the cities considered as a whole or the rest of private industry combined.

The bulk of this agri-business pollution is plain ordinary dirt, sediment from land farmed or forested in disregard of known erosion-prevention techniques. Agricultural lands contribute most of the more than 4 *billion* tons of sediment produced each year in the United States [1] (compared to some 32 *million* tons of organic wastes and suspended or settleable solids produced by all industries and municipalities combined). The soil loss from land in continuous row crops can be as much as 70,000 tons per square mile annually. Soil conservation methods are available which could cut these losses by up to 95%.[2] But no one is required to use them. As a result, suspended solids in our lakes and streams attributable to surface runoff are, it is estimated, more than 700 times as great as those from the discharge of human sewage.[3]

One of the problems with this plain ordinary dirt is that it is almost always highly contaminated, having picked up the residues of pesticides and other chemicals with which it comes in contact. Suspended silt can also destroy a river's capacity to assimilate organic waste. It blocks sunlight needed to keep bottom plants, a precious source of oxygen, alive. Without oxygen, the natural process of organic waste decomposition by aerobic bacteria cannot take place, and the river putrefies. Heavy siltation destroys aquatic life, damages industrial equipment (turbine blades, for example), and can more than triple drinking water treatment costs. When these waste solids settle, they clog harbors and shipping channels, and usurp reservoir storage capacity. (It has been estimated that 20% of the nation's water supply reservoirs, numbering around 2700, will have a useful life of less than 50 years at the present rate of siltation.[4]) In some river basins—Washington, D.C.'s Potomac River, for example—sediment dwarfs the impact of all other pollutants combined.

Much more dangerous than sediment, however, and a close second when measured by mass alone are the wastes generated by America's farm and ranch animal population. With more than 107 million cattle, 3 million horses, 53 million hogs, 26 million sheep, and 490 million chickens presently residing in the U.S., the Department of Agriculture could boast in a 1969 report that "animal wastes in this country [which it estimated at over *1.7 billion tons* annually] probably exceed wastes from any other segment of our agricultural-industrial-commercial-domestic complex."[5] But Old MacDonald has not been providing sanitary facilities for his barnyard population. As a result of years of neglect, the sullying tide of farm and ranch wastes has reached alarming proportions in most river basins. The Interstate

Commission on the Potomac River Basin could have been describing almost any one of the major bodies of water in the country when it found, in a 1966 report, that

> every time it rains . . . enormous amounts of animal wastes are washed from farmyards into the river, rendering it unsafe for swimming. . . . Although only a quarter-of-a-million people live in the river basin above Great Falls [just a few miles north of Washington, D.C.], it has been estimated that the number of farmyard animals . . . is the [waste] equivalent of a human population of 3.5 million. While most of the human population is served by some sort of sewage treatment plant, there is no comparable treatment for the animal wastes.[6]

Developments in the cattle industry within the last 10 years or so have made this people-only policy even more ineffective. Ever since the days of the Open Range, most animal manure had simply fallen to the ground and decayed there through natural processes. Just a few years ago, however, the march toward modernization in business caught up with the ranchers. As an efficiency measure, they began concentrating their herds on smaller and smaller plots of ground. Pork and poultry producers quickly followed suit, and now the Department of Agriculture estimates that as much as half of the national total of animal wastes comes from what are known as animal feedlots—huge pens where the hogs, heifers, or what have you are confined and fattened before slaughter.[7]

In the cattle industry each one of the unfortunate inmates of these overcrowded institutions produces more than 16 times as much fecal matter as the average human being.[*] There they stand, nearly bumper to bumper with their fellows (often numbering 10,000 or more), barely able to move as their own excrement piles higher and deeper around them. Waste control techniques are available (composed of dikes, ditches, and retention ponds). But their use is mandatory in only a few states.

* Cecil H. Wadleigh tells us, in his *Wastes in Relations to Agriculture and Forestry* (p. 41), that the average cow or steer produces the waste of 16.4 humans, the average horse that of 11.3 humans, sheep 2.45 humans, and the average hog 1.9 humans. Poultry brings up the rear with an average waste equivalent of 0.14 humans.

Most of the feedlots' liquid waste evaporates. But that is not, unfortunately, the end of it. As much as 90% of the urinary nitrogen excreted on feedyards is volatized as ammonia. The airborne ammonia is then reabsorbed by nearby bodies of water, where it stimulates eutrophication and increases the danger of nitrate poisoning in drinking water supplies. Most of the remaining 10% percolates down through the pasture to contaminate the groundwater underneath.[8]

The solid waste matter, on the other hand, accumulates until the rains come. Then it runs off or even "landslides" into the nearest lakes and streams, visiting violence upon the homesteaders downriver. Until about 1966 the Department of Agriculture actually *taught* beef producers to locate their feedlots on hillsides near streams precisely so that these concentrated cowpie collections would tumble off into the water rather than remain somewhere on the land.

One reason an avalanche of animal excrement can cause such serious pollution damage is that the manure from these feedlots is a much "stronger" pollutant than human waste. For a given quantity of waste, the BOD of feedlot runoff is from 5 to 100 times greater than that of municipal sewage.[9] Thus, the waste output from a single feedlot can be much more dangerous than the sewage from a major urban center.

The corporate cowmen's barrage of concentrated waste products comes down harder on the Missouri River—which runs through the heart of the cattle country—than on any other major water body in the U.S. There are more than 90,000 commercial feeder operations in the Missouri Basin alone. It has been estimated that the aggregate animal wastes in the Missouri Basin equal the output of a human population of 750 million, or about that of mainland China. Only about 5% finds its way into the Missouri or its tributaries, but this is still the equivalent of 37.5 million people dumping their untreated sewage into the Missouri River system day in and day out.

The chemical content of feedlot runoff may be even greater cause for concern than its "strength," its concentration, and its sheer volume. An estimated three-fourths, or 30 million, of the beef cattle slaughtered annually in the United States have been fed diethylstilbestrol (DES), a hormone additive that produces rapid weight gains in cattle and an estimated $300 million extra in annual profits for the cattle industry. DES also produces cancer in animals and is a suspected cause of cancer in man. (See Chapter 1.) For this reason the Food and Drug Administration banned it in chicken feed 10 years ago. But although at about the same time the governments of France, Switzerland, and the Netherlands outlawed the use of DES in fattening beef, cattlemen in the U.S. have so far staved off a similar ban.

In September 1970 the FDA announced that beef producers would henceforth be permitted to *double* the amount of DES they can add to their cattle feed. At the same time, however, the FDA required that cattlemen stop feeding their animals the hormone two days prior to slaughter. During this 48-hour withdrawal period, according to the FDA, steers would work off the carcinogenic additive, so that no traces would be found in meat sold on the market. What happens to the DES after the cattle "work it off" is, of course, not the FDA's concern. FWQA officials conceded to the Task Force that they had only begun to look at the problem because it was "so new." But they agreed—with great consternation—that if, as many suspect, the hormone *has* been finding its way into cattle excreta, it can henceforth be expected to show up in feedlot waste, in the rivers, and probably in drinking water supplies, in greater quantities than ever before. As one regional FWQA official put it, "Something which looks good at one end may be pretty bad at the other end."

The trend toward concentration and chemicalization in the cattle industry that has made the animal waste problem so serious has been paralleled by similar developments in farming patterns and techniques—developments that have also proved highly destructive of the environment. The use of massive quantities of inorganic chemical fertilizers has made it possible for farmers to "intensify"—to grow more food on less land. But the soil nutrients that support higher crop yields have also made it possible for our lakes and streams to support a much larger and more troublesome crop of algae. It is estimated that, typically, 13–15% of the phosphate pollution in most U.S. waterbodies originates as agricultural runoff.[10] In addition to the phosphates, more than 6 million tons of nitrogen are spread annually on our nation's farms and forests. The most dangerous chemicals in farming's arsenal, however, are pesticides and herbicides. More than 600 million pounds of chemical poisons—including insecticides, fungicides, rodenticides, and fumigants—are used by agri-business annually in the U.S., or about three pounds for every human being in the nation. Use of pesticides increases at a rate of almost 15% a year.[11]

The steady increase in yields per acre from the use of these chemicals has not stopped U.S. agri-business from continuing to expand its acreage as well—despite (or because of) the $3 billion worth of Federal Soil Bank subsidies handed out each year to induce farmers to let their land lie idle. Whenever irrigation is necessary to open up new farmland, the chances are strong that pollution will result. Irrigation water usually leaches solids—notably salt—from the arid soil through which it passes, at the same time losing liquid that evaporates or is soaked up by the crops. By the time the water wends its way back to the river it came from, it carries a much heavier concentration of salt or other solids than when it started out. Salinity, the most important water pollution problem in the Southwest, has been greatly aggravated by irrigation—much of it carried out by the Interior Department's Bureau of Reclamation, which has been "reclaiming" barren desert in the Southwestern states through irrigation since the 1930's (while fertile land throughout the 50 states lies fallow, earning its Soil Bank subsidies). Once a river turns salty, the surrounding area ceases to be attractive to industry, since saline water is too corrosive to be used in industrial processes. Eventually even the farmers cannot use the water any longer. This has happened already in parts of the lower Colorado River, where every five acre-feet of water now contains six tons of salt.

And what has been the Federal government's response to agricultural pollution, the nation's single largest, oldest, and most unmanageable set of pollution problems? It has, for the most part, simply buried its head in the sediment, apparently hoping that the runoff contamination which growers produce may somehow disappear of its own accord. After years of sanitary-engineering professional inbreeding, FWQA has done little more than officially acknowledge the existence of agricultural pollution. Since 1956 the Federal government's enforcement program has zeroed in on thousands of industrial and municipal polluters—but only one token farmer. (The Midwestern Feeding Company of Manley, Nebraska, an animal feedlot enterprise, was notified in the summer of 1970 that it was violating Federal pollution requirements.) And while Federal water quality standards, authorized by a 1965 amendment to the Federal Water Pollution Control Act, set numerical limits on relatively innocuous industrial pollutants, nowhere do the standards make any mention of pesticides. Similarly, Federal standards for the Colorado River—the nation's saltiest—lay down maximum pollution limits on everything but salinity.

No one claims all the answers are easy, of course. But FWQA has hardly begun to look for them. Out of a research budget of $48.5 million in fiscal 1970, the Water Quality Administration allotted only $2.85 million—about 6%—to finding more innovative techniques with which to combat agricultural wastes. While FWQA's budget program plan for fiscal 1971 announces that there will be a new "increased emphasis" on agricultural pollution, that is news to researchers who are working in the area. The 1971 budget request shows a cutback to $2.35 million for research on runoff problems.

Outside the research area, there is virtually nothing going on. In the summer of 1969 FWQA had only one headquarters employee assigned to developing new practical Federal programs for reducing agricultural pollution. Shortly thereafter he was transferred to another post and has not been replaced. The loss is less important than it might seem, however. When the Task Force interviewed FWQA's onetime agricultural specialist—Gary Dietrich of FWQA's Special Controls section —back when he was still on the job, he described his activities in the area as little more than extracurricular anyway. He usually managed to fit in two solid hours each week on the subject, the Task Force was informed. In his short-lived search for new solutions Dietrich apparently left one small stone unturned. He could not recall ever having made contact with anyone at the U.S. Department of Agriculture.

This is unfortunate, because USDA is much more than the logical home for most of the Federal government's expertise on the various techniques for tilling the soil. The Agriculture Department has always had most of what little practical authority there is over the polluting activities of the American farmer. Since USDA's self-proclaimed mission has always been to promote the economic well-being of agri-business, its record in exercising its authority to reduce pollution has been predictably dismal.

In the area of pesticides, for example, USDA was given responsibility under the Federal Insecticide, Fungicide, and Rodenticide Act of 1947 to make sure pest poisons sold in interstate commerce were both effective and safe. Day-to-day responsibility for regulating pesticide sales under that act fell to the Agricultural Research Service's Pesticide Regulation Division, which quickly developed a close working relationship with the pesticide chemical industry. Heavily lobbied by the National Agricultural Chemical Association, trade organization for the industry's two giants—Shell and Dow Chemical—and over 100 smaller pesticide manufacturers and distributors, Pesticide Regulation has always taken a militantly restrictive view of its regulatory role. From its inception the Division regularly approved the sale and use of the most persistent killing compounds, asking only whether there was any *short-run* danger in their use, and whether the manufacturers could demonstrate that the agents did in fact eradicate pests.

The fact that the pesticide might eradicate higher forms of life as well through its *cumulative* impact was not a matter for official concern until 1964. In that year, concern over the long-range dangers of pesticides that Rachel Carson's *Silent Spring* had documented over a year earlier prompted an interdepartmental agreement at the Federal level under which pesticide applications were to be reviewed by the Department of Interior's Bureau of Fish and Wildlife. The shadow Pesticide Regulation Division within the Fish and Wildlife Bureau which had been designated to carry out Interior's responsibility for review had no actual authority under the agreement. Its assignment was simply to advise its pesticide regulation counterpart in the Department of Agriculture on the danger to aquatic life presented by the various toxicants that the USDA division had been asked to approve.

The Department of Interior pesticide group soon found out that when Fish and Wildlife spoke, Farming rarely listened. Between 1965 and 1969 Interior's Pesticide Division made adverse comments on 25.6% of the 32,724 pesticide registration applications submitted to it by the Department of Agriculture. Fish and Wildlife staff members told the Task Force that, except for a brief period at the beginning when Agriculture officials volunteered the information, the Interior division has never been able to find out what became of its recommendations. If what the Interior staffers found out during the brief period of feedback was any indication, however, most of their comments were rejected by USDA out of hand.

While USDA's regulatory authority outside the pesticide area is more limited, it still has immense educational and promotional influence over the activities of its farming clientele. Despite this vast potential for controlling farm pollution, however, most of the Department's efforts in the environmental area appear to have been aimed at the public rather than the polluter. In 1967, for instance, USDA established a high-level eight-man Executive Committee on the Environment. According to the memorandum that set up this blue-ribbon panel, its mission was to "contribute significantly to Department-wide communication and planning on all facets of USDA activities and responsibilities relating to the quality of the environment." In the following year Agriculture solicited the help of other Federal agencies in preparing *A Report to the President: Control of Agriculture-Related Pollution,* which it released in January

1969. Referred to by Agriculture officials as the "last word" in pollution control, the "greenbook," as it has been dubbed, is in fact an impressive study. It makes a comprehensive sweep of the entire field and, at least for some types of pollution, outlines methods of control that are not only presently workable but ready for universal implementation. The greenbook was a good start; but the Executive Committee stopped right there.

In July of 1969 the Task Force sought out the USDA Executive Committee's Assistant Chairman, Dr. T. C. Byerly, and its Executive Secretary, David Ward. Our question: based on the ideas discussed in the greenbook, what concrete programs has your committee promoted or is it now promoting to enhance the quality of American waters? The answer was straightforward: "None." USDA's Executive Committee on the Environment had not made a single recommendation, directive, or effort of any sort in support of new programs in the more than two years it had been in existence, either before or after the release of the greenbook. Their only job, Byerly and Ward explained, was to coordinate whatever was going on inside or outside the Department in the evirontmental field. They were not very busy.

The Department of Agriculture has had powerful help in keeping other Federal agencies, and particularly FWQA, off its turf. In order to get its operating money every year, FWQA has had to go before the House Appropriations Committee's Public Works Subcommittee. One of the Subcommittee's key members is Congressman Jamie Whitten, a Democrat from the Mississippi Delta with 29 years on Capitol Hill. The Mississippi Congressman has also been since 1946 the Chairman of the House Appropriations Committee's Subcommittee on Agriculture. Jamie Whitten is the virtual czar of the Federal farm bureaucracy. He is also an intimate of the Agriculture Department's regular corporate clientele. Whitten is an honored guest and an occasional keynote speaker at the annual banquet of the Limestone Institute, a major lobby for the fertilizer industry. The National Agricultural Chemical Association (NACA), powerful trade organization for the chemical pesticide industry, also counts itself among the Congressman's most ardent supporters. In fact, it was Parke Brinkley, president of NACA, who suggested to Representative Whitten that he write a pro-industry rebuttal to Rachel Carson's *Silent Spring*. Whitten's book, *That We May Live* (which was turned down for publication by the Public Affairs Press because, as editor M. B. Schnapper said, "it seemed substantively weak"), was printed only after three major pesticide manufacturers (Velsicol Corp., Shell Oil, and Geigy Agricultural Chemicals) agreed to subsidize sales.[12]

Whitten has become famous around Washington for the meticulous attention he gives to even the smallest, most insignificant items in the Department of Agriculture's budget. He brings most of that same attention to detail to his annual review of FWQA's request for financing. But Whitten makes no secret of the fact that in his role as the arbiter of FWQA's budget, his heart remains with the farmers. Thus, the price of FWQA's encroachment into Agriculture's carefully preserved sphere of influence might well be a serious cut in the support for this pollution control program or that.

One of the major goals of President Nixon's December 1970 reorganization of the Federal pollution control bureaucracy was to consolidate under one roof the government's fragmented responsibility for controlling agricultural pollution. How much or how little can the new Environmental Protection Agency be expected to accomplish toward cleaning up "the farm mess"? Under the reorganization plan, Pesticides Regulation came out of the Department of Agriculture to join, at least formally, the pollution control camp. It is possible the new atmosphere may bode a favorable turn in pesticide policies. The most significant change, in the long run, may turn out to be FWQA's removal from the Department of Interior. Under Interior, whatever disagreements the water pollution agency had with irrigation projects carried out by Interior's Bureau of Reclamation had to be sacrificed all too frequently to the goal of departmental harmony.

For the most part, however, prospects for improvement are dim. FWQA and its sanitary engineers are no better equipped, from a personnel standpoint, to press for pollution control on the farm than they were before the transformation. And it is difficult to predict what will happen now that FWQA has officially linked arms with Pesticides Regulation, since few observers expect that the former USDA division will alter its allegiance to the chemical industry overnight. In trading its very limited license to criticize from a distance for a chance to influence the pesticide group at close range, FWQA may still have made a poor bargain. Spokesmen for NACA have observed quite frankly that their job will be easier now that the reorganization has taken place. Whereas responsibility for pollution control used to be scattered throughout the Federal government, the trade associations can now focus all their lobbying resources on a single target.

Should the trade lobbies not find EPA malleable, they will have Congressman Jamie Whitten to insure that the new environmental agency does not step out of line. In the reshuffling of Congressional committee assignments to accommodate the organizational change, the powerful Mississippian tightened his choke hold on the pollution control program. Whereas prior to the shift the water program got its money from the Appropriations Committee's Public Works Subcommittee (to which Whitten belongs), in February 1971 Appropriations Committee Chairman George Mahon (D.-Tex.) assigned control over EPA's budget to the committee's Agricultural Subcommittee, which Whitten chairs.

But, most important, the agencies that make up EPA have no better laws to work with than they had before. Preventing soil erosion is still purely voluntary. Pollution control authorities must still rely on exhortation to get growers to restrict fertilizer use or take measures to control runoff. No one but the Food and Drug Administration has any power to determine what may or may not go into animal feed. The pesticide laws still place the burden of proving damage on the citizen (rather than placing on the industry the burden of proving safety). EPA's "coordinated attack" on farm polluters will be easily repelled if the new super-agency has no legal weapons to use. Until the Federal government gets the basic legislative authority it needs to control agricultural wastes, it will continue to do little more than flail help-lessly away at the nation's largest pollution problem.

The Developers

Poor land management is not the special province of rural America. Today, uncontrolled real estate develop-ment threatens to strip the nation of its natural resources just as agriculture has been doing for so long. Experi-ence and the economics of farming have fostered at least a rudimentary notion of agricultural conservation. But the public has yet to connect the real estate and con-struction industries' haphazard methods of clearing, digging, and building on the topsoil with the widespread water pollution they cause.

Some of the worst sources of sediment pollution, for example, are highway and building construction sites and bulldozed urban areas. While clearing and cultivat-ing land can increase land erosion rates four to nine times over the rates for land under natural cover, con-struction may increase erosion a hundredfold or more. It has been estimated that the average sediment yield during a rainstorm at highway construction sites is about 10 times that of cultivated land, 200 times that of grass-covered areas, and 2000 times that of forest areas.[13]

By definition, construction disturbs the stability of the soil. But most suburban sediment pollution results from expedient and often irresponsible building practices. In real estate developments throughout the country, con-struction goes on in blithe disregard of some of the fundamental rules of intelligent siting and building—e.g., do not build on flood plains or tidal lands, do not disturb inclines greater than 30 or 40 degrees, do not build on watersheds. Most new subdivisions are stripped of all tree and plant cover before construction begins, thereby removing the natural restraints on topsoil movement during rains or spring thaws. Excavations and leveling operations often are marked by huge mounds of earth that lie unprotected for up to a year in some projects while awaiting removal.

Rapid real estate development often creates pollution problems that linger long after the sodded lawns have been laid. Development interests usually obtain official approval to tie into overloaded public sewage systems, adding to municipal pollution loads. It is also common for real estate interests to get first priority on state and Federal grant funds for sewer and treatment plant construction.* Sewer lines to outlying districts built to accommodate would-be developers sometimes lie empty for years while overburdened facilities in older parts of town are causing sewage to back up or are bypassing it into the rivers untreated.

In addition to distorting pollution control priorities, the real estate developers who get new sewer lines built for themselves rarely pay the full cost of the extensions, either in fees or in "contributions." The International Business Machines Corporation and a shopping and industrial center in Montgomery County, Maryland, for example, together contributed only $500,000 of the $2.4 million cost of laying sewer lines from their complex to the Washington suburban sewage system, according to a 1965 study by Francis X. Tannian cited in James Ridgeway's *The Politics of Ecology.*

But what confounds pollution control authorities the most is when construction outstrips the ability of public sewer systems to keep up, as it frequently does. Where that happens, individual septic tanks (holding tanks that release sewage into the ground at a controlled rate) for new homes are the well-known stopgap measure. Septic tanks are designed to take advantage of the natural filtration capacity of the earth. But where that capacity is exceeded—as it often is—the best septic tanks money can buy won't prevent pollution of the groundwater. Fecal contamination of groundwater supplies for the Potomac River, which provides drinking water for the Washington area's 3 million people, was traced in 1967 to septic tanks in Potomac, Maryland, a wealthy bedroom suburb with homes in the $150,000 range. And in Suffolk

* The State of Ohio offers a typical example. There, according to February 1969 testimony by Representative Charles Vanik (D.-Ohio) before the House Public Works Committee, Federal grant funds totaling $28.6 million had "been generally allocated to permit the economic growth and development of communities rather than emphasizing the need for eliminating existing pollution." Vanik attributed this misallocation to the fact that "the industrial land-use promoter and the tract-development operator have directed the course of priorities in the distribution of Federal funds."

County, Long Island, where few of the county's 1.1 million residents are served by sewers, tap water has become so foamy and foul-smelling that the county legislature voted in November 1970 to ban the sale of detergents, which have been seeping out of septic tanks into groundwater supplies.

The danger of water pollution from a tract development naturally increases when that project is close to a lake or a stream. For that reason, the most destructive construction work of all is carried out by the resort and recreation spot developers, who like to build right on the water's edge. The effects of construction activities on Lake Tahoe, on the California-Nevada border, provide a revealing case study of both the resort promoters' *modus operandi* and the government's inability to go beyond traditional sanitary engineering solutions in dealing with this aggressive branch of the construction industry.

A deep, glacial lake of crystal-blue water, Tahoe has a reputation as one of the most beautiful lakes in the world. Biologically, it is still relatively free from nutrient infiltration, so that fish and other oxygen-based life forms flourish with uncommon ease. Unlike most of our rivers and lakes, which are so polluted that one cannot see more than a few inches or a few feet into them, Lake Tahoe is so clear that a 9.5-inch dinner plate can reportedly be seen submerged to a depth of more than 100 feet.

Elaborate precautions have been taken to preserve Tahoe's beauty against the threat of pollution from human sewage. At FWQA's recommendation, more than $7.5 million in Federal and local funds have been spent to construct and operate one of the world's most sophisticated sewage treatment plants, to serve the booming Tahoe resort population. The sewage is treated so well, in fact, that it has met Public Health Service standards for drinking water.

But Tahoe is threatened by organic pollution nonetheless. How? Earth, unless it is extremely sandy, contains natural organic nutrients similar to those in artificial fertilizers—phosphorus, nitrogen, and carbon products —which encourage excessive algal blooms and speed the eutrophication process. It seems that all the while Tahoe's exceptional sewage treatment system was being planned and put into operation, not a thought was given by Federal or state planners to protecting the lake from sediment pollution caused by unrestricted land development around it. Real estate developers have been busily putting up housing developments, shopping centers, recreational areas, and roads, tearing up the land, cutting down trees, leveling banks, and digging holes in the process, usually without using diking or other basic erosion-control techniques. Federal and state authorities have carefully padlocked the front door against pollution while leaving the back door wide open.*

While profit-minded developers have, on the one hand, torn up the land without a thought for the environmental consequences of soil runoff, they have, on the other, taken to filling up or draining submerged lands for purposes of development, often with equally dire environmental consequences. The filling of San Francisco Bay is probably the best-known instance where controversy has flared over landfills. The bay and its adjacent marshlands are already about 40% smaller than in 1850, and modern developers are reportedly eager to fill in at least another 11 or 12% of the remaining tidal lands. Furthermore, a California official has warned that pressures "for a helluva lot more filling" are likely to grow in the next 50 years.[14]

The developers are interested because bay areas are easy to fill, cheap to develop, and ideally situated for residential and industrial uses. What doesn't enter their cash-flow projections is the fact that these estuarine areas, in their natural state, also support vital organisms in the food chain and operate to reduce pollution (water picks up oxygen as it passes through shallow, reedy areas). Already there have been sharp declines in clam and shrimp fishing in San Francisco Bay. The 2 to 3 million waterfowl that used to nest and feed in the bay area have been reduced in number to fewer than 600,000.[15] Nationwide, about 7% of the estuarine habitat

---

* According to Dr. Charles R. Goldman, director of the Department of Ecology at the Davis campus of the University of California, the "fertility" of Lake Tahoe has increased by 72% in the past 10 years.

—important to many species of fish and shellfish—was destroyed by dredging and filling operations between 1947 and 1967.[16]

Of all the harmful types of development activity, land-fill is probably the easiest for the Federal government to control—when it wants to. Under the River and Harbor Act of 1899 the U.S. Army Corps of Engineers must approve any proposed structural modification of navigable waters. In the past, unfortunately, the Corps not only routinely licensed private landfill except where it actually threatened to obstruct navigation, but often-times did the dirty work itself, dredging and filling land to be sold to private developers or expanding munic-ipalities.

The Engineers' friendship with the real estate industry has cooled somewhat—at least formally—since May 1970, when the Army Corps adopted regulations per-mitting landfill by private fillers only where the applicant affirmatively proves the proposed work is consistent with ecological considerations.[17] It's too early, however, to tell what practical effect the new regulations will have. And as far as the larger problem of polluted sediment runoff from unrestricted development projects goes, FWQA (along with state and local agencies) is still stalled in its tracks. As FWQA Commissioner David Dominick explained to the Task Force, the water pollu-tion agency is "beginning to do some real thinking about the problem." No one, unfortunately, is beginning to act.

The Miners

Even before mining operations were begun in the U.S., streams were, to some extent, naturally polluted by acid water draining from areas of coal deposits. In the pres-ence of air and water, the sulfuric materials present in coal deposits form sulfuric acid and acid salts. The familiar yellow or reddish colors of naturally occurring "sulfur water" led to the initial discovery of coal in the United States more than 270 years ago. In 1698 Gabriel Thomas observed, "I have reason to believe there are good coals also, for . . . The Runs of Water have the same colour as that which proceeds from the Coal-Mines in Wales." [18] By 1803, however, the mining of coal had exacerbated the natural acid pollution to the point where a sojourner in Pittsburgh complained, "But the spring water, issuing through fissures in the hills, . . . is so impregnated with bituminous and sulphurous particles as to be frequently nauseous to the taste and prejudicial to the health." [19]

Coal-mining operations speed up natural acid drainage by exposing more sulfur-bearing materials to the air (which causes the sulfur to oxidize, and to form acid in the presence of water) and by increasing the contact of acid material with the water that seeps in and out of most mines. A study by the U.S. Bureau of Mines completed in 1969 showed that in two-thirds of all active mines, water that has drained into the mine is pumped back up to the surface and discharged directly into an adjacent stream. At the time of the Bureau's investigation, only 16% of this discharge water was being given treatment. Acid water in the remaining active mines and in those that have been abandoned also reaches streams eventually, typically through groundwater.[20]

More than 20,000 acres of lakes and 9000 miles of streams in 31 states already suffer pollution damage from mine discharge or drainage, and mine pollution is increasing. Coal is not the only culprit. All told, industries that mine and process some 19 different minerals have been implicated. But the coal industry is the worst. The Bureau of Mines reports than 93% of the stream mileage and 63% of the reported lake acreage affected results from the underground mining of coal.[21]

The Ohio River Basin states are hardest hit. On countless streams throughout Ohio, West Virginia, Pennsylvania, Kentucky, and Tennessee, the approximately 4 million tons of acid discharged annually in the Ohio Basin has been too much for the fish and other aquatic life to take. On just one of the Ohio River's tributaries, the Monongahela, acid corrosion damage to boats, barges, bridges, locks, dams, and industrial and municipal water supplies was estimated by the Public Health Service in 1961 to cost more than $2.2 million annually in Pennsylvania alone. FWQA has estimated that Appalachia loses more than $7.5 million each year in recreation income because of acid mine drainage.[22]

Solving the problem would not be nearly so difficult if not for the fact that 60–80% of the coal acid pollution comes from abandoned mines (numbering around 90,000, according to a "conservative" Bureau of Mines estimate).[23] For years state and Federal legislators were persuaded by the coal industry to exempt acid mine water from anti-pollution statutes, claiming that it was technologically impossible to control the drainage. Only within the last two decades have legislators disabused themselves of the notion that it is impossible to have coal-mining and clean water too.

Federal-state pollution abatement programs for the past 50 years have been piecemeal, underfunded efforts that have produced more failures than successes. The extensive Works Progress Administration (WPA) program, conducted during the Depression, to seal up abandoned mines (to prevent water from getting in) was no exception. Money was not allocated for the continued maintenance of the completed seals, and many of them have since deteriorated. When World War II broke out, most of the remaining seals were removed so that the mines could be reopened to help meet the increased demand for coal. After the war the seals were not replaced.

It took 25 years to get the next project off the ground. A Federal-state water pollution enforcement conference was convened in 1963 to bite off a much smaller piece of the overall problem—West Virginia's Monongahela River. The 1963 conference set up a technical committee to compile an inventory of the mines draining acid into the river and to make recommendations for abating the drainage. Seven years later, in October 1970, the committee was still mulling the question and had not yet come up with any official answers. The Task Force was informed, however, in the summer of 1969, again in the summer of 1970, and a third time in October 1970, that most of the details had been worked out on what would eventually be the committee report's major proposal—a Federal-state cooperative mine-sealing program. Shades of WPA!

In 1961 an amendment was added to the Federal Water Pollution Control Act which directed the water pollution unit to make a study of the acid mine drainage problem. After completing the required study, FWQA initiated a number of pollution control demonstration projects at abandoned mines in Appalachia from 1964 through 1967. The largest demonstration project, in Elkins, West Virginia, was a dismal and expensive failure.

The project ran into trouble from the beginning. Anxious to make up for its embarrassingly languorous record in acid pollution control, FWQA hastily chose the demonstration site at Elkins without bothering to ask the Bureau of Mines for advice. At the outset of the Elkins project, water pollution workers found themselves bogged down in legal negotiations over property rights. Their efforts were plagued continually by the threat that owners of the mineral rights would reopen their abandoned mines and thereby cancel out all of FWQA's work. In 1967 the project ended—before it was completed—as abruptly as it had begun. The official explanation for the premature burial went like this:

In January 1967 the Director of [FWQA's Technical Services] Division recommended that the Acid Mine Drainage Project #1 at Elkins, West Virginia, be terminated at the earliest practicable date. This recommendation was based upon difficulties encountered in obtaining easements and the high probability that similar difficulties would be common to any similar project undertaken in any of the coal regions. After extensive evaluations . . . ., it was decided to close the project as expeditiously as possible. . . .

Unofficially, FWQA's officials confess they would have had little to show for their efforts had the project continued. They had tried to examine too many control techniques at one time over too large an area, and most of the sub-projects were languishing for lack of supervision. When the Elkins project was scuttled, FWQA knew little more about mine drainage control than it had when the project began. The explanation in FWQA's 1967 Technical Services manual (quoted above) concluded by announcing that "all acid mine drainage control activities were reassigned to the Research and Development program." And, except for the Monongahela mine-sealing project, that's where they are today.

Further research, while important,* will not curb acid mine drainage. Available techniques that are said to reduce mine drainage by about 70% [24]—improved mine design, plugging holes in mine surfaces, diverting underground and surface watercourses around instead of through the mines, pumping water out and neutralizing it with chemicals before dumping, flooding or plugging inactive mines—are not being used. The Bureau of Mines, closely tied to the coal industry and hardly a champion of pollution control, said as much in its 1969 report on "Environmental Effects of Underground Mining and of Mineral Processing." † Pollution from mining activities can be attributed in part, the report concludes, to the fact that the industry has made "*minimal* application of *available* preventative and control measures . . ." (emphasis ours).[25]

Use of available techniques is, unfortunately, still optional; pollution control laws have never shed their point-source bias. Typically, the laws do not give control authorities any right to intervene unless they can trace a pollutant in the river back to the polluter who dumped it in. Many state laws prohibit *dumping* mine wastes directly into a stream. But there is nothing illegal about just letting them *drain away*.‡ Research will never cure that problem. As Assistant Secretary of Interior Frank DiLuzio testified in 1967 before the Senate's Subcommittee on Air and Water Pollution, "Some of the most important preventive techniques are not scientific in nature. They are legal, sociological, and economic." [26]

The same is true of the other non-point source pol-

---

* According to the 1969 Bureau of Mines Report, even if known methods of control had been universally applied during the period 1965–1975, the mine drainage problem would have continued to grow. "Only the development of control technology which approaches 95 percent effectiveness and its total application between 1970 and 2000 will reverse the present pollution trend."

† The report was obtained and released to the press by Ralph Nader, who charged that it was being suppressed. As of February 1971, the report had still not been published, and the recommendations to Congress, envisioned by the Johnson Administration in ordering the study, had not been made.

‡ Pennsylvania is the only state whose laws require that mines be sealed upon abandonment to prevent water entry and drainage. Even Pennsylvania's law has no provision requiring maintenance of the seals or providing for the assignment of damages which might occur from the failure of a seal. There are no laws regulating drainage from active mines.

lutants as well. For the most part, stopping farm or real estate runoff is no more a question of new scientific techniques than is bringing a halt to acid mine drainage. In each case, it will take laws that require the polluter to control his drainage before it starts, rather than laws that require pollution control authorities to prove, after the stream has been polluted, where the offending wastes originated.

Part Three

POLITICS, ACTION, AND INACTION

6

Discretion and Dereliction

Unless pollution carries a penalty, the individual interest is to pollute. State and local governing bodies first began to acknowledge this fact officially in the middle 1800's when the first pollution control laws were passed. Sometime before the turn of the century most states had empowered their Public Health Services to step in and force cleanup when pollution became a serious health threat. It gradually became apparent, however, that state regulation alone would not suffice to protect the public from pollution damage. Like individual industries and cities, the states showed little interest in what went downstream when downstream was in another state.

The Federal Water Pollution Control Act of 1956 embodies the first primitive recognition of that fact at the Federal level.* Although several revisions of the law since 1956 have expanded Federal jurisdiction and powers, the keystone of Federal authority has always been the "enforcement conference." The word "enforcement" has rarely been so grossly misused. The conference is the classic embodiment of the currently fashionable notion of "consensus" or "partnership" government.

A Federal enforcement conference may be called at the initiative of the Administrator of EPA when water pollution crosses state lines, or by the Administrator in response to a request by a state governor when a pollution problem is intrastate. (The statutory authority for enforcement has always belonged to the man with overall administrative responsibility for the Federal pollution

---

* The 1956 law (P.L. 84–660) was actually a permanent set of amendments to the original Federal Water Pollution Control Act (P.L. 80–845), a temporary piece of legislation enacted in 1948, which was never used. Hereinafter, when we speak of the Federal Water Pollution Control Act, it is the permanent 1956 amendments to which we are referring.

program—i.e., the Surgeon General until 1961, the Secretary of HEW until 1966, and the Secretary of Interior until the December 1970 reorganization that created EPA.)

In addition, the "shellfish clause" allows the Administrator to call an enforcement conference, without regard to whether the pollution is interstate or confined within one state, if he finds that the interstate sale of shellfish suffers as a result of the pollution.

The language of the Pollution Control Act abounds with imperative phrases like "the [Administrator] *shall* . . . call . . . a conference" (emphasis added). At first glance these phrases might suggest that the Federal enforcement power is what is known as a "mandatory" power. A law enforcement official subject to a mandatory duty is actually a lawbreaker himself when he fails to carry it out. Persons to whom such a duty is owed— in a pollution law, for instance, citizens offended by pollution—are entitled to obtain a court order compelling the official's performance. In point of fact, however, it is in the nature of the essentially political function which high-ranking political appointees are expected to perform that they are rarely given mandatory responsibilities. The Pollution Control Act is no exception.

There is, for practical purposes, only one circumstance in which he is legally compelled to call an enforcement conference—when a governor or state agency requests a conference to deal with a pollution problem that is claimed to be interstate. In all other cases, the EPA Administrator is obliged to call a conference only if he has "reason to believe," on the basis of reports or studies, that the water pollution exists. The catch is that the Administrator is under no obligation to make these studies or look at the reports. So unless state officials ask the Federal government to step in, no interstate pollution problem, no matter how ominous— no request from a governor regarding *intra*-state pollution, no matter how urgent—no damage to the shellfish industry, no matter how pervasive—can force the EPA Administrator to convene an enforcement conference. The where, the when, and the why of a decision to convene a conference are almost entirely within the realm of the Administrator's discretion—and good will.

The conference itself is exactly what the word implies, a convening, a coming together to discuss mutual problems. The only participants who get official invitations from the Administrator of EPA are the so-called conferees: one representative from EPA's Water Quality Office, one from as many state water pollution control agencies as are affected by the pollution, and one from any interstate agency involved (the Delaware River Basin Commission, for example). But with a scrupulous concern for political etiquette, the Act permits these official parties to bring all their interested friends. Polluters and their victims all get a chance to speak, some to equivocate and apologize and others to call loudly for reform. All are heard, but some are listened to more attentively than others.

No testimony is offered under oath, no formal rules of evidence apply, no judge, jury, or administrative tribunal sits at these conferences. These are purely informal convocations, marketplaces of pollution control ideas, from which the conferees are supposed to issue advisory recommendations to the Administrator of EPA. When "official" cleanup recommendations are adopted by the Administrator, who may or may not heed the suggestions of the conferees, they are purely exhortatory and have no force as law.

After a minimum of six months has elapsed, the Administrator—again at his complete discretion (if he determines that satisfactory progress has not been made) —may go on to the next stage of the procedure by convening a Hearing Board. The deliberations of the Hearing Board are formal, but nevertheless result in yet another set of "recommendations." Another six months must elapse before the Administrator may—once again, this is discretionary—invoke the third and final stage of the conference procedure. He may ask the U.S. Attorney General (who presumably could exercise his own discretion and refuse) to initiate injunctive proceedings in Federal court to abate the pollution.

If at the end of this lengthy process, which may include several appeals, the government finally succeeds in obtaining a court order to abate the pollution, the penalties for violating that order under the Pollution Control Act are nebulous. They depend entirely on the

discretion of the judge. In other words, a polluter weigh-
ing the costs against the benefits of continued pollution
will find in the Act no visible penalty—no fine or criminal sentence—that might alter his calculations.

As one might suppose, the idea of holding enforcement conferences was not strongly opposed by industry.
In fact, it was included in the Act partly on their behalf,
"probably because they thought it would serve as a
delaying measure," recalls Jerry Sonosky, former aide to
Congressman John Blatnik (Dem.-Minn.), who sponsored the Pollution Control Act. The record would suggest that industry was right. Since 1956 the Federal government has initiated official discussions on 51 different
pollution trouble spots,* and none of these conferences
is listed as having been officially "closed." Many have
been reconvened several times in the years since their
initial session to evaluate new information on the problem, see if any progress has been made, and resume the
conversation where it left off. In only four instances
has the government proceeded to the Hearing Board
stage. And except for a single uncharacteristic case in
which the Federal government's monumental patience
wore thin—against the city of St. Joseph, Missouri, in
1960—the final court stage has gathered dust in the
back of the statute books.

Murray Stein, currently the Water Quality Office's
Assistant Commissioner for Enforcement, has been the
architect of the real-life conference format, since he has
chaired most of these meetings since 1956. The Act
contains few procedural rules regarding the conduct of
the conference, and Stein made up his own as he went
along. Stein's Brooklyn accent and straight-talking demeanor make him a colorful figure and center of attrac-
tion at these convocations. But this does not mask the
fact that he is a shrewd, knowledgeable, and, when the
occasion requires, appropriately diplomatic man who
is widely respected in the water pollution fraternity.

Stein has spent the better part of his professional life
trying to squeeze what little cleanup he can out of this
essentially unworkable procedure and, by focusing public pressure on polluters and local control agencies, has

* As of February 1971. A list of the government's enforcement conference actions is included as Appendix A.

usually had some effect. But from the very outset of the enforcement program, the strictures of the statutory form have proved a serious handicap.

If the Federal government's anti-pollution weapons have been unreliable, the targets have at least been easy to find. By the time the enforcement program was inaugurated in 1956, water pollution was everywhere, just as it is now. Every single one of our major urban rivers has been grossly polluted for so long that data collection for enforcement rarely results in an authentic discovery —only a confirmation of one's worst suspicions.

Two distinct geographical areas can be singled out as having been most drastically affected. The waters adjoining our major Eastern urban-industrial centers have carried a heavy pollution load for more than 100 years. And economic expansion through the upper Midwest carved out a swath of territory from western Pennsylvania and New York through Illinois and up to Minnesota so thoroughly contaminated that it became in a very real sense the nation's "pollution belt." The most critical pollution problems facing the Federal government's new enforcement program in 1956 were to be found in these two regions.

Among the highest of these problems on any rational list of priorities would have been the contamination of Lake Erie, the world's most heavily polluted large body of water. The lake's death rattles were audible as early as 1953, when the local Mayfly population—a nuisance to lakeside residents but a staple food for its fish—mysteriously disappeared. Shortly thereafter a search party of curious scientists found the Mayflies in a mass grave —millions of them—at the bottom of the lake, where they had choked for lack of oxygen. Their extermination was only the most dramatic sign of Erie's gradual demise. An inexplicable decline in the lake's herring catch in the early 1920's had been followed, in 1926, by a sudden shocking drop in the catch of cisco fish from an annual average of 25 million pounds down to 6 million pounds. Decreases in the critical dissolved-oxygen levels were first noted in 1929. And in 1965 it was found that the oxygen deficit was 2.6 times as great as the 1933 measurement.[1]

But well before Erie became a *cause célèbre* a number of pollution-belt rivers were also being openly and flagrantly fouled. The upper Mississippi, particularly around Minneapolis-St. Paul and farther down at St. Louis; the Calumet River and her many brothers and sisters (the Grand Calumet, the Little Calumet, etc.), which join the southern end of Lake Michigan near Gary, Indiana; Youngstown's Mahoning River; Pittsburgh's Monongahela; Charleston, West Virginia's Kanawha—in fact, the entire network of rivers and streams emptying into the Ohio River and, of course, the Ohio itself: all were terribly polluted.

Given the fairly well-defined geographic scope of these pollution problems—problems more serious by orders of magnitude than those facing other sections of the country—one might expect to find that the Federal government had concentrated its enforcement efforts primarily in this area. But one need only scan the enforcement program's roster of past abatement actions to discover that these are, for the most part, the very places that the government has tried hardest to avoid.* Not a single case was brought in the heavily industrialized Great Lakes region during the first six years of the Federal government's enforcement program. Long-standing Eastern trouble spots such as the Delaware River (at Philadelphia) and the Hudson (which runs through New York City) were similarly neglected. The Surgeon General of the Public Health Service (who was then vested with the authority to initiate enforcement actions) did commence proceedings at St. Louis, Missouri (the largest Mississippi River polluter), Kansas City, Missouri (on the Missouri), and Raritan Bay in New Jersey, three big industrial-population centers whose heavy output of organic waste pollution posed serious dangers to human health. (The Raritan Bay conference, for example, came in 1961—five years after the outset of the enforcement program—immediately upon the heels of an outbreak of infectious hepatitis traced to clams from the bay.) But with those few exceptions, the worst places were ignored altogether.

A review of the history of Federal enforcement reveals that the government fell early into a pattern of

* See Appendix A.

reluctance to proceed where its presence was most needed. It has, to the present day, rarely moved to abate concentrated pollution except when an official state request has made it legally obligatory or in belated response to a burst of public pressure. In the meantime, while places like the Hudson River and Lake Erie languished, the government stalked pollution in the nation's creeks—the North Fork of the Holston River, Big Blue River, the Corney Creek drainage ditch, and so on.

As pollution control became more popular in local politics in the 1960's, the Federal enforcement program began to back into a number of the places it had previously avoided. An enforcement conference on the Detroit River, the major source of Lake Erie's pollution, was called in 1962 at the insistence of Michigan Governor John Swainson. It took prodding by Senator Robert F. Kennedy and invitations from Governors Rockefeller and Hughes to produce the 1965 enforcement conference on the Hudson River, where 540 of the 575 miles of waterfront had been declared dangerous for recreation since 1915. And in the same year, more than eight years after the inception of the enforcement program, the Federal government's enforcers finally held a conference covering the whole of Lake Erie—but only after Governor Rhodes of Ohio asked for it.

About this time Interior Secretary Stewart Udall was campaigning vigorously at the White House and in Congress to bring the pollution control program out of the Public Health Service in HEW and into Interior, where, Udall claimed, it would receive the attention it had long deserved. As the likelihood that HEW would lose out to Interior increased, HEW Secretary Anthony Celebrezze put on the only real display of energy the enforcement program has seen before or since. The Federal government actually initiated enforcement actions, entirely on its own, in two major interstate industrial trouble spots—the Calumet River in Chicago and the Mahoning River in Youngstown, Ohio—and in a number of less notoriously polluted locations as well.

The 1966 executive reorganization came in spite of this brief flurry of excitement, and it quickly became apparent to clean-water enthusiasts who had pushed for the move to Interior that the new aggressive enforcement policy they had hoped for would not materialize. The Federal government's enforcement program went into virtual retirement for three years. (Not only did FWQA shy away from those heavily polluted interstate waters which still had not seen Federal enforcement, but the government also stopped reconvening and checking up on conferences it had opened earlier.) The Federal enforcement program's only assault on an important industrial-pollution center in the years immediately following the shift to Interior was the 1968 Lake Michigan conference, called at the request of Illinois Governor Otto Kerner.

The Nixon Administration boasts (in its June 1970 pollution control status report, *Clean Water for the 1970's*) of "stepped up" conference activity since the Republican takeover in January 1969.[2] That claim is not wholly unsubstantiated. The government has at least begun reconvening old conferences (eight of them at last count) to check on progress in places that FWQA had visited earlier and then simply forgotten about. But of the five *new* enforcement conferences that the Department of Interior points to as evidence of its accelerated pace over the past two years, only one—in Mobile Bay, Alabama—was initiated by the Nixon Administration. Three of them (Perdido Bay in Alabama, Escambia River and Biscayne Bay in Florida) were state-requested. The fourth, at Lake Superior, came under the January 1969 signature of Interior Secretary Stewart Udall, his last act before the Democrats relinquished their posts.

How did the government manage to ride out the spring 1970 storm of national political pressure for environmental protection without ever taking the Federal enforcement program out of low gear? One might imagine that Washington had, after all these years, eventually been embarrassed and dragged into so many serious interstate pollution spots that there are not many left to ignore. That is, unfortunately, not the case. There are still a vast number of cases of acute—and distressingly obvious—pollution that the Federal government has left entirely undisturbed.

The government has had its eye on most of these prospective enforcement locations since at least 1963. Enforcement chief Murray Stein submitted for the rec-

ord that year during hearings of the Special Subcommittee on Air and Water Pollution a list of some 220
polluted water areas whose dossiers were on file in the
enforcement office, 90 of which "really need some looking into right now." [3] Or, as Senator Muskie rephrased it,
"which could conceivably result in enforcement action
beginning with the conference stage." [4]

When Senator Jack Miller (R.-Iowa) asked Stein if
the Enforcement Office planned to be more active in the
future than it had been over the previous six years, the
enforcement chief replied cryptically that this "would
depend on the policy."

SENATOR MILLER: You indicated that the possibility of having these cases grow out of the 90 instances that you
   mentioned would depend somewhat on policy. Would you
   elaborate on that, please?
MR. STEIN: Well, obviously this depends on how active a
   program, how active an enforcement program we are
   going to have. . . .
SENATOR MILLER: Well, is your policy likely to change from
   the way it has been? Do you have any changes in policy
   in mind?
MR. STEIN: I don't have any indication that there will be a
   change.[5]

Seven years after Stein made that prediction, an
examination of his timeless list of 90 interstate pollution
spots suggests that there has indeed been little "change"
in the "policy." Sixty-eight of the 90 polluted waterways
have still not seen a Federal enforcement conference.
While some of the 68 are smaller rivers (or separately
designated sections of larger rivers), the inventory also
includes such indisputably major bodies of water as the
Arkansas River, the upper Connecticut River, the Mississippi estuary, the upper Potomac, the Rio Grande, the
Susquehanna, the Ohio, and the Delaware.*

Moreover, the potential enforcement targets on that
1963 list are not the only polluted places in which the
Federal government has voluntarily stayed its hand. In
1963 the Federal government had jurisdiction to initiate
action only when pollution from one state injured persons in another state, and so the list included only cases

* The 1963 list is included as Appendix B.

of interstate pollution. In 1966 the government got the "shellfish clause"; Federal jurisdiction under the Pollution Control Act was expanded to take in those places where "substantial economic injury results from the inability to market shellfish or shellfish products in interstate commerce, because of pollution. . . ." This boost in authority meant that in situations where pollution remained wholly within one state, FWQA could now intervene without first receiving a request from the state's governor, provided that marketable shellfish were being damaged. But given the Federal program's policy of peaceful coexistence with pollution, the shellfish clause has served only to lengthen the list of polluted places that could have had Federal enforcement pressure for cleanup—but never got it.

The shellfish clause could have been invoked in 427 areas, where, according to the Public Health Service, 1,750,900 acres of shellfish beds have been completely closed to shellfishing,[6] as well as in an untold additional number of areas where marketable shellfish *would* be found had they not been eliminated from their old habitats by pollution. The shellfish clause could have been used effectively against almost all types of polluters, since discharges of toxic metals, bacteria, and most other pollutants including simple settleable solid matter either contaminate shellfish or destroy the environment they need. But FWQA has invoked the shellfish clause to conduct enforcement only five times since its enactment.*

The places where the government has allowed shellfish damage to continue uncontested are to be found in virtually every coastal state in the country. In the Northeast, for example, a January 1968 Shellfish Sanitation Technical Report of the Public Health Service suggested that FWQA initiate enforcement conferences under the shellfish clause for 17 areas in New Jersey, four in Maine, two in Connecticut, and one in New York.[7] The coasts of the Carolinas, Georgia, Florida, Alabama, Mississippi, Louisiana, and Texas have scores of places

* At Maine's Moriches Bay (1966–67) and Penobscot River (1967), eastern New Jersey (1967), Boston Harbor (1968–69), and Mobile Bay, Alabama (1970).

which have been closed to shellfishing. In the West, pollution has affected the oyster industry since the early 1900's. Oyster production in California was 3.4 million pounds in 1899. It had dropped to 700,000 pounds by 1925, due in part to heavy pollution of the bays of San Francisco and San Diego.[8] Today in the San Francisco Bay Area the Association of Bay Area Governments has reported that some 3 million tons of refuse a year are dumped by 83 collection agencies at 77 sites.[9] California's Bolinas Bay is also too filthy to support shellfish, as are several bays in Oregon and Washington.

If one were to tour the hundreds of polluted water-bodies that have never been admitted to the select circle of Federal enforcement targets, he would find a few places—very few—where state and local control officials have been taking rapid strides on their own all these years. He would find a few—again very few—where the pollution problems are realistically just not quite so serious as they are on many of the Federal government's favored 51. But he would find a disconcertingly large number of places with severe pollution problems that the Federal enforcement program has chosen to ignore —like the ones described below.

The Delaware River

The Delaware River flows southward out of New York State's Catskill Mountains to form the border between New Jersey on the east and Pennsylvania and then Delaware to the west. It hosts one of the largest petro-chemical complexes in the East. Traveling northward upriver from Delaware Bay, one sees huge refineries and chemical-processing plants clustered along the banks from DuPont's home base in Wilmington, Delaware, through the giant urban-industrial center at Camden, New Jersey, and on to Philadelphia—the largest of a number of serious municipal polluters.

"The Delaware is a dirty river," begins FWQA's 1965 *Delaware Estuary Comprehensive Study*. The pollution, the study notes, has been "self-evident" for "three generations." [10] Water sports are out of the question, since bacteria levels are high and toxic industrial discharges are a regular occurrence; the water in many places is a dark broth of industrial acid and oil, laced with fresh human feces from Philadelphia's perpetual sewer overflow.

The lower Delaware is alive with fish once a year when the plucky shad set out for spawning grounds upstream. The journey is treacherous at best, but if they time their try to coincide with the yearly spring flush of runoff melting snow, most make it. If the shad start late, however, the polluters are sure to win; straddling the Delaware-Pennsylvania state line is a 20-mile zone of certain death where, during the hottest summer months, the water's oxygen content dips to zero. If the shad run up against the summer Delaware, local residents say, the event always ends with tens of thousands of shad, belly up, floating slowly to the ocean, putrefying as they go. The hardy eel and an occasional carp are the only fish that survive for several miles south of Philadelphia. FWQA's 1965 study of the Delaware reported that fishing along the entire estuary was at only 8% of its potential.[11]

The Delaware can probably look forward to eventual redemption despite its undeserved immunity from Federal enforcement to date. The Delaware River Basin Commission (DRBC), a political hybrid comprised of representatives of the four basin states (New York, Pennsylvania, New Jersey, and Delaware) and a representative of the Federal government (who is appointed by the U.S. President) has set cleanup requirements that, on paper, promise to make the Delaware one of the cleanest urban rivers in the country. But the paper dreams are far from being realized. The timetables for abatement are generous—Philadelphia, the river's largest discharger, has until October 1977 to complete construction of its new treatment facilities. And the DRBC's enforcement powers, unused so far, are so weak that a Federal boost may still be needed. Fines range from $50 to $1000, less than the daily treatment costs of many large polluters.

Assuming the day of improvement finally does arrive, it will still have been a long time in coming. The annual shad catch on the Delaware has decreased from 19 million pounds in 1896 to no more than 50,000 pounds in recent years. The river was already so foul by 1900 that ferry boats traversing the channel churned up enough hydrogen sulfide gas to peel the paint off shoreside buildings.[12] The last noticeable change for the better was in 1954, when Philadelphia installed primary treatment. It has been all downhill since then. Until 1968 the DRBC—to whom the Federal government has deferred —had not set a single cleanup requirement.

The Federal government has limited its activity on the Delaware to beginning a study of the problem in 1957 (when the water program was under the Public Health Service) at the request of the Army Corps of Engineers. FWQA has been studying the Delaware ever since. Since the beginning of the Federal enforcement program in 1956, municipal and industrial wasteloads on the river have approximately doubled.[13] No one but the polluters benefited while the Federal enforcement program waited this one out.

The Ohio River

Most rivers start out as a babbling brook and become polluted later in their journeys to the sea. The Ohio River has never been given that fighting chance; it starts out already grossly polluted with steel mill wastes and acid mine drainage at Pittsburgh, Pennsylvania—where the Allegheny River from New York State joins the Monongahela from West Virginia to form the Ohio— and stays polluted from there on. The Allegheny and the Monongahela are only two of more than a dozen major rivers—most of them highly industrialized—that flow into the Ohio as it winds its way westward out of Appalachia. Among them are the Beaver in Pennsylvania; the Kanawha in West Virginia; the Miami, Scioto, and Muskingum in Ohio; the Big Sandy and the Tennessee in Kentucky; and the Wabash in Indiana. By the time it enters the Mississippi River at Cairo, Illinois, the Ohio has picked up pollution from 12 different states.*

* The Ohio River Basin drains the states of Indiana, West Virginia, Ohio, New York, Illinois, Kentucky, Pennsylvania, Virginia, Tennessee, Alabama, North Carolina, and Georgia.

The Ohio River Valley is the birthplace of the nation's steel, coal, and chemical industries; the river itself was a menace to health before the turn of the century. Stepped-up production during two World Wars forced Ohio Valley steel plants to expand without heed to pollution. By the end of World War II the Ohio was virtually dead for most of its 981 miles.

In 1948 a major effort to clean up the Ohio and its tributary rivers was initiated by the newly formed Ohio River Valley Water Sanitation Commission (ORSANCO), set up under a compact among eight of the basin states. ORSANCO drafted a set of anti-pollution standards and encouraged industrial and municipal polluters to follow them. The response was truly impressive. Over 1400 municipalities and over 1500 private companies installed primary treatment—a number that nearly matches the Federal enforcement program's abatement record for the entire nation over its 14-year existence—at a cost of more than $1 billion, and the river began to breathe again. There was pleasure boating on the Ohio and, in some places along the stream, water skiing and even swimming for the intrepid, where there had been none before. But only the most obvious contamination had been eliminated. Since that time the Ohio River has seen little but failure.

There is still boating on the Ohio in 1970, but in a number of places a strong stomach is recommended equipment. The powerful stench of raw sewage drifts along the Ohio and its tributaries near the 295 communities that still have no sewage plants. Few of the Ohio Valley communities that do treat their wastes provide more than the obsolete primary treatment they installed in the 1950's, including Pittsburgh, Cincinnati, Louisville, Evansville, Indianapolis, and Dayton.

The massive municipal pollution in the Ohio Basin cannot match the even greater industrial violence. The Ohio is the nation's industrial river, and throughout the entire Ohio Valley region the waterborne wastes from one manufacturing site run into and mingle with the wastes from the next ones down the line. The Monongahela, for example, comes in at Pittsburgh still reeling from a losing bout with one of the largest concentrated steel-making complexes in the world—U.S. Steel, Bethlehem, Jones & Laughlin, and National Steel. Five miles later the river takes on an enterprising middleweight, Shenango Inc. (another steel-maker), and readies itself for the blow from Jones & Laughlin at Alliquipa, Pennsylvania. Just a few miles farther on at Midland, Pennsylvania, pipes from a Crucible Steel Corporation titanium plant pour a bright green, poisonous fluid into the Ohio along with huge quantities of iron oxide and other solids. The bottom of the river near Crucible Steel is covered with the company's soluble oil, visible at the surface near the outfall before it coalesces and slowly sinks. At Steubenville, Ohio, the Wheeling-Pittsburgh Steel Corporation wields a powerful one-two combination punch: the company's blast furnace on the Ohio side of the river spews out suspended solids, oils, cyanides, and iron oxide, staining the river a reddish brown; across the river at Follensbee, West Virginia, the Wheeling-Pittsburgh coke plant dumps in high concentrations of phenols, coppers, and—like its brother plant on the other side—cyanides.

Interspersed among the large steel companies in the upper Ohio region are scores of smaller steel fabricators, coal tar refiners, and chemical firms. Most of them leave their marks on the Ohio too. At Potter Township, Pennsylvania, Sinclair-Koppers Company has completely covered the bottom of the river with plastic beads from its styrene-manufacturing operations. Pollution control authorities have not yet analyzed the chemical content of the river's new synthetic bottom, but they do know that the plastic pellets make their way at least 10 miles downstream from the plant, where there are still 100–1000 on each square foot of river bottom. The Ohio River runs a gantlet of industrial polluters like these all the way to the Mississippi. Around 240 of the industrial plants on the Ohio and its battered tributaries have yet to comply with the minimum control requirements set by ORSANCO in the 1950's.[14]

The Federal government knows the Ohio River is polluted. The Ohio held one of the places of honor on a list of the nation's 10 most polluted rivers which FWQA made available to the media in the spring of 1970.* So

* The entire list is reprinted in Chapter 9.

far, however, the government has limited its enforcement activity in Ohio Valley territory to conferences on two of the Ohio's tributaries. The enforcement program tackled acid mine drainage problems on West Virginia's portion of the Monongahela River in 1963; and in 1965 the government held a conference on Ohio's Mahoning River.

By 1968 some of the Ohio Basin states had begun trying to push their polluters to higher treatment requirements, a delayed reaction to Federal legislation passed in 1965. (See Chapter 14.) The effort had become more general by 1970. But the majority of polluters have not yet moved beyond the "planning" phase of cleanup, and completion dates in some places are as far away as 1977. Federal authorities have learned to be cynical about Ohio Basin industries and suspect that most companies won't be ready by their deadlines. For the last few years, says Bill West of the Federal Water Quality Office's Ohio Basin Regional Office, "progress has been *very* slow—industry-wise." That's a sign of improvement, nevertheless. Before that, progress was virtually nonexistent. For more than a decade it has been business as usual on the Ohio and its poisoned tributaries.

The Houston Ship Channel

Water pollution control officials in Houston, Texas, worry as much about fire as they do about water. "At Houston we have the most dangerous port in the nation, if not the world," says W. L. Farnsworth, who heads the National Cargo Bureau—the inspecting agent for insurance companies that write policies on the ships and cargoes moving through the port. "Seventy percent of the cargo which moves through the port is classified as dangerous." [15] It is perhaps fitting that all this explosive cargo, including 90% of the poisons moved by water in the U.S., must traverse the nation's most poisoned, and potentially the most explosive, body of water: the Houston Ship Channel.

The Ship Channel was constructed in 1914. It twists and turns for 52 miles from the Gulf of Mexico, through Galveston Bay, and deep into the Texas coastal prairie before it ends five miles from downtown Houston. The Ship Channel was relatively unpolluted 40 years ago, before industrial growth spurred by the easy access to shipping tripled the population of metropolitan Houston. Now the last 12 miles of the 400-foot-wide channel play host to the partially treated and untreated industrial effluent of more than 1000 factories and the domestic sewage of 2 million people.

Fire can erupt anywhere along the crowded channel: in the chemical plants and oil refineries that line its banks, in the ships that ply its waters, and—most frightening—in the waters themselves. Oil, chemical, and refuse discharges place the Houston Ship Channel high among the nation's aquatic fire hazards. The oil- and chemical-saturated water could quickly transform even a small fire into a major catastrophe. What have the local authorities provided to meet the obvious dangers? One 20-year-old fireboat that can make a maximum of eight knots with the throttle jammed full speed ahead—and nothing else.

In comparison with governmental efforts taken to guard against the other more subtle dangers presented by the Houston Ship Channel's pollution, the little fireboat represents a major investment. In addition to being the most polluted waterway in the U.S. and probably in the world, the Ship Channel may also be the most talked about. Cabinet Secretaries and their assistants, Presidential commissions, and local politicians vie with one another in lamenting the sorry state of the water. To date, however, no one has seen fit to apply the sanctions necessary to clean it up.

The channel receives a daily wasteload equivalent to the raw sewage of 2 to 3 million people. It would remain contaminated for years to come even if all the discharges were to dry up instantaneously, because the channel bottom is covered by a blanket of polluted silt and sludge that Texas A&M University environmental engineering professor Dr. Roy Hann estimates is now two feet thick.[16] The filth that does not settle on the bottom floats into Galveston Bay, a 533-square-mile estuary now threatened with the loss of its valuable fishing and recreational resources. *Science* magazine reports:

The bay still supports major commercial and sports fisheries, and oysters, shrimp, crabs, and redfish, sea trout, and other finfish are plentiful. However, nearly half the bay is now closed to oyster harvesting because of pollution. . . . Fishery biologists are worried that, given the degradation and numerous man-made changes in the bay environment, the bay's productivity for marine life will decline. And since the bay is an important nursery for shrimp and certain fish . . . which spend part of their life cycle in the Gulf of Mexico, conditions which hurt fishing in the bay will hurt fishing in the Gulf also.[17]

Enforcement efforts on the channel have been confined to a few hit-or-miss prosecutions by the Harris County air and water pollution control agency during the 1950's and early 1960's. In 1967 the Texas Water Quality Board for all practical purposes pre-empted most of the activities and powers of the Harris County control agency. At that time the state board had only four full-time employees of its own. Today the largest state in the lower forty-eight can boast of 40 water pollution enforcement officers for an area larger than all of Illinois, Indiana, Iowa, Wisconsin, and Michigan combined. The Lone Star State enforcement officers do not enforce. Most of their time is spent persuading industries and municipalities to apply for permits to pollute, a full-time job even though the board has shown itself a willing accomplice to almost any kind of environmental mayhem. The ostensible aim of the permit system is to insure that certain water quality goals or standards will eventually be met. There are only three problems:

1. The state standards themselves are so permissive as to guarantee the continuing pollution of Galveston Bay.[18] At the turning basin in Houston where the Ship Channel ends, the dissolved-oxygen requirement is fixed at 1.5 parts per million—well below the 4.0 ppm needed for fish to survive and the 2.0 ppm recommended as the lower limit for industrial water supply, and not much more oxygen than can be found in an open sewer.

2. The permits held by channel industries allow for such wholesale degradation that even these minimal water quality standards cannot be met.

3. Violations of permit conditions are frequent.

The result of the state control program is predictable. According to Dr. Walter A. Quebedeaux, Jr., director of the county control agency, pollution is a great deal worse on the channel than it was 10 years ago and is likely to be much worse 10 years from now. Federal authorities agree. The most hopeful prognosis came from William B. Halliday, environmental director for the Atlantic-Richfield Corporation. "You used to be able to walk on the water," Halliday maintains. "Now the channel is getting a little spongy."

A few years ago the Texas Water Quality Board began to show what, for it, was a lively interest in the subject of pollution from oil-field brines, a recurrent problem in the Houston area. Minions of the petroleum lobby in state government responded quickly by transferring regulatory jurisdiction over oil brine to the Texas Railroad Commission, whose primary function is to fix prices for the oil industry. This was the last anyone heard about oil pollution.

Federal interest in the Ship Channel peaked in 1967, when President Johnson's Water Pollution Control Advisory Board—a high-level fact-finding body composed of civic, business, and conservation leaders—wound up a whirlwind tour of Texas by declaring that the Houston Ship Channel is "overwhelmingly polluted." [19] The highlight of the Advisory Board's visit had been the ride around the channel on the *Sam Houston,* the city's inspection boat, in midsummer when the fragrance is especially acute. James Quigley, then Commissioner of FWQA, led the tour and had harsh words for Houston's polluters. "Look at that outfall off the starboard beam," he said, manning the boat's public address system. "That's pure Champion water." Quigley did not need to point. Everyone present could see the gushing torrent of orange-brown water coming from a huge pipe just below the Champion Paper Company. "There's only one Champion," the Commissioner told his fellow passengers, "not only for water pollution but for air pollution." [20]

Two and a half years later the Advisory Board returned. Once again the group was squired around the channel, this time with Assistant Secretary of Interior Carl Klein at the microphone. (Commissioner Quigley had long since left the government to become director of environmental affairs for Champion Paper.) Once again board members deplored the gross pollution of the channel. Once again most agreed that there was only one Champion, although Diamond Alkali Corporation, Shell Chemical, Sinclair Refining, Signal Oil, and Humble Oil and Refining were gaining.

Having twice dispatched a Presidential Advisory Board to Houston, Federal officials seem to feel that they've gone about as far as they can go. When the Task Force asked FWQA's South Central Region Enforcement Officer Edward Lee about the possibility of Federal enforcement, he dispelled all doubt on that matter: "There has never been, nor do we anticipate, a Federal enforcement action on the Houston Ship Channel."

Were the Federal government not so determined to let Houston stew in its own juice, the Ship Channel would appear to be an ideal place to invoke the "shellfish clause." The Administrator of EPA "*shall* call . . . [an enforcement] conference," the Pollution Control Act declares, "*whenever . . . he finds* that substantial economic injury results from the inability to market shellfish . . . in interstate commerce, because of pollution . . ." (emphasis added). Apparently neither the Secretary of Interior nor the Administrator of EPA, during his brief spell as head of the pollution control program, has been able to "find" the fact that 48% of the Galveston Bay complex has been declared off-limits for commercial production of oysters due to pollution. How "substantial" is the economic injury? Noting that the Houston oyster crop is currently $11.6 million each year, longtime pollution foe Congressman Bob Eckhardt (D.-Tex.) figures that, roughly speaking, Houston is "losing approximately $10 million annually in protein supply of oyster meat." [21] In keeping with its tradition of straining out gnats while swallowing camels, the Federal government last used the shellfish clause in January 1970 at Mobile Bay, Alabama. Secretary of Interior Walter Hickel found that pollution at Mobile Bay was causing a "substantial economic hardship" of $227,000 annually in lost retail sales of shellfish.

Perhaps when the Houston Ship Channel goes up in flames and takes a goodly portion of Houston with it, someone in EPA's Water Quality Office will realize that his agency could have done something about that problem.

7

Better Late Than Never: The Lake Superior Story

Lake Superior is the largest fresh-water lake in the entire world.* It is also one of the most unspoiled, a treasure of vast beauty fully 31,820 square miles in surface area. This is the great North Woods country, and civilization's only serious encroachment has been in the city of Duluth, Minnesota, at the lake's southwestern corner. The rest of its nearly 3000 miles of shoreline is a primeval wilderness, much of it sheer rock cliffs, uninhabited except for scattered small towns in Michigan and Wisconsin to the south and along the northern shore in Minnesota and Canada. All lakes are mortal, of course. But the lines of age have barely begun to show on Lake Superior. The water is pure and clear—so clear that phantom trout more than 30 feet below can be seen easily from the surface. Because it is "thousands of years behind the other Great Lakes," an FWQA report explains, the "lake nearly resembles its pristine condition as created eons ago." [1]

Lake Superior has been the least studied of all the Great Lakes. The only information the Federal government can provide about the "deepest portions of the middle of the lake" is that "essentially nothing is known about bottom organisms, bottom character, and fish species." [2] But the lake is believed to be, like most creatures that have evaded captivity, extremely sensitive to man's unfamiliar influence. FWQA's report cautions that the lake is "delicate":

* Superior, 350 miles long and 160 miles wide, has a greater surface area than any other lake in the world. Lake Baikal, in Russia, has more water volume.

Increases normally considered insignificant or acceptable in most lakes will dramatically alter this lake. For example, an increase in 5 units in turbidity will result in a reduction of many feet in light penetration and significant loss of fish food organisms. . . . The quality of Lake Superior water is so high compared to other lakes that the early signs of damage may go undetected or may be excused as being insignificant. Using standards of clean water normally considered appropriate in pollution control programs, Lake Superior could be degraded considerably and changed significantly. . . .[3]

Superior's slow rate of drainage makes it even more vulnerable than other lakes to human insensitivity. Lake Erie, for example—which saw a life expectancy of more than 15,000 years stolen away in less than 50 [4]—drains into Lake Ontario more than 10 times as fast as Superior empties into Lake Huron.[5] Once Lake Superior begins to deteriorate, the end may come even more quickly than it did for Lake Erie.

Lake Superior is far removed from the nation's main commercial centers and from the earliest well-developed transportation routes. Until the 1940's the lake had attracted almost none of the industry that settled so much earlier on the other Great Lakes. During the 1950's a few mining, paper, and chemical concerns began to locate in small company towns along the shore, and with the completion of the St. Lawrence Seaway in 1959 the area was ripe for industrial development. Duluth Harbor had already been handling most of the iron ore bound for the steel mills of the lower lakes for a number of years. When the seaway opened, Duluth began to look attractive to other commercial traffic as well and soon became the fifth busiest port (by tonnage handled) in the U.S.

By the early 1960's the all too familiar pattern was beginning to emerge. Fishermen started reporting drops in their catches and algae masses in their nets similar to those which had gone almost unnoticed on Lake Erie 50 years earlier. Senator Gaylord Nelson of Wisconsin asked about the possibility of Federal action as early as 1963. And by 1967 he was calling for a Federal enforcement conference to put an end to what he described as "provable interstate pollution" coming from two small Wisconsin cities—Ashland and Superior—and Duluth, Minnesota.[6]

Nelson was particularly concerned about the activities of the Reserve Mining Company of Silver Bay, Minnesota (about 60 miles north of Duluth on Lake Superior's northern shore). Reserve Mining is a joint subsidiary of Armco Steel and Republic Steel and their principal supplier of iron ore. Reserve mines a low-grade ore called taconite, and uses upward of 350,000 gallons of Lake Superior water every minute in its refining operation. In 1948 Reserve negotiated with the Army Corps of Engineers (which is charged with preventing obstructions to navigation) for a permit to dump the unusable waste "tailings" from its taconite into the lake. Since the company went into full-time operation in 1955, it has deposited more than 190 million long tons of tailings in Lake Superior, and each day—from two gargantuan chutes that extend out from the lakeside plant—it adds another 60,000 tons to the total.

Reserve has always described its discard as "harmless sand" that sinks rapidly to the "Great Trough," the deepest part of Lake Superior, and remains there inert on the bottom. But a Federal investigation conducted in 1967 and 1968 told a very different story. The Army Corps of Engineers and the Department of Interior had agreed in 1967 that all dumping permits should be reviewed with environmental considerations in mind. When the Corps (which still retained the final permit authority) sought Interior's guidance in dealing with Reserve's permit, Interior Secretary Stewart Udall assigned a task force of five Interior agencies to investigate and make a recommendation. Interior's regional director Charles Stoddard was put in charge, and on December 31, 1968, he sent a copy of the study group's findings to the Corps, to Reserve Mining, to Minnesota officials, and to Washington.[7]

The report's conclusions were shocking. Each day's refuse from Reserve, the study showed, contains over a ton of nickel, more than 2 tons of copper, a ton of zinc, 3 tons each of lead and chromium, 25 tons of phosphorus, and 310 tons of manganese, in addition to silica, arsenic, and substantial quantities of iron. Over 5000 tons of the daily deposit has been ground so fine as to remain in suspension for an indefinite period—perhaps inert, perhaps not—traveling around the lake with the currents at rates greater than nine miles per day. While the coarsest particles (approximately 45% of the waste) settle quickly, forming an artificial beach that now protrudes half a mile from shore, the finer tailings travel suspended in discharge water over the Reserve "beach" and out into the main body of the lake. These fine tailings, the study concluded, are in all probability accelerating the eutrophication of the lake. Over one-third of the total amount Reserve has discharged since operations began, the Stoddard Report also claimed, is unaccounted for somewhere in Lake Superior. The report's recommendation was that the mining company be given a maximum of three years to find another location for its taconite wastes and to discontinue dumping into Lake Superior entirely.

One of the first persons to learn of the Stoddard Report's findings was another resident of the Duluth–Silver Bay area, the 8th Minnesota Congressional District's Democratic Congressman, John Blatnik. On the national scene Blatnik is often referred to as "Mr. Water Pollution Control," the founding father of the Federal pollution control program and author on the House side of every piece of water pollution legislation enacted since 1956. Representative Blatnik has long been the chairman of the powerful House Public Works Committee's Rivers and Harbors Subcommittee, to which FWQA must turn for any changes in its legislative authority. (In 1971 he acceded to the chairmanship of the full Public Works Committee.) In his 8th Congressional District, on the other hand, Blatnik is better known as a booster of the taconite industry—backbone of northeastern Minnesota's economy and his district's largest employer.

It was only after visiting Blatnik's Washington office that the Task Force began to appreciate the real dangers facing Lake Superior. A Task Force member who asked the Congressman's staff about taconite was enthusiastically loaded down with pamphlets and publications on mining and related industries. Among them were several promotional brochures put out by the Northeastern Minnesota Development Association (an organization of businessmen formed to attract new industry to the Duluth area), chock full of persuasive come-ons to investment capital in search of a home. A sample:

In the past couple of years Dow Chemical, DuPont, Mallet Minerals, Hercules Powder, Union Carbide, Monsanto Chemical, Spencer Chemical, American Brake Shoe Corp., plus a dozen additional firms, are now operating taconite related plants. But *there is plenty of room for others who can envision the ramifications of a 100 million ton taconite industry in future years.* . . . Much less known but nonetheless potentially highly profitable are Northeastern Minnesota's water . . . resources. Over most of the nation *good industrial water in huge amounts* is becoming a scarce commodity. Not in Northeastern Minnesota. Our watershed contains 41 million tons of water and this doesn't even include *Lake Superior representing the world's greatest single supply of fresh water.* . . . With the entire region boom-bound, area manufacturers are expanding operations. [Emphasis ours.] [8]

If northeastern Minnesota is indeed "boom-bound," the trend is a recent one. In the mid-1950's the high-grade natural ores that had made the region the iron capital of the world for more than 60 years were beginning to be mined out and steel-makers started to search for cheaper foreign sources. When Reserve Mining commenced the first commercial taconite operations in 1955, northeastern Minnesota had been on its way to becoming another Appalachia. Congressman Blatnik is understandably close to Reserve, which put an end to the soaring unemployment in his district with more than 3000 jobs and an annual payroll of $31.5 million (1969 figures). He is a good friend of Reserve Mining's president, Edward Furness. According to a Blatnik staff aide, the Congressman and Reserve Mining Company officials worked "shoulder to shoulder" to obtain the passage of the 1964 Taconite Amendment to the Minnesota Constitution, providing for a special tax policy designed to attract new taconite industry to the state (and protect

old companies like Reserve as well).* Such close cooperation cemented "a very strong relationship with the
Reserve people," the aide explained. "They have a real
rapport."

So it was only natural when the Stoddard Report came
out on December 31, 1968, with its recommendation
that Reserve's dumping permit be terminated in three
years that Ed Schmid, Assistant to the President of
Reserve Mining for Public Relations, should telephone
Blatnik's Washington office immediately to express his
outrage at the findings. Schmid's call signaled the beginning of an all-out attempt by Reserve to quash or at
least discredit Stoddard's work. Blatnik was caught in
the middle—torn between his businessmen friends in
Minnesota and his national reputation as a conservationist.

What happened next has been a matter of controversy
ever since. Maurice Tobin, longtime Chief Counsel to
Blatnik's Rivers and Harbors Subcommittee, told the
Task Force he was in the office when the Reserve call
came in. The best compromise the Congressman could
strike, according to Tobin, was to get in touch with
Stewart Udall and ask him to "get ahold of this report"
before it got out, at least long enough to "let the dust
settle a little bit." Blatnik remembers it differently. He
told the Task Force that he did talk with Udall about
the report, once in the Interior Secretary's office and
once in his own. But the Congressman was, in his
words, "just trying to get the facts." And besides, Blatnik added, "Why the hell should the Federal government go into Minnesota to abate pollution in *their*
waters?" Stewart Udall could not recall Blatnik ever
having asked him to hold up the report, but he said he
"certainly knew [Blatnik] wasn't happy with it." Udall
assured the Task Force that he had nothing to do with
the Stoddard study's later suppression at Interior.

Reserve's Ed Schmid also asked Maurice Tobin to

* The Taconite Amendment provided that state taxes could not be
raised on taconite companies for 25 years unless taxes were raised on
other in-state businesses proportionately at the same time. This measure was intended primarily to reassure potential investors that Minnesota would not follow what had been its traditional policy of taxing
mining companies at a higher level than other industries.

lobby on the company's behalf. The Blatnik assistant was just wrapping up his government work at the time, preparatory to becoming an industrial consultant, but he thought better of the Reserve assignment—since anything he did for the company would make Blatnik look bad—and declined. "And besides," Tobin told the Task Force, "I didn't need the money."

Another government official contacted by Reserve was Max Edwards, the Assistant Secretary of Interior for Water Research and Pollution Control, FWQA's administrative superior in the Department of Interior. Like Tobin, Edwards was leaving government to become an industrial pollution control consultant and presumably wouldn't have minded lining up a future customer—Reserve Mining.

Edwards went right to work. He ordered all Interior copies of the Stoddard Report held in his office for "review" and refused to release the study or its findings to inquiring newsmen.* Fortunately, Stoddard had taken precautions to guard against suppression by "leaking" a copy of his report to the *Minneapolis Tribune*.[9] But when asked by newsmen about the Stoddard study, Edwards described it as "not an official document" and "full of inaccuracies" (none of which he specified).[10] FWQA Commissioner Joe Moore did not share his departmental superior's critical opinion of the report. He stated publicly that there were no "great differences of views" between the water pollution agency and Stoddard's findings and defended Stoddard's procedures in preparing the document and sending it directly to the parties involved.[11]

In the meantime the outside pressures on Stewart Udall to take enforcement action had been rapidly increasing in the few weeks since the Stoddard Report first

* Edwards later told the Task Force that he did not withhold the report from circulation entirely, but maintained he "only kept it from members of the press." The Task Force asked him who he *had* released copies of the report to, and he mentioned a Congressman whose name he couldn't remember. He told the Task Force that handling the report in that manner was his decision, but that he kept Secretary Udall "generally informed." Former Secretary Udall agreed with the Task Force that Edwards should have released the report to the public under the Freedom of Information Act. "That's what we normally do in these cases," Udall said. "He should have released it." Udall told the Task Force he had not been aware the report was being held up.

began to circulate in official channels. Minnesota conservation groups had begun to charge that the Secretary was "owned by Blatnik." Wisconsin's aggressive Attorney General Bronson La Follette was demanding Federal action, claiming that the tailings were crossing over into his state. Senator Gaylord Nelson was urging, as he had for several years, that Udall step in. And the Interior Secretary was, he later told us, genuinely impressed by Stoddard's findings. Still, when Udall received a draft of orders for an enforcement conference early in January 1969, prepared by FWQA's Enforcement Office, the Secretary let it remain on his desk unsigned.

The situation was still in limbo a few days later, according to Department of Interior officials close to the Secretary, when Udall received a nudge from a wholly unexpected source. The Interior Secretary had been pushing for a long time to have more Federal land protected from future commercial development by placing it in the national park system, and as the Democratic Administration drew to a close it looked as if he had President Johnson convinced. As a dramatic parting conservation measure, LBJ was going to set aside an additional 7 million acres of parkland—an increase of almost one-third in the total parks system at the time. The Department of Interior had press releases all prepared to go out announcing the increase when Udall learned the plan had fallen through. Reportedly miffed at Udall over a number of personal and political matters, the President had slashed the number of acres in the order from 7 million to a paltry 384,500 acres. Apocryphal or not, the rest of the story points up how much a matter of personal whim enforcement decisions can be. When Udall found out what the President had done, he was reportedly so furious that he went back to his office and signed every pending conservation measure on his desk. The order calling for the Lake Superior conference was one of them. The Secretary's move had come none too soon; Udall handed over the reins to Walter Hickel less than a week later.

The fact that Secretary Udall considered the Stoddard Report the basis for his conference call was still not sufficient to clear the report's name in some quarters, however. In mid-January Congressman Blatnik, who had been in touch with Assistant Secretary Edwards (for "fact finding") as well as Udall, echoed for the press what Edwards was saying about the report. The study, according to Blatnik, had "no official status" and was only a "preliminary report." [12] At the same time the new Administration's Interior appointees had more reason to feel beholden to their Congressional overseers than Udall had in his final devil-may-care days in office. So while preparation for the upcoming enforcement conference proceeded apace, the Stoddard Report stayed buried. The "official" government report, so the story went, was to be the one FWQA was busy preparing for the enforcement conference. (Max Edwards' replacement as Assistant Secretary, Carl Klein, later waved away the Task Force's question about the Stoddard Report with one of his own: "How could I have suppressed the report? I never even *saw* it. But Max Edwards thought it was bad.") And Representative Blatnik's public statements grew bolder still.

The May 11, 1969, *Duluth News Tribune* quotes Blatnik as saying the Stoddard Report on Reserve Mining's tailings "was completely false." * By this time Max Edwards, the first public official to criticize the report from his post as Assistant Secretary of Interior, was out of government and on retainer as a consultant to Reserve Mining.

By the time the enforcement conference opened in Duluth on May 13, 1969, the government was still walking a shaky tightrope between Congressman Blatnik and Lake Superior. The political sensitivity of the proceedings was underscored by Secretary Hickel's unusual choice for conference chairman. Assistant Secretary of Interior Carl Klein headed the gathering, the first time in 46 Federal enforcement actions that FWQA's Murray Stein had not been in charge. If Klein's performance at the conference is any indication, he had been brought there for one reason: to repudiate the Stoddard Report.

---

* In response to inquiries, Blatnik's aides claimed that the Congressman never made that remark. We were unable to find any public denial by Blatnik, however. And the comment the *News Tribune* reported was consistent with remarks Blatnik was making to others at the time. Around March 20, 1969, for example, he called from Washington to tell Minneapolis lawyer Grant Merritt, an activist in Minnesota conservation causes, that the Stoddard report was "a phony."

The Assistant Secretary stayed only one day, just long enough to run through what appeared to be a well-rehearsed routine with Congressman Blatnik. Their dialogue might well have been borrowed from an old Edgar Bergen and Charlie McCarthy script; no one who viewed the scene they acted out had any doubts that it was Congressman Blatnik who was pulling the strings.

The stage was set for the exchange between Blatnik and Klein when the prepared statement of Wisconsin Governor Warren Knowles was read into the conference record. Knowles recommended that the "conference . . . evaluate reports in the news media alleging that a member of Congress has attempted to interfere with a Federal report identifying a major source of industrial pollution in Lake Superior." [13] At the end of the first morning's technical testimony, Congressman Blatnik rose to ask "just one question" of conference Chairman Klein:

MR. BLATNIK: . . . I ask you for a brief comment at this point, Mr. Secretary. Do you or any of your administrators or officials under your jurisdiction to your knowledge, know of any Federal report that has been suppressed?

MR. KLEIN: Congressman Blatnik, you give me a chance to lay the ghost to rest. . . . the official report and the only official report of the Department of the Interior . . . was issued about a week ago. There has been no attempt at suppression by any Congressman or any other Federal official. There is in existence a report put out by an individual who used to be employed by the Department of the Interior shortly before he left and that is his report, despite the fact it bears the words "Department of the Interior." *The Department of the Interior did not authorize it* and is not bound by the report. The only report that was put out officially by the Department of the Interior is this one put out a week ago. [Emphasis ours.] [14]

In laying the ghost to rest, Klein had given him an indecent burial. The Task Force had heard and seen the Stoddard Report dismissed so many times as "unofficial" and only a "preliminary draft" that the explanation had almost begun to sound convincing—until we spoke to former Secretary Udall. The Secretary himself had been the one who had directed Stoddard to prepare the report that according to Klein "the Department of Interior

did not authorize." And under routine Interior procedures that he had established, the former Secretary told
the Task Force, the report did not need to come to
Washington for clearance or final review. It was a final
document when completed by the Regional Director—
Stoddard—and expressed Interior's official position on
the question of Reserve Mining's dumping permit. It
was properly sent by Stoddard directly to the Corps of
Engineers.

The ghost came back to haunt the conference in person on the second day. Testifying as a witness for the
State of Wisconsin, Stoddard explained to the conferees
essentially what Udall later told the Task Force about
his report and charged that Assistant Secretary Klein—
who was no longer at the conference—had "not done
his homework." [15] The remainder of the conference session was characterized by an unconcealed bias in favor
of industry, not only on the part of state officials but—
more unusual—from the Federal government as well.
Minnesota's conferees John Badalich and Robert Tuveson (Executive Director and Chairman, respectively, of
the Minnesota Pollution Control Agency) constantly
manipulated the agenda, which was never distributed,
to comply with the wishes of Reserve Mining's representatives; FWQA's new Commissioner David Dominick,
the conference chairman after Assistant Secretary Klein
departed, exercised his legal responsibility for scheduling
by endorsing the Minnesota conferees' every move. To
one conservationist who complained, when his presentation was postponed for the second time to allow Reserve
witnesses prime time, that after three days of waiting to
speak he had to get back to his business and earn a living, Minnesota's Tuveson replied contemptuously (and
incorrectly), "You're getting paid by *somebody*."

It was clear to those who were in attendance, however, that the witnesses for Reserve Mining had more
going for them than just preferential treatment. They
also had, to all appearances, the balance of expertise at
the first conference meeting. Because the Federal Water
Pollution Control Act requires that pollution be interstate (i.e., "endangering the health or welfare of persons
in a State other than that in which the discharges . . .

originate") before the Federal government has jurisdiction to issue recommendations for cleanup, the battle was necessarily fought on technical grounds. The Stoddard Report's real weakness was not that the evidence of pollution it presented contained "inaccuracies," as Max Edwards had claimed; the Stoddard technical data had been incorporated largely intact into FWQA's later report and formed the basis of the government's conference presentation. The problem, according to FWQA scientists the Task Force interviewed, was rather that the Stoddard team had not gathered *enough* evidence to *prove conclusively,* against all possible technical objections, its claim that Reserve was damaging the lake— let alone to show that the company was causing damage in Wisconsin waters across the state line.

It is important to remember that the question which the Stoddard investigators had been assigned to answer was not the same question that faced the enforcement conference. The Stoddard group's job had been to make an advisory recommendation to the Army Corps concerning Reserve's license to dump, a privilege that the Corps can withdraw for good cause without satisfying the complicated legal requirements in the Pollution Control Act. One thing the earlier Interior study *had* demonstrated was that the mining company was discharging waste particles so microscopic that, according to the best scientific theory, they would be carried in suspension throughout the lake, and this was damning evidence in its own right. But the conference took on a much more difficult task: proving that the tailings actually were traveling as far as science predicted they should. In order to do this the government had to find a tailing somewhere in Wisconsin waters that it could convince the conferees had started its journey in Reserve Mining's discharge chute.

On this question Reserve's battery of hired scientists perforated the government's case with one technical objection after another. While the conference did show interstate pollution from other sources in Michigan, Wisconsin, and Minnesota, the indisputable case of interstate pollution against Reserve was not developed. To be sure, some of the evidence offered by Reserve witnesses at this opening session was gimmickery pure and simple.* But most of their scientific presentation could not be challenged on the basis of the data available at the time. And all of it was convincing to the conferees. The best they felt they could do under the circumstances was scale down Stoddard's original suggestion that the dumping be terminated to an official conference recommendation that Reserve's wastes be kept "under continuing surveillance."

In succeeding skirmishes Reserve gave ground only grudgingly, dropping back a single step at a time. Six months later, in the face of new evidence gathered by FWQA, Reserve had dropped its contention that the tailings were not crossing the state line and had fallen back on the defense that, at any rate, no damage could be shown. By April 1970 Reserve's argument was that, in any event, the damage the government could prove was very slight.† As its position eroded, Reserve saw the writing on the wall and began to propose modest

---

* For example, Reserve's scientists sampled copper concentrations both in the center of the lake and around the company's discharge point. All of the readings for metal concentration in the middle of the lake (about 29 readings) were zero except for about three or four. The three or four which were *not* zero, however, were larger on the average than the average metal concentrations right next to Reserve Mining. Reserve's scientists disregarded all the zero readings in the middle of the lake and simply averaged the few high ones. (The concentrations in the middle might have been more than zero but still too small to be detected. The very least Reserve could have done would have been to average the "minimum detectable level" rather than throw these readings out entirely.) The Reserve scientists used this jimmied average to "prove" that metal concentrations were higher in the center of the lake than they were right next to Reserve's plant, evidence they used to support their claim that Reserve's discharge did not increase metal concentrations in the water. The details of this deception did not, of course, emerge until after the session was over.

† FWQA, having shown that the tailings enter Wisconsin waters, went on to demonstrate that in traveling this far the metal constituents in the tailings enter into solution and become toxic. While Reserve's initial claim was that these metallic solutions were not toxic to living organisms in concentrations less than 1 part per 1000, the government has shown that concentrations of these metals can be lethal to several aquatic organisms at 200 parts per million (only one-fifth the concentration Reserve claimed was necessary for lethality). At the time we checked on the government's progress (March 1970), FWQA did not yet know over what area of the lake this concentration existed. The government also was demonstrating that the tailings precipitate algal growth, particularly at higher summer temperatures. Any pollution of Superior's waters would of course eventually flow into the other Great Lakes, reducing the rehabilitative effects of the dilution flow from Lake Superior to the more polluted waters.

alternatives to Stoddard's original recommendation that the dumping be stopped entirely.

The State of Minnesota may eventually demand of Reserve what the Federal government has not demanded so far—that the dumping be halted. The Democratic Farmer-Labor Party of Minnesota has had a plank in its campaign platform since 1968 (which Congressman Blatnik tried unsuccessfully to block) declaring Lake Superior off limits as a "dumping ground for mining or industrial wastes." Persistent pressure from Minnesota's aggressive conservation groups—the Minnesota Environmental Control Citizens Association (MECCA), the Save Lake Superior Association, and Clear Air Clear Water Unlimited—has sparked the once-hesitant Minnesota Pollution Control Agency into taking a stronger stand against Reserve Mining's pollution.* In fact, one of the latest rounds in the battle is a suit by Reserve against the state challenging the applicable state water quality standards (pollution limits for Lake Superior), which Reserve exceeds by some 800%.

All the while the U.S. Army Corps of Engineers has been following the proceedings against Reserve—but that is about all. The long-standing Federal license to dump is still in effect. On March 3, 1971, Minnesota's new Governor, Wendell Anderson, telegramed General Frederick Clare of the Corps of Engineers to request "immediate revocation" of Reserve's permit from the Corps. Governor Anderson asked Clare to grant Reserve a conditional permit to dump for two years, provided that the company made plans to shift to a state-approved land disposal method within the two-year period. Clare's March 10 reply to the Governor indicated only that the matter was under staff review. It has now been under "review" by the Corps since Charles Stoddard's study was first begun in late 1967.

As of March 1971 the Lake Superior conference has been reconvened five times. Every Lake Superior polluter has been given a cleanup deadline to meet but one: Reserve Mining. Reserve's 60,000 tons per day of dense

---

* Former MECCA leader Grant Merritt was named the Pollution Control Agency's new head by Governor Wendell Anderson when he assumed office in 1971.

black taconite continue to pour into the lake, with no end in sight. Persons in the pollution control profession have had to learn to be content with small improvements, however, and Reserve's pollution has at least been brought out into the open. And all but the most pessimistic observers now believe that in the end either the state or the Federal government will, one way or another, force the mining company to modify its old dumping habits substantially.

In many ways, the most significant aspect of the Lake Superior experience is not the fact that government scientists were eventually able, with persistence, to overcome Reserve's initial advantage, but rather that they floundered so badly in the beginning. FWQA scientists at the National Water Quality Laboratory in Duluth who performed the technical investigations for the Lake Superior conference blamed their poor preparation on lack of notice. "Conferences are called without anyone checking with the technical people to see if they have enough time to get ready," * they told the Task Force.

This is a satisfactory explanation only if one accepts what was obviously FWQA's assumption: that preparations for enforcement should begin only after the politicians inform the engineers and scientists that the water is polluted. It's not as though FWQA's technical people had no advance warning—no clues prior to the Stoddard Report's release to tell them something might be wrong. There had been reports of green water around Reserve's plant for several years. The running debate among old-timers in the area was whether this was algal growth precipitated by Reserve's tailings or simply an optical effect from the tailings themselves. And concern over Lake Superior need not have depended on evidence of Reserve's pollution. There are 61 other industries and 93 municipalities discharging into the lake (1969 figures), many of which have never treated their wastes at all.[16] While the main body of the lake had not yet been affected, serious oxygen depletion of the water from

---

* The conference was called in January 1969. Winter is not a good time to collect data on Lake Superior, and by the time the ice had melted, the initial May 1969 conference session was rapidly approaching.

both municipal and industrial pollutants had been observed in Duluth-Superior Harbor and on the interstate Montreal River downstream from Hurley, Wisconsin, and Ironwood, Michigan.[17] And there were the fishermen's reports of algae and lower catches, complaints which—while not necessarily related to Reserve Mining's discharge—provided signals that need not have been ignored. Senator Gaylord Nelson, not a scientist at all, had spotted these clues more than seven years before the enforcement conference was held and in 1963 had even called the Reserve Mining problem to FWQA enforcement chief Murray Stein's attention during a hearing of the Senate Air and Water Pollution Subcommittee:

SENATOR NELSON: What about the pollution in Lake Superior caused by mining, pouring pollution into Silver Bay from the big taconite mines in Minnesota?

MR. STEIN: We have looked at that. Now, is that interstate in nature? I believe it is largely intrastate [i.e., remaining wholly within one state], isn't it?

SENATOR NELSON: The Lake is interstate waters.

MR. STEIN: Yes. But does that pollution endanger the health and welfare of persons in Wisconsin? We have looked at that taconite discharge and I think you can see it from an airplane. Generally it stays relatively close to shore and drops out. . . . I don't know that we have had any clear evidence that that affects the other State.[18]

For more than five years after Stein's Subcommittee appearance the Federal government considered that view from a spotter aircraft sufficient proof that none of Reserve's taconite tailings were wandering into Wisconsin waters. In so doing, FWQA disregarded not only the signs of growing pollution in the lake already available in 1963 but warnings that came later from one of FWQA's own scientists at the lakeside National Water Quality Laboratory in Duluth.

The Duluth National Water Quality Laboratory has always been known to persons familiar with the pollution control program simply as "Blatnik's Lab." In the early 1960's the only way funds for such superfluous items as pollution control research laboratories could be obtained from Congress was to build them in the home districts of ranking members of the House and Senate Public

Works and Appropriations Committees, probably the most sophisticated "pork barrel" projects ever dreamed up by vote-counting legislators. Senator Robert Kerr of Oklahoma got his lab put in Ada. Rhode Island Congressman John Fogerty got one in Narragansett, and Congressman Blatnik got one right on the shore of Lake Superior in Duluth. Blatnik is proud of the Federal laboratory and mentions it frequently in his public appearances. At the first session of the Lake Superior enforcement conference, for example, he described the research lab as a "watchful, protective beacon" standing ready to "preserve Lake Superior." For the first several years of its existence, however, Blatnik's beacon was carefully trained away from the Reserve Mining Company.

The Duluth laboratory's senior researcher when it opened in late 1965 was aquatic biologist Dr. Louis Williams. Dr. Williams' initial assignment was to determine the properties of Lake Superior water, and he found almost immediately that the water was not as pure as everyone had believed. His experiments showed that Reserve Mining's taconite discharges were increasing the trace metal content of the lake and stimulating blue-green algal growths, particularly at higher water temperatures. (At 78° F., samples of raw Lake Superior water—which Williams found contained traces of taconite—gave rise to algal growths within a week.) And in the Duluth water supply Williams discovered algal conditions that suggested a gradual trend toward Lake Erie conditions.

Williams made no attempt to keep what he was doing a secret; he was broadcasting his findings all over town. In the spring of 1966 he was the guest speaker at a luncheon meeting of a local civic group. He discussed the taconite pollution problem. Word travels fast in northeastern Minnesota, and the story of Williams' experiments quickly made its way up to Reserve Mining's hometown, Silver Bay. The mining company's indefatigable public relations man, Ed Schmid, got in touch with Williams to tell him, as Williams remembers it, not only that his findings were wrong but that he was wasting his time, since Reserve Mining's scientists had already been paid to do the same kind of water research that Williams was doing.

The tremors of excitement that Williams' Duluth speech caused in Silver Bay reached Washington not long afterward, and Williams received a surprise visit from his research supervisor, Dr. Clarence Tarzwell. Tarzwell was the full-time Director at the Narragansett, Rhode Island, laboratory at the time, and most of his previous communications with Williams in Duluth had been by long-distance telephone or by letter. Dr. Tarzwell explained, as Williams remembers it, that the work he had been doing was very "sensitive"; pollution control requirements imposed upon Reserve Mining might bring unemployment back to this previously depressed area. What Congressman Blatnik had given to the pollution control program, Congressman Blatnik could presumably also take away, and FWQA headquarters was concerned that funds for the new laboratory might be cut off before it could ever get a real start if its activities threatened local economic interests. According to Williams, Tarzwell directed him to break off the work he had been doing on Reserve's taconite tailings and their effects on the lake and focus instead on the more theoretical long-range laboratory water studies that the Duluth facility had been assigned to carry out. As Williams remembers it, Tarzwell implied that he could continue his experiments with taconite on the side as long as he kept them under wraps. "And besides, there was no way I could have avoided studying the tailings," Williams recalls, "because the water is full of them." But, above all, Williams was not to write or speak publicly about the controversial question again without clearance from Washington headquarters.

Williams tried to get the clearance, but it never came. He was forced to turn down numerous speaking invitations. An article on the tailings problem that he wrote for a scientific journal, *Limnology and Oceanography,* was never cleared for publication. Believing himself blocked at every turn, Williams began to resort to bureaucratic guerilla warfare. He started writing letters of complaint directly to Washington headquarters without going through his immediate superiors and "leaking" information to newspapers in Wisconsin. By June 28, 1967, Williams had become so frustrated that he wrote a letter describing the studies he had been "ordered to stop" to Wisconsin Senator Gaylord Nelson, the main advocate of a Lake Superior enforcement conference:

Since I am already in trouble for letter-writing—one more might be a public service and probably cannot hurt me more, since I have about decided to return to university teaching and research. . . .

The question that I would like to raise concerns the pollution of Lake Superior (eutrophication) with taconite wastes from Silver Bay, Minnesota. . . . The question arises . . . how much restraint should a professional water-quality scientist exercise who is a federal civil employee?

I am aware that the power structure desperately wants taconite business (and I am for it, too) and they are now acting within the permitted law; thus my suggesting massive Lake Superior pollution would mark me as unfriendly. On the other hand, good water quality in Lake Superior is an important resource for many users, present and in the future, so it must be protected, while this is still possible. We must not allow Lake Superior to become another Lake Erie.

In August 1967 the Duluth laboratory's maverick taconite hunter left government service for a teaching position at the University of Alabama, submitting a parting memorandum that recommended "further research" on the taconite question and urged that Duluth lab scientists "be encouraged to make public their findings."

Were Federal pollution control officials actually subjected to political pressure, as Williams suspected they had been, to quash his research on a politically sensitive subject—the pollution from a large Minnesota industry with close ties to a powerful Congressman? Why was Williams not permitted to speak or publish on taconite pollution? In attempting to piece together the other side of Williams' story, the Task Force came away with more questions than answers, since FWQA officials professed not to remember most of the details. But Dr. Leon Weinberger, head of FWQA's entire research effort at the time when Dr. Williams had been making public pronouncements about taconite pollution, and now out of government doing work as a private pollution control consultant for Reserve Mining, did tell the Task Force emphatically that neither he nor any of his subordinates in the research program had ever been subjected to any political pressure whatsoever. What Weinberger had been concerned about, the former research head said, was that his government scientists should not make unwarranted positive statements about matters that still might be, in scientific terms, a matter of conjecture. Dr. Williams had been a good scientist— hard-working, sincere, and extremely idealistic, Weinberger said. But Williams' associates suspected that he might on occasion jump to final conclusions a bit too hastily. And he had shown he could not be counted upon to measure his words appropriately when speaking in public. The news of Dr. Williams' taconite speech in the spring of 1966 had caught FWQA headquarters completely unawares. "He had no clearance," Weinberger recalls. "It's upsetting to have someone give information like this to an after-dinner group when you haven't checked it out to see that it's accurate." The paper Williams had hoped to present to his fellow limnologists and oceanographers, Weinberger assured the Task Force, was held up—as are many articles written by aspiring authors throughout the bureaucracy—for purely technical reasons.

Had Williams actually been ordered to stop his research on the taconite problem, and if so, why? Weinberger explained that Williams' specific work assignment would have been a matter for his immediate supervisors to decide and not a headquarters problem. But, he added, there had never been enough researchers available to carry out all the long-term research tasks, let alone the short-term investigations necessary to prepare for enforcement activity. If Williams' enforcement-oriented research had been of interest to FWQA's enforcement people, Weinberger observed, they could have made their interest known.

Dr. Donald Mount, who became Director of the Duluth lab near the end of Williams' tour there and who has headed all government studies on Reserve's pollution since the 1969 enforcement conference was called, explained to the Task Force that Williams was supposed to have been working on other research assignments at the time he was carrying out his experiments on Reserve Mining's pollution. According to Mount, Dr. Tarzwell, Williams' supervisor, thought Williams was just "farting around," although "it has since turned out otherwise." But, Mount reminded the Task Force, "we have to keep people on their mission."

What had Williams' mission really been? Dr. Clarence Tarzwell described it the way Williams had: to determine the properties of Lake Superior water, presumably including its tendency to age more rapidly as a result of high taconite content. Tarzwell could not, on the other hand, recall ever having told Williams to stop his work. He had not been particularly impressed with it, however, Tarzwell said, because "Williams didn't ever suggest that his findings had anything to do with enforcement." Tarzwell did remember ordering Williams to spend more time "in the laboratory" and less time "at his typewriter writing letters." (This rebuke came long after the original order to stop his experiments, according to Williams.) "I understand Don Mount's investigations have later [in 1970] come up with the same thing Williams did," Tarzwell added.

Both Tarzwell and Mount have impeccable scientific reputations (though not in areas closely related to Williams' highly specialized field), and it was difficult to know what to believe. But for whatever combination of scientific, bureaucratic, and political reasons, it is clear that the Duluth laboratory was, during this early period, a far cry from the "watchful, protective beacon" that Congressman Blatnik says he established to "determine in a scientific manner the best preventive methods to avoid pollution of Lake Superior." If the government had really been interested in finding out the facts about Lake Superior's pollution, FWQA could have ordered Williams to step up his damning inquiry into Reserve's destruction of the lake or could have rushed in reinforcements to conduct a more detailed investigation until it considered the case against the mining company absolutely unassailable. An enforcement conference could even have been convened on the basis of the information already available—more than enough to give the most skeptical Secretary the "reason to believe" there is interstate pollution, which, under the Pollution Control Act, is supposed to trigger Federal enforcement. FWQA could have, at the very least, subjected Williams' results to independent scientific scrutiny. Instead, the water pollution agency chose only to try to shut Williams up.

Several months prior to Williams' August 1967 departure a personnel appointment was made at the Duluth lab which could only have served to reinforce Williams' belief that Reserve Mining's pollution was a sensitive subject within the water pollution bureaucracy. Mike Lubratovich, a Blatnik acquaintance from Duluth, was named the laboratory's Assistant Director. As everyone well knew, Lubratovich obtained his position of power at the lab through Congressman Blatnik's good graces, his principal qualification for the job reportedly being unswerving loyalty to the Congressman and to the Congressman's political friends. Although the Assistant Director's formal duties include control over the laboratory's administrative and personnel matters, FWQA senior research officials in Washington assured the Task Force that Lubratovich's authority has never extended to the supervision of scientific work or to the hiring and firing of scientific personnel. Be that as it may, researchers at the lab have been well aware since the day Lubratovich reported aboard in 1967 that an agent for the district's influential Congressman is advantageously situated in their midst.

As an example of the way Lubratovich has been relied on to use his inside position in the pollution control hierarchy to protect Congressman Blatnik's industrial benefactors, consider the events of August 11, 1967. Planned for that day was the formal dedication of the National Water Quality Laboratory after two years of construction (during part of which time the Duluth researchers operated out of rented quarters in downtown Duluth). Preparation for the festivities began in early July when Maurice Tobin, Congressman Blatnik's legislative aide on the Rivers and Harbors Subcommittee, telephoned Reserve Mining's public relations man, Ed Schmid, to give him advance notice of the upcoming occasion. The story of the events that followed is best told in a confidential internal communiqué that Schmid sent the next day to Reserve President Edward Furness.

Because the Schmid memorandum illuminates not only the role of FWQA's Lubratovich but, more generally, the intricate web of mutual interest that grants large industrial polluters favored access to the government officials ostensibly charged with regulating them, the Task Force has included it here in its entirety.

MEMORANDUM

Confidential

To   E. M. Furness   Date   July 7, 1967
From   Ed Schmid

Subject   Secretary Udall to inspect Reserve's use of Lake Superior

A few days ago you told me you had received a confidential tip that Secretary Udall was going to fly over Silver Bay to "see for himself" Reserve's use of Lake Superior. He would do this while he is in Duluth to dedicate the new Federal Water Quality Research Laboratory on August 11th. This memo will furnish additional information.

Late yesterday I received a phone call from Maurice Tobin (Counsel, House Public Works Committee) telling me the Lab would be dedicated August 11th; Representative John Blatnik will be in charge of the affair. Blatnik, Lud Andolsek (Civil Service Commissioner and best friend to Blatnik), James Quigley (Commissioner, Water Pollution Control), Tobin and Secretary Udall will be present.

Knowing (from you) that Udall was going to fly up the North Shore to "inspect" us I asked Tobin if there was any way we could arrange for Udall and Quigley to come to Silver Bay to see our operations. If so, they'd never be concerned about our use of Lake Superior.

Tobin said, "We've got to. We have no choice. The Secretary insists on flying over to see for himself. I've already arranged for a helicopter from the Army Air Force Base in Duluth. Quigley, John [Blatnik] and I will fly up there with Udall . . . maybe a few others, too."

I pointed out that John might be embarrassed if he cannot identify from the air, the trout streams, villages and settlements, state parks, etc., in his own district . . . he might even have trouble identifying our plant buildings. I suggested they needed a "guide" and offered my services. He was delighted at the idea and said he would arrange it. *After this was accomplished, I suggested Mike Lubratovich (formerly manager of Duluth's water department and now program manager of the Federal Water Quality Laboratory) be included in the flying trip, too. As you know, he's one of our strongest and best informed supporters.* I did this because I feared Udall might say "no" to a Reserve man accompanying him. *With Mike along, we'll have a spokesman —and he can be our ears and eyes, too.* I can make sure Mike has the plane fly along the shore all the way—not simply go across country direct to Silver Bay. [Emphasis ours.]

When I suggested driving to Silver Bay and perhaps spending the night at our Guest House, he said no; felt that this might be "suspect" and leave the Secretary open to criticism. However, Tobin didn't hesitate to ask me if I would "gracefully and quietly" pay for three motel rooms for August 10th and a breakfast of Bloody Marys at the Kitchi Gammi * the next morning for about 25. I agreed. He told me he had reserved 10 single rooms at the Edgewater Motel, Duluth, but seven of the party of ten "had their own resources" so I'd only have to pick up the tab for three. We plan to talk again soon about details, travel schedules, the Kitchi Gammi breakfast, etc. At that time I think I should suggest that Reserve should not openly "host" the breakfast. We can arrange for someone else to do it—the University, the Chamber, perhaps—if you think this prudent.

Tobin sketched this tentative schedule: Party will leave (commercial air line) Washington about 6 PM, arriving Minneapolis at 8 PM, where a press conference will be held. "3M Company,† or somebody like that will fly us to Duluth on one of their planes," he said. "We ought to get there about 10 PM." Bloody Mary type breakfast at the Kitchi Gammi Club about 9:30 AM. . . . dedicate the Duluth Lab at 11 AM . . . lunch at the Arena-Auditorium at 12 noon . . . dedicate Northwest Paper's new water-air treatment works at Cloquet at 2 PM . . . board helicopter to fly to Silver Bay and return to Duluth . . . "Jeno Paulucci (Chun King Foods) ‡ will fly us to his lodge on Lake Kabetogama for the weekend . . . Sunday, Chun King plane will take us to Minneapolis to catch a commercial plane back to Washington."

I asked him if there were any new developments on HR25 Estuarine bill, any hearings, informal discussions or anything. "No," he said. "They're too smart to do anything now." I did not ask what he meant. It is possible, of course, that Udall's determination to take a look at Reserve is related to HR25—though I think the more likely explanation

* According to those who know, "everybody who is anybody who's got money belongs to the Kitchi Gammi Club" in Duluth.
† The chairman of the board of the 3M Company, Bert Cross, is now chairman of the quasi-official National Industrial Pollution Control Council housed in the Commerce Department.
‡ Paulucci, former owner of Chun King Foods, is now owner of Jeno's Pizza Rolls.

is that the "fly-over" is related to pollution and the 1965 Water Quality Act.

For your information, after I learned from you about Udall's intended fly-over and before my talk with Tobin yesterday, I arranged that on August 11th:

a. A commercial photographer would take a series of air-view colored slides—one every sixty seconds—of the North Shore of Lake Superior and the Wisconsin side, too. (If Udall should see "green illusion water" in front of our tailings delta on that date I would *hope* our photos would show that he would have seen areas of apparent discoloration elsewhere, too. You'll recall that August last year was our worst month for occurrences of "green water days" in the area of our tailings discharge. We had no occurrences of the phenomenon in six of the months last year, but in March we had 3; April, 3; May, 2; June, 2; August, 7; September, 6 and October 5.)

b. R & D will take turbidity and temperature readings in and out of our permit area—especially in areas of "green illusion water," if any exist.

Among other things we should do promptly:

a. Plant the rye, sedges or "grasses" on the tailings delta, as we have planned for some time. And we should plant it wherever possible, even if some of it *does* wash away. The cost is negligible, all things considered. In the month remaining, much of that delta can be green, attractive and obviously solid and stable.

b. Talk with Lee Vann, Vice President of Chun King. When he was executive secretary of the Arrowhead Association I arranged for him to speak on Reserve's behalf at several of our water pollution hearings; he's reasonably well informed and is "committed" on our side. He likes Reserve for other reasons—even applied for the job of assistant director of public relations of Reserve. I want to show him the record of commercial fishing on Lake Superior and several other things, because he'll have the whole weekend to talk to Quigley and others.

c. Show Mike Lubratovich our existing series of "sixty second slides" so he'll know how frequently huge areas of apparent discoloration exist in the Lake obviously *not* associated with Reserve. This can be done easily since his top technical man, Armond Lemke, has asked me to show them to his staff and Mike will be delighted to sit in.

ES/bh   Signed   Ed Schmid

When the big day finally arrived, everything came off according to plan—Bloody Mary breakfast hosted by

Reserve Mining, dedication ceremony with Congressman Blatnik presiding, and so on—with one possible ironic exception. The party may never have had time to make the aerial inspection of Reserve Mining's lovely green tailings delta. While most of the celebrants thought they remembered flying over Silver Bay in somebody's private plane,* Reserve's master strategist, Ed Schmid, told the Task Force he was sure the Udall reconnaisance mission never made it that far. "I would remember better than anyone else," Schmid said. "Because I very badly *wanted* them to fly over Silver Bay." In any event, the pollution control laboratory's dedication day had been a smashing success thanks to the special efforts of the lake's largest industrial polluter. Former Blatnik aide Maurice Tobin was careful to point out to the Task Force that these arrangements constituted nothing unusual. It was, as Tobin put it, "strictly an above-board relationship with a good corporate citizen."

The remaining pieces of the Lake Superior story fall into the familiar pattern of reluctant response by FWQA to outside pressures for pollution abatement. Secretary Udall's interest in the Reserve Mining problem and the Army Corps' November 1967 request to Interior for guidance on the Reserve dumping permit reopened what had been a dead question at the water pollution agency. Shortly thereafter a small FWQA group from the agency's Chicago Regional Office joined representatives of four other Interior agencies to begin the study from which the Stoddard Report later came.

Ironically, Reserve Mining still had some reason to believe it had little to worry about from the 1967 and 1968 investigation. The man in charge, Interior's Regional Coordinator Charles Stoddard, was himself a Blatnik confidant. Both Stoddard and his brother were old friends of the Congressman and his wife, and Blatnik had reportedly interceded on behalf of Stoddard on occasion in scraps that the Duluth Regional Coordinator had had with the Interior bureaucracy. Thus, many people in the Duluth area were taken by surprise when

---

* Ed Schmid told the Task Force he was "sure [the plane] wasn't Jeno's [Jeno Paulucci of Chun King Foods]" but "didn't know who arranged to get it." Some of the others thought it was Reserve's company plane.

Stoddard's report turned out to be so uncompromising. Mike Lubratovich, Assistant Director at the Duluth lab, called Blatnik's office immediately to report the disaster and find out what he should tell the press. As Blatnik's former aide Maurice Tobin explained it to the Task Force, "Lubratovich was grateful to us for his job and knew what he should do." And as for Stoddard's betrayal? "How could he have done such a thing? He should have at least had the grace to check with Congressman Blatnik first."

8

With a Little Help from Their Friends

I think that some have been quite correct in saying that water pollution is one part water and one part politics.
—Interior Secretary Stewart Udall

Why has the Federal government always given the most serious pollution problems such a wide berth? Not because of a shortage of suspects. The polluters in most locations have been all too conspicuous. It is the nation's largest cities and best-known industries that are polluting on such dumping grounds as the Houston Ship Channel, the Delaware River, and the Ohio River. Lake Erie's worst municipal polluters are Cleveland and mighty Detroit. And when the names of Erie's corporate polluters are added to the roster, it becomes truly formidable.

The commercial community of greater Detroit includes such notorious enemies of the environment as the Ford Motor Company, Scott Paper, Firestone Tire and Rubber, Great Lakes Steel, Mobil Oil in outlying Trenton, and an impressive battalion of chemicals—Wyandotte Chemical, DuPont, Pennsalt Chemical, Allied Chemical, and Monsanto. Clustered around the streams that drain into Lake Erie from Michigan, Indiana, Ohio, Pennsylvania, and New York, and ringing the shores of the lake itself, are other pillars of the business community: Interlake Iron; U.S., Bethlehem, Republic, and Jones & Laughlin Steel; General Motors; Swift and Company; International Harvester; General Electric; the B&O Railroad; General Tire and Rubber Company; Gulf and Sun Oil; Union Camp Corporation; Campbell Soups; and Olin Mathieson, to name but a few.

Just to scan this list of corporate and metropolitan polluters is to sense why their transgressions have gone so long unchallenged. Pollution is wedded to political power. Wherever water is bad the industries and cities that made it that way are ready to use their influence and economic leverage on every level of government to stave off abatement. And the Federal Water Pollution Control Act's enforcement scheme could not have been better designed to make the Federal program vulnerable to these influences. Enforcement decisions are made by a handful of men located far from the water's edge in Washington, D.C.—currently Water Quality Office Commissioner David Dominick, then Deputy and Assistant EPA Administrators, and ultimately (since December 1970) EPA Administrator William Ruckelshaus, the only man who can give an enforcement action the final go-ahead. Any enforcement recommendation that Assistant Commissioner Murray Stein's office sends up the EPA chain without the imprimatur of a state request for Federal intervention can, for all practical purposes, be tabled by any one of these men for any reason—personal whim, inertia, timidity, the press of other business, political loyalty, or political pressure.

The enforcement effort has been responsive to occasional bursts of pressure for cleanup from the public and the press as well as to the everyday obstructionist tactics of polluters and their friends in office. But during the long stretches when Federal officials have no one at their backs pushing for pollution control, they have generally responded to the only pressures operating on them—pressure to ignore pollution problems entirely or suppress what information comes to light.

Too Little Too Late: Potlatch Paper

Industrial polluters generally prefer not to fight their own battles with the Federal enforcement establishment, because they usually don't have to: while their political allies take the hard stand against Federal encroachment, the corporations themselves are left free to ooze concern for the environment. But on occasion Federal enforcers have been subjected to a crude frontal assault by a panicky polluter unschooled in Washington politics—sometimes late in the game, after the decision to enforce has already been publicized. These polluters usually get burned for their efforts, but their blatant displays of resistance provide some clues to the way the more successful—and less conspicuous—polluters make their influence felt also.

An announcement late in 1963 by the Secretary of HEW that an enforcement conference would be held on the Snake River set the stage for one such attempt. The Snake wriggles its way harmlessly north along the Idaho-Washington border until it reaches Lewiston, Idaho, and Clarkston, Washington, on the opposite bank. There the Snake turns poisonous, and stays that way for the rest of its journey northwest through Washington to the Columbia River. Clarkston, Washington, with its raw municipal sewage and meat-packing wastes, had for many years tried—and failed—to drop as much into the Snake as its rival city directly across the stream. Lewiston owes its superiority to the efforts of two prominent corporate citizens, a slovenly pea and potato cannery, and the Potlatch Forests paper mill—whose dark, smelly residues slither along with the Snake through the lowlands of eastern Washington, while its tiny wood fibers settle to the bottom near the mill, leaving hometown Lewiston a legacy of putrid sludge.

Three weeks before the scheduled January 1964 opening of the Snake River conference, one of Murray Stein's assistants on the headquarters enforcement staff, who was manning the office in Stein's absence, received an unexpected visit. Two Potlatch vice presidents came in and said they just thought he would be interested to know that if the conference wasn't called off, Idaho Senator Frank Church's "possibility of receiving campaign funds" in the future would be cut "down to zero." They asked Stein's assistant to cancel the enforcement action. Since the enforcement office has been the scene of many such visitations by polluters, Stein's assistant was well prepared to show his visitors the door.

Phase two of the Potlatch offensive showed a little more imagination. When Stein's staff telephoned to reserve hotel rooms and a hall for the conference, they were informed that there were "no accommodations in Lewiston." Having heard of no major conventions being held in Lewiston, Idaho, at that time of the year—a blizzard had closed the roads and the town was reportedly almost deserted—the enforcement staff was suspicious. At their request, the public relations director of the American Pulp and Paper Association got in touch with Lewiston to find out whose wood fiber was causing the logjam. When the Lewis and Clark Hotel phoned the next morning to report "unexpected vacancies," Stein's men were sure that Potlatch had deferred to the pulp and paper industry's wish to preserve its national image. But more zealous efforts to protect the Potlatch part of that image were still in store.

From somewhere in the HEW organization (although all memories have conveniently become fogged and no one recalls just where) the word came down that there was to be "minimum crew" and "no public information office personnel" at the conference. In accordance with the decree, the public information gang stayed home—but to no avail as far as Potlatch was concerned. Murray Stein was snowed in at the Lewis and Clark Hotel for several days with no one to talk to but the newspapermen staying there too, and the paper company came out of the affair with a terrible black eye from the press.

Since 1965 Potlatch has greatly refined its political techniques, exchanging the frontier antics it put on in Lewiston for more sophisticated—and commonplace—methods of persuasion. It is easy to see why one of the first places Potlatch instinctively turned was to Frank Church, the Senator from its own state and, by its account, the recipient of its campaign contributions. The pressure on Church was not only heavy-handed but misplaced: first because, as far as we could tell, he refused to help them; moreover, because he is not one of the water pollution "insiders" in the Senate. He sits on neither the Public Works Committee nor the Appropriations Committee—the only two with direct power over FWQA's legislative and budgetary fate. While Church, a Westerner, has his fair share of friends in the Department of Interior—John Carver, his former Administrative Assistant, for example, was Stewart Udall's Assistant Secretary of Interior for Public Land Management—the pollution program was still in HEW at that time. Especially because the conference had already been announced, Church couldn't have done much on the company's behalf even if he had wanted to.

Potlatch, no slouch, has learned a lot since then. On April 18, 1969, Senator Frank Moss (D.-Utah), a leading Senate spokesman on conservation matters, read into the *Congressional Record* a paper presented at an International Convention and Exhibition on Water Pollution Control in Rome, Italy, earlier that year, which gave Potlatch favorable mention. The paper had been prepared by Maurice Tobin while he was still counsel to the House Congressional Committee on Public Works (which handles air and water pollution matters) and staff aide to Congressman Blatnik, sponsor of the Federal Water Pollution Control Act. In the paper Tobin first explains how the positive efforts of American industry in pollution control "are most encouraging and reflect the new, refreshing and enlightened point of view and spirit in the business community." He goes on to say:

And, I think, they deserve greater publicity. The business community produces many a civic-minded industrial leader. One of the fast-growing forest producing companies in America, Potlatch Forests, Inc., of San Francisco has a dynamic president, Ben Cancell, who was one of the founders and former president of the Clean Streams Improvement Council.[1]

This was the kind of publicity Potlatch liked. Since leaving Congressman Blatnik's staff, Tobin has been retained by the company—in his new capacity as a "legislative listening post" in Washington—to represent the Potlatch point of view on a more permanent basis among the many friends he made while working in the House of Representatives.* With Tobin's finely tuned ear to the ground, the new "civic-minded" Potlatch is better equipped to catch the first faint signals of impending enforcement long before the news breaks. And if it re-

---

* A "legislative listening post"—Tobin's term—appears to be someone who uses his access to political figures and familiarity with the ways of Congress to find out how existing and impending legislation will affect his clients, and advises his clients on what to do about it. He would almost certainly not, in this capacity, engage in formal and open lobbying activity. But the line between gathering information and getting one's client's point of view across is a fine one.

ceives these signals in time, its portfolio of pressure plays will no doubt be put together more artfully than in 1963.

Polluters with Class: The Mahoning River

While the Federal government has never enforced on the highly industrialized Ohio River, Washington has put in enforcement appearances on two of its many polluted tributaries—the Monongahela and the Mahoning. These were both courageous but essentially futile expeditions deep into enemy territory; the massive resistance the government encountered in these small segments of the economic and political muscle in the Ohio Basin is an indication of the greater opposition that the Federal enforcement program would face were it ever to take on the Ohio itself.

It was at the Mahoning River conference, held in 1965, that the spirit of the Ohio Basin received its most untrammeled expression. The Mahoning is typical of the Ohio's tributary rivers—little known outside the local area, very small, and grossly polluted. It runs from the industrial town of Alliance, Ohio, north through Warren and then southeast down to Youngstown, the site of the conference. Before rudimentary control efforts were made in the 1950's, hot steel effluent pouring into the Mahoning from plants between Warren and Youngstown turned the river into a steaming caldron of chemicals; the Mahoning used to bubble out of Youngstown at 140°F.[2] The river's temperature was down by the 1960's (occasionally exceeding 100°F.),[3] but nine miles downstream from Youngstown at the Pennsylvania state line the Mahoning was still, in the words of the Federal conference report, "biologically dead."[4] Twenty-three miles farther downriver is Beaver Falls, Pennsylvania (1960 population: 16,240), with one of the worst raw water supplies in the nation. According to Dr. Graham Walton, drinking water specialist of the U.S. Public Health Service, only 13 other communities in the country had more bacteria at their water intake points when he last took a national survey in 1955.[5] Walton testified at the 1965 conference that the bacteria problem was still of "the same order of magnitude." Beaver Falls' water also comes enriched with phenols, iron, manganese, oil, and other industrial by-products, all far in excess of PHS raw drinking water limits.

Local witnesses at the conference quickly made it clear that, as far as they were concerned, the government was treading on industry's turf. In describing their vast economic power, the industrialists who testified provided a simple but revealing picture of the force that could be marshaled against cleaning up the Mahoning, or any other small industrial river in the Ohio Basin. R. F. Doolittle, for example, Vice President and General Counsel for Youngstown Sheet and Tube, started off his presentation with an intimidating list of the companies he was speaking for:

Republic Steel Corporation, Copperweld Steel Company, Jones & Laughlin Steel Corporation, Pittsburgh Steel Company, Sharon Steel Corporation, United States Steel Corporation and, of course, my own company, the Youngstown Sheet and Tube Company.[6]

The Mahoning Valley he vividly described was a valley of steel producers.

As you must know, the Mahoning Valley from Warren down to the Pennsylvania State line, supporting a population of some 509,000, is primarily a center of steel production and steel fabricating. There are nine steel companies with 15 major plants, including 63 rolling mills, along the river. . . . The giant mills that line the banks of this small stream for a number of miles account for nearly seven percent of all the steel produced in the Nation, and nearly every family in this community is vitally concerned, directly or indirectly, with whether steel in this valley is doing well or poorly. . . .

The existence of the steel industry in this valley, and the employment that goes with it, is utterly dependent on the Mahoning River. The industry relies on this river, first, for enormous quantities of water needed for cooling purposes, and second, for discharge of certain of its wastes, and this is why the steel plants are all located right along the river's banks. Small as this little stream is, without it, and without its use as a workhorse, there would be no steel industry in either Youngstown, Niles or Warren.[7]

It was puzzling at first to hear local representatives at the conference claim the river was not "polluted" and Federal officials claim that it was, but gradually it became clear that to the local officials the river was not polluted if it was serving what they called its "highest and best use." J. P. Richley, Commissioner of the Waterworks in Youngstown, explained how such an unsavory river as the Mahoning could have retained such a favorable local reputation:

For many years, the Mahoning River has served admirably —but not completely—its highest and best economic use— that of providing industry along its banks with their primary water needs and that of providing a place of discharge for the liquid wastes of the valley.[8]

And when Murray Stein suggested he might hesitate to endorse the construction of a proposed Lake Erie–Mahoning canal fed by tainted Mahoning water, Mayor Flask of Youngstown exploded with his most deeply held convictions:

MAYOR FLASK: What do we care about the river as long as it brings the barges down, the iron and coal into the steel mills?

CHAIRMAN STEIN: Sir, I recognize what you want, but I—

MAYOR FLASK: And what you are going to get.

CHAIRMAN STEIN: Bless you. I certainly am not going to stand in your way. We have statutory obligation under our law to . . . see that the water quality is protected.

MAYOR FLASK: . . . As long as the water is clean and cool enough for a canal and for industry, business and commerce, I think that's the big motive behind the Mahoning River. I think that's the only thing we hope to look at, because no plants or no industries are going to build in Youngstown or in any part of the Mahoning and Trumbull Valley unless they have a water supply for industry. . . .

I don't care what kind of treatment they have . . when you see the barges come down there and the smoke bellowing out of these stacks and thriving business for economy's sake. . . . That is the important part and we have a most pleasant site in this valley of ours today because it is thriving. Business is here and people are working much more than ever before, and I think that's the primary purpose.

If you make people happy, you have got good citizens, you have got good Americans, and when they aren't working, they aren't happy, and, oftentimes, we may be led astray by other groups that we are not interested in and, so help me, I believe all of us are here for that one aim in life, good Americans, happy times, and thank you very much.[9]

The fact that these "outsiders" from Washington wanted to impose some new treatment requirements on the City of Youngstown added to the Mayor's agitation:

MAYOR FLASK: . . . We have attempted in all instances— well, we were compelled to, should I say, go along with the State Pollution Board, and each year we had to go forward and get a certificate and show our intent on this elimination of the pollution of the Mahoning River. And, oftentimes, we had the most difficult time with the state authorities in getting the certificate to be able to discharge our sewage into the river. And now we have gotten to a point where I thought we could ignore the State Pollution Board, to be exact, I mean, we have accomplished what they have been asking for in all these years, and all of a sudden now someone else comes into the picture.[10]

The government apparently got the message. Federal enforcement has never gone back to Youngstown—not even for the usual follow-up conference session—despite the fact that the Federal recommendations for treatment have been completely and openly ignored for more than five years. None of the municipalities has the secondary treatment they are supposed to have had by October 1969; two of them, Niles and Struthers, had not even submitted satisfactory plans by September 1970. Most of the Mahoning steel-makers sat out the four-year abatement schedule they were given without making a perceptible move toward meeting the Federal demands. And as the following table shows, the water at the intake for the Beaver Falls' drinking water supply downstream was more contaminated than ever before.

The only really surprising thing about the Mahoning River experience is that the government went there in the first place. The Mahoning's polluters have been calling the shots in the small valley they inhabit for over 100 years and would represent a formidable opponent even if they were left entirely to their own devices. But during that time they have also garnered some powerful support. The Mahoning Valley runs like a well-oiled machine, with industry—the valley's lifeblood—at the controls. And fronting for the Mahoning's polluters in

CONCENTRATIONS OF MAJOR POLLUTANTS IN RAW WATER SUPPLY
FOR BEAVER FALLS' DRINKING WATER
1964 AND 1969[a]

	1964 AVG.	1964 MAX.	1969 AVG.	1969 MAX.	% change in avg. from 1964-1969	Recommended "permissible" safe limit[b]
Coliform bacteria[c]	280,000/100ml	750,000	401,500	1,100,100	+44%	10,000/100ml
Iron[d]	870 ppb	[e]	1,270	9,000	+46%	300 ppb
Manganese[d]	390 ppb	800	540	1,500	+38%	50 ppb
Fluorides[d]	[e]	700	850	1,580	[f]	1700 ppb
Phenols[d]	14.5 ppb	14.5[g]	10	20	[f]	1 ppb

[a] Data is from Beaver Falls Municipal Water Works, the State of Pennsylvania, and EPA's Water Quality Office; the drinking water treatment plant manager at Beaver Falls informed the Task Force that all these pollutants are reduced in treatment to safe levels for drinking (the city doesn't measure phenol content after treatment, but believes it "low"). Treatment levels are considered excessive by PHS, however (e.g., chlorination in amounts up to six times the PHS recommended limit).

[b] Limits established by the National Technical Advisory Committee on Water Quality Criteria, Subcommittee on Public Water Supplies.

[c] Per 100 ml. of water.

[d] In parts per billion (ppb).

[e] No data available to Task Force.

[f] Data insufficient to make calculation.

[g] Only one observation.

any conflict with the Federal government are the machine's smoothly functioning parts—local Congressmen, regional control authorities, and the regulatory officials who represent the sovereign state of Ohio.

The State of Ohio

"Profit is not a dirty word in Ohio." Every state manages to put that message out in one form or another, but Ohio's full-page ads are bigger, bolder, and more to the point. Ohio, the copy explains, is the businessman's dream—the complete gracious host to corporations that decide to set up shop in the state. ("Write or call collect Governor James A. Rhodes" for questions.) Industrialists who read on into the smaller print beneath the eye-catching caption will learn that in Ohio they will find cheap and plentiful natural resources and a haven from taxes. ("Ohio's state and local taxes are the lowest in the nation!") As anyone who lives near one of the Buckeye State's many contaminated rivers and streams can testify, Ohio has also always tried its best to provide industry with a safe haven from the tough pollution control requirements that would otherwise be part of the cost of settling in a heavily industrialized state.* The temptation to attract new industry and placate old industry by keeping control requirements low is one that every state faces and one to which most of them have succumbed. But among the highly industrial urban states, Ohio has a reputation for hospitality that none can match.

Much of the credit for Ohio's impeccable reputation in this regard must go to Ohio State Health Director Emmett Arnold. Arnold is in overall charge of Ohio's six-man Water Pollution Control Board,† best known for its unyielding devotion to Ohio's industrial interests.

---

* As one Ohio ad (published in the *Wall Street Journal* on May 14, 1968) says, "Pollution control . . . is a cost of production—and Ohio leads the nation in its progressive programs to provide tax incentives and direct assistance to industry to insure clean air and clean water. . . . We'll show you . . . how you can benefit from our understanding of the cost of production." Another popular Ohio ad notes that "harassment of industry by public officials adds to the cost of doing business. . . . When these costs of doing business become too great industry moves out."

† Arnold is scheduled to be replaced in January 1971 when the new Gilligan administration will take over for Governor Rhodes and his appointees.

Lest the Board members occasionally forget where their loyalties lie, they have Barton Holl to remind them. Holl, an executive of the Logan Clay Products Company of Logan, Ohio, fills the slot on the Board formally allocated to "industry" by Ohio's lobby-ridden state legislature. (A representative of municipal polluters also has a reserved seat on the Ohio Board.) Not that this arrangement is unusual; 32 of the 50 state control agencies are manned by one or more representatives of polluting special interest groups,* and on many of them the strength of industrial representation reveals unconcealed contempt for the public. (Alabama's 14-member Water Improvement Commission includes six "industry" seats, for example; all six are occupied by executives of polluting corporations.) But Arnold, Holl, and company on the Ohio Water Pollution Control Board have always carried out their prescribed task of "regulating"—and accommodating—the state's polluters with a special efficiency not many other state agencies possess.

Dr. Arnold has bitterly opposed Federal enforcement, and the reason is plain. Having established a comfortable working relationship with Ohio polluters, the Ohio Board can think of nothing worse than having the Federal government come in to disturb the equilibrium by making higher demands. The natural alliance between the polluters and their state regulators is cemented, ironically, by the enforcement procedures of the Federal Water Pollution Control Act. At enforcement conferences the Federal government is pitted, in effect, against the state—not the polluters. The encounter is one most state agencies have good reason to fend off; they are first compelled to listen to a public recital of their failures in pollution control and then saddled with an onerous cleanup program—the Federal "recommendations"—which they have to carry out under Washington's supervision.

State representatives at Federal enforcement conferences frequently make plain their agony over this arrangement. But Ohio State Health Director Emmett Arnold's distress is in a class by itself. As Arnold scornfully observed at the Mahoning River conference:

* A list of state agencies with special interest membership is included as Appendix C.

In this case . . . what the Federal government seems to have chosen, instead of reasonable approaches and contacts, is a cumbersome device of the Federal Water Pollution Control Act by which . . . the Federal Water Pollution Enforcement unit . . . may make itself a party to somebody else's accomplishments.[11]

But it was also the Federal government's belated and seemingly capricious assertion of authority that the State Health Director resented so much—and not entirely without cause. The Federal enforcement program had not moved on the Mahoning River until 1965, nearly nine years after the Federal program began and long after the worst pollution had been eliminated:

DR. ARNOLD: Now we are suddenly brought into a conference on pollution of the Mahoning River by Federal action which implies by the nature of the call that health and welfare may be endangered and that some sort of emergency exists. If an emergency existed, it was 10 years ago, and the Federal government is considerably late.[12]

The fact that the government was considerably late is a tribute to Ohio's—and Arnold's—effectiveness in keeping it out. State hostility to Federal enforcement has been so consistent that the gun-shy Administration officials empowered to make final enforcement decisions routinely flinch when they contemplate initiating an action without a state invitation. As Murray Stein explained to the Senate Subcommittee on Air and Water Pollution in 1963:

Generally, in developing [a potential] enforcement case, we do two things. One is we set out the facts as we see them. And second, what we anticipate would be the reaction of the States . . . involved. . . .

There are many, many factors to be considered in Federal-State relations. We have many, many programs dealing with the States. And there is a reluctance on a good part of the States to accept Federal enforcement and a resistance to it.[13]

All the while Stein's superiors in the Federal water program have been debating the political pros and cons of ever going back to the Mahoning River for another conference session, the citizens of Beaver Falls, Pennsylvania, are being driven not to drink by the industrial invasion from across the border. And what has the Ohio

Water Pollution Control Board been doing on its own? The best commentary on the State of Ohio's contribution to Beaver Falls' plight was made, quite disingenuously, by the late Congressman from Youngstown, Michael Kirwan, in a prepared statement presented by one of his aides at the Mahoning River conference:

Permit me to make this observation, based on my own long experience: what has been done with and to water in the Mahoning Valley is truly an outstanding example of what can be done when local people work with local—and I include state in that term—authorities on their problem.[14]

Boy, you can say that again.

The Local Congressman

Prior to his death in July 1970, Youngstown's Congressman Mike Kirwan (D.-Ohio) could virtually guarantee that the Federal government would never again enforce on the Mahoning River. Kirwan was Chairman of the Subcommittee on Public Works of the House Appropriations Committee—the committee to which FWQA has always had to turn for its operating money. Although the Congressman did not attend the 1965 enforcement conference, every witness, even the few from conservation groups, paid tribute to "our beloved Michael Kirwan." Kirwan was described at the conference by Youngstown Mayor Flask as "probably the leader in our Nation in protecting natural resources such as water," but he was well known in Washington as the foremost booster of the filling, dredging, and canal-building projects of the Army Corps of Engineers —projects that have proved conducive to industrial growth but highly destructive of the environment. The Congressman made it amply clear in the prepared statement his aide read to the conference that his allegiance lay with the region's industrial polluters. With somewhat tortured logic, the Congressman first observed that "Youngstown district people who want to fish are certainly not deterred from the pursuit of their favorite hobby [by pollution] because there are no fish in the Mahoning River [as the result of pollution]." Kirwan then went on to spell out in more detail his views on conservation:

I leave you with this thought.

There was no doubt a day on the Mahoning when the Indians fished the stream in pristine cleanliness. I don't know how long ago that might have been. But I do know this. The Indians had no television to watch nor jet airplanes on which to ride. And I ask you now, who was better off, the Indians or you. . . ? [15]

The Basin Organization: ORSANCO

The most consistently effective voice on behalf of the polluters in opposing Federal cleanup—on the Mahoning and throughout the Ohio River Basin—has always been the eight-state Ohio River Valley Water Sanitation Commission (ORSANCO). When the Commission's compact was drawn up in the 1930's, citizen interest was invisible, rampant pollution was destroying industrial water supplies, and the threat of strong Federal control legislation hung as heavy in the valley air as the river's perpetual stench. Ohio Valley businessmen were justifiably worried. The product of their concern was a regional alternative to a Federal presence: ORSANCO. The regional compact's bylaws (finally ratified by all the participating states in 1948) insured its dependence on industry by making enforcement virtually impossible. Each member state becomes party to a reciprocal pollution agreement: you may pollute us provided we may pollute you. In legalese, the key clause reads:

no . . . order [to abate pollution] upon a municipality, corporation, person, or entity in any state shall go into effect unless and until it receives the assent of not less than a majority [two out of three] of the commissioners of such state.

In other words, each member state's Commissioners have veto power over any proposed abatement action in their home state. ORSANCO's signatory states have almost always appointed Commissioners who can be counted upon to exercise this veto with corporate beneficiaries in mind. Hailing from Ohio, for example, are State Health Director Emmett Arnold and Barton Holl of Logan Clay Products Company and the Ohio Water Pollution Control Board.

The West Virginia delegation's credentials are also in order, with Luther Dickenson from Union Carbide—

the state's number one polluting chemical concern—and Dr. N. H. Dyer, West Virginia's State Health Commissioner. Dyer combines expertise on his state's most pervasive pollution problem—acid mine drainage—with an abundance of sympathy for the coal companies that create it. Before he went into the pollution control business full time, the State Health Commissioner put in 23 years as company physician for the Pond Creek Pocahontas Coal Company (now a subsidiary of Island Creek Coal Company, the nation's third largest coal producer).

New York State, known for its independent and forward-looking pollution control program, knows whom to send to Ohio Valley deliberations; New York's Joseph Shaw is head of Associated Industries of New York State, Inc.; his colleague on the Commission, Lyle Hornbeck, is an attorney with Bond, Schoeneck and King, a Syracuse law firm that represents Allied Chemical Corporation. Indiana sends along Joseph Quinn of the Hulman Company.

ORSANCO makes its insulation from the public interest complete by surrounding itself with advisory committees, one each for the Ohio Basin's seven major polluting industries: chemicals, coal, metal-finishing, petroleum, power, pulp and paper, and steel. (An eighth committee represents Ohio Valley water users—mostly industrial—and a ninth gives technical advice on aquatic life.) The general public is nowhere represented among the Commission's advisory groups, and if an ordinary citizen turned up at an ORSANCO meeting, the industrial representatives gathered there would be nonplussed. Industry's domination of ORSANCO has paid a rich dividend in continued pollution: in over 22 years the Commission has not initiated (or even threatened to initiate) an abatement order against an industrial polluter.*

ORSANCO's total impotence guaranteed, any enforcement of its low water standards must be done by the member states, but those ORSANCO signatories most heavily responsible for the Ohio Valley's blight

---

* ORSANCO issued the only abatement order in its history to Middleport, Ohio (population: 3000), in 1964. It has, at the request of its member states, threatened to serve abatement orders on eight other small municipalities, all of which then complied voluntarily.

are hardly paragons of pollution control virtue. Ohio, Pennsylvania, and West Virginia have always been weak when it came to controlling industrial emissions. But the weakest of the weak among the ORSANCO states is Kentucky.

The Bluegrass State has always been the Ohio Basin's strongest bastion against Federal control requirements. When ORSANCO's pact was drawn up in the 1930's, two powerful Kentuckians carried the Ohio Valley region's fight to Washington. Kentucky's Democratic Senator Alben Barkley (later Truman's Vice President) and Democratic Congressman Fred Vinson, both closely allied with valley coal interests, were the main Congressional opponents of strong Federal pollution control laws for over a decade.* Kentucky still takes good care of the coal industry today. Fred W. Luigart, currently a member of the Kentucky Water Pollution Control Commission, is also the president of the Kentucky Coal Operators' Association, trade lobby for the state's mining companies.

The Bluegrass boys are partial to coal, but they have a soft spot for other industries, too; sitting beside Luigart on the Kentucky control board is Roy Sharp of West Virginia Pulp and Paper. Kentucky Commission executive director Ralph Pickard is not known for his sense of urgency about industrial pollution either. The most specific statement Pickard normally provides Federal authorities about his state's industrial polluters is that their controls are "adequate." In the midst of the mercury pollution furor of 1970 Pickard told reporters that there was "no problem" in Kentucky just two weeks before the Justice Department sued Calvert City's Pennwalt Chemical, one of the nation's largest and most recalcitrant mercury dischargers. Pickard was mystified by the excitement over mercury: "No one has yet shown up at the mortuary." [16]

With agencies like Pickard's providing ORSANCO's only enforcement punch, pollution control progress in the Ohio Valley has been all but imperceptible since

---

* In 1948 Barkley put Kentucky's lasting imprint on the Federal control effort. As an alternative to stronger measures, he sponsored the first Federal Water Pollution Control Act, feeble forerunner of today's law.

the cooperative cleanup in the 1950's. But it is characteristic of organizations that they do not die abruptly when their jobs are done, and ORSANCO is no exception; the Commission grinds mechanically on, legally still alive, dying only by degrees. The Commissioners —many of whom have held the same posts since ORSANCO first began operating—have begun to lose interest and are drifting away one by one. Prior to their retirements in 1970, Kentucky Commissioner Russell Teague and New York Commissioner Hollis Ingraham usually sent substitutes to ORSANCO meetings.[17] According to ORSANCO's executive director, Robert Horton, Indiana's current Commissioner Joseph Quinn "hasn't been here in years." [18] He isn't missing much. At ORSANCO's public meetings, unanimous resolutions are quickly passed and the rest of the time is spent swapping stories about stupendous cleanup accomplishments in days gone by.

Each year the Commission justifies its existence in a glossy "yearbook" of accomplishments, skillfully designed to sell its defective river to an unwary citizenry. The yearbook's "Tally for the Valley" is always comforting and impressive; it shows that some 95% (always a fraction of a percentage point gain over the year before) of the municipal population and some 87% (always up by a similar minuscule amount from the previous year) of the valley's industries are now "complying"—with a primary treatment requirement that has not been revised since the 1930's.

Once the dynamic architect of Ohio River Valley cleanup, ORSANCO is currently a viable organization for only two reasons: (1) the Commission's small and highly competent permanent technical staff is doing pioneer work developing sophisticated automatic techniques for monitoring and analyzing the Ohio River's growing pollution load; and (2) ORSANCO continues to be the Ohio Valley polluters' most effective device for influencing Federal enforcement policy.

A regional pollution control commission like ORSANCO is uniquely equipped to be a mouthpiece for special interests. It is responsible to no electorate, impervious to public pressure, and blessed with a compact that is as difficult to amend as the U.S. Constitution.

(Each state legislature plus the U.S. Congress must ratify any change before it becomes effective.) ORSANCO, like any other union, turns the grievances of individuals into group concerns and thus can multiply the pressure on the Federal government to back down from an enforcement action in member states. Most important, ORSANCO provides its constituent industrialists with a smooth and regular conduit through which they can gain the ear of ranking Federal pollution control officials, since three of its 27 Commissioners represent the Federal government.

Ohio Valley polluters also get attractive fringe benefits from their states' membership in a club to which the Federal government belongs. The states can accuse the Federal government of not being a partner in good faith if it "penalizes" them (i.e., conducts Federal enforcement) for making decisions in which Federal representatives to ORSANCO took part—decisions not to use the Commission's enforcement powers against polluters in a member state, for example, or to set water quality standards too low. Even if the Federal Commissioners voted against the action in question and lost, they can be accused of not playing fair if they turn around and call the big boys in to do what they couldn't do on their own. Federal officials in the Departments of Interior and HEW told the Task Force that the various Secretaries and Assistant Secretaries in charge of the water program had always been extremely careful to avoid provoking this kind of criticism.

Federal representatives in a basin compact are faced with a difficult dilemma. If they push from within, perhaps even threatening to recommend Federal enforcement if movement is not forthcoming, they not only invoke charges of bad faith but provide the region's polluters with a convenient early warning signal, a cue to descend upon Washington *en masse*. If Federal representatives take a passive role, they have acquiesced in continuing pollution.

As far as the Task Force could determine, the last time the Federal government seriously considered enforcing on the Ohio was in late 1962. When the suggestion leaked out, it stirred up a hornets' nest of resistance. Washington was bombarded with hostile missives from the states involved, and, predictably, ORSANCO came in person. Ed Cleary, then the Commission's executive director and now one of its staff consultants, appeared in the office of Assistant Secretary of HEW James Quigley to demand that the idea be shelved. Quigley explained to the Task Force that Cleary had friends not only in all the states but in the U.S. Public Health Service as well (including Gordon McCallum, Murray Stein's boss as head of the Division of Water Supply and Pollution Control), and if the government had gone ahead, there would have been "trouble." Quigley concluded that "the better part of valor" was "to let it ride for a while." The government has "let it ride" ever since.

Near the beginning of our investigation in the summer of 1969 the Task Force naïvely asked Assistant Secretary of Interior Carl Klein why the Federal government had never enforced on the Ohio River. His answer, though the Task Force did not know it at the time, could as easily have come from any of his predecessors in charge of the pollution control program during the past 13 years: "We'll be going to ORSANCO in the next ten days, and we're going to take a real close look at the situation."

9

The Squeaky Wheel Gets the Grease: Public Pressure and Pollution Control

"We are in an area where we need to be very, very careful that we do not strike a match and set something on fire that gets away from us," Congressman Jamie Whitten (D.-Miss.) warned FWQA officials at the House Public Works Appropriations Subcommittee hearing on the water pollution agency's 1970 budget.[1] The Congressman was not describing the Cuyahoga River, the Houston Ship Channel, the River Rouge, or any other highly combustible body of water in the U.S. What disturbed Whitten—who has always been one of the pollution control program's principal antagonists in the Appropriations Committee—was the fact that the water pollution agency had asked for $46,000 more for its tiny Office of Public Information than it had received the year before.

Whitten's fears were not unfounded. Both opponents and proponents of pollution control know the power of informed public pressure to shake lethargic bureaucracies into action against polluters and pry increased funding out of public coffers. For this reason, an active government information program inevitably faces heavy political and administrative constraints. These constraints are especially serious in pollution control precisely because the role of public opinion is so vital. It has usually taken an outside force pushing for cleanup to induce Federal officials to initiate enforcement actions. Some of the effective pressure has come from political figures— Senator Gaylord Nelson at Lake Superior, for example. But in most of the places where enforcement conferences have been held, public pressure eventually made the difference. This is what finally happened at Lake Erie, for example, long after any apolitical appraisal of the lake's condition would have sent an activist anti-pollution program racing in with sirens wailing.

In 1964, Lake Erie began putting out her loudest cries for help. Scientists from the Federal pollution program's Chicago-based Great Lakes Regional Office were crisscrossing the lake in a small boat that summer, taking water samples to analyze for planning purposes. When they got the data plotted on a grid chart of the lake late in the fall, the result was startling—about 2600 square miles of lake bottom (approximately a quarter of its entire area) were almost completely devoid of oxygen and unable to support any but the few toughest forms of aquatic life. Although the study had still not been completed, Washington headquarters was notified immediately.

Chicago Regional Director H. W. Poston was particularly disturbed. With the assistance of the public information office at Washington headquarters, Poston arranged to sound the first alarm in a speech in Cleveland—the second largest source of the lake's pollution. He was scheduled to address the elite membership of the Cleveland City Club, with full radio and press coverage, and force Cleveland's most prestigious citizens to face the question in the title of his speech: "Is Lake Erie Dying?"

Poston was never allowed to tell. A few days after the date for the speech had been set, headquarters Public Information Officer Mort Lebow was informed that it would have to be canceled. The man who had ordered the cancellation, Lebow was told, was the Assistant Secretary of HEW, James Quigley. The reason, Lebow claims Quigley told him later, was that information on how badly polluted Lake Erie was would embarrass Quigley's boss, Secretary of HEW Anthony Celebrezze —a former Mayor of Cleveland and a well-known friend of Lake Erie's largest industrial polluters.* For his own

* Quigley probably decided to cancel the speech without showing Celebrezze the Lake Erie data. He was accustomed to handling most water pollution matters for the Secretary. Quigley told us, "The Lake Erie report was quite unsatisfactory. My suggestion was that it was not a

part, Quigley was known to feel that Lake Erie was too big a problem to take on; it would inevitably stir up too many powerful interests. He not only canceled Poston's speech, but held up a brochure on Lake Erie that the HEW Public Information Office had prepared.

Fortunately, the preliminary findings of the Lake Erie study seeped out through informal channels. Cleveland City Club manager William Sanborn complained to the *Cleveland Press* that Quigley had canceled Poston's presentation—because the issue was, as Sanborn claimed Quigley had put it, "too political." [2] *Press* environmental editor Betty Klaric began to badger Federal officials to release their findings. Ohio Congressman Charles Vanik got wind of the Erie report and wrote to Celebrezze on March 2, 1965, urging him to schedule an enforcement conference immediately.[3] Quigley was still reluctant to get involved without a request for Federal intervention from a state governor, however. He wrote Congressman Vanik that the Secretary did not have the evidence of interstate pollution he needed to call a conference on his own.[4]

Meanwhile the Federal officials whose initial efforts to inform the public had been thwarted had not yet given up. "To be frank," HEW Public Information Officer Mort Lebow told the Task Force, "what we did then was not very loyal." Chuck Northington, one of the technical people who had helped plot the disturbing Lake Erie data the previous fall, was asked to speak at a League of Women Voters seminar on pollution control in Cleveland. Lebow helped Northington prepare his speech. But Betty Klaric of the *Cleveland Press* beat the League to the news. A few days before the scheduled seminar she finally got Northington and William Kehr, another pollution control official who had worked on the study, to grant her an interview about the Erie findings. The next day—March 25, 1965—the story made

---

very sound strategy to discuss it when the report was not yet published, in smooth form—after all, it was right in the Secretary's hometown and alluding to a report which was so far so poorly prepared and not really completed would have embarrassed him. . . . I thought until it was firmed up, the 'better part of valor' was not to discuss it." Quigley was occasionally considered overzealous in guarding his boss's interests. According to one former HEW official, those who tried to be "Celebrezze's best friends [i.e., Quigley] were his worst enemies."

the front pages under the headline, "Critical Pollution Is Found in Lake." The Cleveland newspapers also covered the League seminar on March 30, where additional information about the Federal findings came out. The *Washington Post* and *Evening Star* sent reporters to the pollution program's headquarters a few days later to interview people for major stories, and *The New York Times* ran an article under the byline of Gladwin Hill, one of its major environmental reporters.

The publicity was Murray Stein's signal to begin preparing an enforcement case, working to document what he later said was indisputable interstate pollution. Federal officials were still hesitant to act on their own, however, and tried to persuade Ohio Governor James Rhodes to request a conference. The Cleveland papers also kept up the pressure on Rhodes, and the hostility of the public gradually began to outweigh the insistence of Ohio industry that the Governor look the other way. On June 11, 1965—less than three months after the story had hit the front pages—Governor Rhodes succumbed, requesting Secretary Celebrezze to convene a Federal enforcement conference. On August 3 the reluctant Federal and state authorities finally arrived in Cleveland for the first Lake Erie conference session. What they found there was a clear-cut case of attempted murder. Erie was in many places nearly lifeless, tortured almost beyond recognition.

There are still some unsolved mysteries surrounding this Kitty Genovese case in water pollution control. The victims of Erie's polluters may have included not only the lake itself, but as many as 50 Cleveland citizens whom one doctor believes to have died of a "paratyphoid" epidemic from drinking Lake Erie water in 1965, and many others taken ill since then.[5] And although Lake Erie's present condition is hardly a subject of dispute, the eventual outcome is still unclear. Official predictions remain guardedly optimistic. But Federal pollution authorities who have tried to revive Lake Erie since the 1965 conference lament, privately, that the government acted so late. At least a few fear that—as one former Public Health Service official expressed it— "even if the Federal recommendations are fully complied with, Lake Erie may have been lost."

Despite these eloquent sentiments, however, the Lake Erie experience has produced little more at the Federal level than copy for environmental speeches. The lessons have gone unheeded. It still usually takes an obvious crisis or a public outcry to spark Federal enforcement— long after government scientists have, or should have, seen signs of serious danger.

While the public should not have to rely on its own eternal vigilance to see that anti-pollution laws get enforced, it is true that an awakened local citizenry will always be needed to support the tough stands officials will have to take to get the water clean. Thus the Federal government has a clear-cut responsibility to generate local pressure for cleanup by letting people know how bad their water is.

The enforcement program, not FWQA's public information office, has always performed this function best. In scanning old publications for news on water pollution, the Task Force found a remarkably consistent pattern. Apart from anti-pollution campaigns waged by local newspapers, Murray Stein's enforcement conferences generated almost all the coverage of the subject until recently. Conferences focused public indignation on specific polluters, brought out local bigwigs to take public stands against pollution, and sparked bond drives for municipal treatment plants with all their attendant publicity. Thus, where the enforcement program has gone, it has left in its wake a by-product potentially more valuable than individual cleanup victories: a national constituency concerned about water pollution. In far too many of these places, of course, enforcement might never have begun had not a frustrated official alerted the public surreptitiously—by "leaking" suppressed documents or alarming findings.

Since about 1966 FWQA's Office of Public Information has become an aggressive instrument in its own right, despite the budgetary constraints that key Congressmen like Jamie Whitten put on it. Running his office on a shoestring budget of only $500,000 annually (compared to around $27 million for the Department of Defense public information budget), FWQA's Information Officer Charles Rogers has focused on graphic mass media messages designed to stimulate local concern. His office has put out a hard-hitting series of radio and television spots that record the frustrated reactions of citizens to local pollution problems. Back in 1968, these candid anti-pollution ads so offended the mining company constituents of Idaho Senator Len Jordan that Jordan asked the Federal Trade Commission to investigate FWQA's "unfair advertising practices." In the spring of 1970 FWQA took an unusual step for a public agency: it conducted environmental teach-ins for students in key cities around the country and, out of these meetings, set up an organization—Students Concerned about Pollution of the Environment (SCOPE)—to funnel student opinions to the agency's leadership. (What will happen to these views once they have been funneled remains to be seen.) *The Gifts,* a 30-minute color documentary produced by the Information Office in 1970, is probably the best film on environmental destruction ever made by anyone inside or outside of government.* But FWQA's strong information effort still falls short of the mark when it comes to telling the American public what it most needs to know: exactly who is doing the polluting, specifically where, and how much.

Concerned citizens seeking access to specific information about polluters and pollution damage have always run up against a rigid tradition in the sanitary engineering fraternity against "naming names." As *Chicago Tribune* environmental editor Casey Bukro told the Task Force, control officials always give him "the excuse that they have to work with the polluters." The states used to conceal the names of lawbreakers not only from the public but from the Federal government as well. Compliance lists that FWQA received from states sometimes had polluters' names encoded, taped over, or omitted entirely. (There are still states—Ohio and Kentucky, for example—that normally will not release information on individual polluters' effluents to the Federal government.)

Unlike the states, FWQA has never consciously concealed the identity of polluters against whom it has initiated enforcement proceedings; Murray Stein boasts

* *The Gifts* may be borrowed free from the EPA's regional public information offices.

that his "files are always open," and they are. But the Federal government rarely volunteers information about polluters between enforcement sessions. Which polluters, for example, have met their cleanup deadlines and which are in violation? The public may never learn until a follow-up conference five years after the first. Nor is the public likely to find out which companies have the worst overall records in pollution control. The government does not tell.

Outside of announcements in the context of enforcement actions, the identities of polluters and polluted locations have been squeezed out of the Federal government only rarely—typically, in belated response to high-level criticism or in the midst of an obvious and well-known pollution crisis. In May 1970, for example, FWQA published a list of detergents, their manufacturers, and their phosphate content that was picked up by virtually every major newspaper in the country—five months after Congressman Henry Reuss (D.-Wisc.) had published a similar list compiled by a private consulting company and had publicly reprimanded the agency for its equivocation on phosphates. And in September 1970 FWQA released the names and locations of 40 serious mercury polluters against whom the government had decided *not* to initiate enforcement action—information that the government had gathered more than two months earlier when those companies were discharging dangerous levels of mercury. By the time the information was published, mercury pollution had long been a front-page item; Senator Philip Hart (D.-Mich.) and his Subcommittee on Energy, Natural Resources, and the Environment had been hounding the agency to release the information since late July; and charges had been leveled publicly that Interior officials had warned the agency against allowing "leaks" of any kind.

FWQA put out one information release in spring 1970 that set a healthy precedent, although again not on its own initiative. The *Washington Evening Star* asked FWQA to name the nation's 10 most polluted rivers for a series on water pollution; after an internal gnashing of teeth, the agency came up with an unranked list of 10. *Newsweek* magazine later published the FWQA honor roll:

Ohio River	Passaic River (N.J.)
Houston Ship Channel	Arthur Kill (N.Y.-N.J.)
Cuyahoga River (Cleveland)	Merrimack River (N.H.-Mass.)
River Rouge (Detroit)	Androscoggin River (N.H.-Me.)
Buffalo River (Buffalo, N.Y.)	Escambia River (Ala.-Fla.)

FWQA is only a decade or so late in putting out this list. By contrast, the National Air Pollution Control Administration (now in EPA with the water program, but previously in HEW) has been distributing—of its own accord—an "Air Pollution Hit Parade," a comparative ranking of the most polluted cities in the country, more or less regularly since the early 1960's. The list has been the air pollution agency's most effective device for drumming up public support for air pollution control. Interior Secretary Udall began to wonder as far back as 1966 "why we didn't do the same thing in water," as he later told the Task Force. After all, Udall explained, you "need [the list] even more in water pollution because you can't see it. You don't know it's there unless someone tells you, because the discharge pipes are under the water." At a press conference in 1966 the Secretary spoke of bringing the "pitiless glare of publicity" to bear on polluters and hinted that Interior would publish an annual list of the "ten most wanted" cities. Udall then sent a memorandum to FWQA Commissioner James Quigley proposing that the water pollution agency get to work on the idea. "There was dead silence for a couple of weeks," Udall recalls. And then:

When I asked again, [FWQA] people—particularly Quigley —sounded like they were horrified. "You'll make enemies in Congress and everywhere else," Quigley said. There was so much negativism on the idea from [FWQA] that I didn't do it. After all, they are the experts on the problem.

"But why," the Secretary concluded his account, "aren't we all entitled to know this information?"

A good question. It is obvious why public information is unpopular with almost everyone but the public and the Public Information Office—it embarrasses, tarnishes carefully preserved images, reveals unsolved problems, points up governmental failure, keeps things in a constant state of unrest. But, self-serving preferences aside, is there any good reason why the government should not be required to provide the public with the hard facts about pollution problems—particularly the worst pollution problems?

Company representatives typically told the Task Force they would not much mind public information if it didn't lead people to make "comparisons" between one firm's effort and another's. The notion, which sounds strange coming from businessmen, seems to be that competition is inherently unhealthy. It is—unhealthy for pollution and unhealthy for apathy toward fighting it. Polluters know that, and, consequently, the information they dislike most is the kind that comes with the comparisons explicitly drawn—as in the annual list Secretary Udall proposed in 1966, for example, or the list of dirty rivers FWQA gave to the *Washington Star* in 1970.

The bias against such comparisons is usually shared by local control officials for many of the same reasons. In speaking to the Task Force, state and local officials frequently invoked the "apples and oranges" metaphor. How, for example, can the Federal government maintain that 2000 factories dumping waste into the Ohio River, nearly 1000 miles long, is in the same league with 1000 industries pouring pollutants into 12 miles of the Houston Ship Channel? Which is more serious, fecal contamination near a recreation area or dangerous chemicals in a large city's water supply? These are difficult questions to answer. But the Task Force maintains that the government has an obligation not only to make those hard decisions but also to make explicit the criteria by which they are being made. Government is in the business of making comparisons and cannot avoid it; it must decide which problems to attack now and which to put off until later—which ones to fund heavily and which to pass over. Every citizen whose life, health, and security depends on these decisions being made responsibly— that is, everyone who drinks, uses, or simply enjoys water —has an absolute right to know what the decisions are and how they were arrived at.

FWQA's "ten dirtiest" list of 1970 represents a crude first acknowledgment by the Federal government that it owes this kind of information to the public. The government's effort to discover and expose specific pollution problems should be broadened and carried out on a regular basis. Up to now polluters have been forced to stand in the "pitiless glare of publicity" only infrequently —and then, typically, only in response to widespread public outcry for more information. The rest of the time they have rested comfortably in the shadow cast by government suppression of the facts.

Part Four

LAW AND ORDER

10

Law and Order

Our objective in enforcement is to get clean water—for the public's health, public water supplies, agriculture, industrial use, recreation, and propagation of fish and wildlife. In each area, we try to get the maximum number of these uses. The question is can you swim and catch fish where you couldn't do it before?
>   —Assistant Commissioner for Enforcement Murray Stein at FWQA's briefing for the Task Force, July 1969

How much polluted water has been restored to any of its former uses as a result of the Federal water quality program? In July 1965 a special Presidential committee of three—Council of Economic Advisors Chairman Gardner Ackley, Secretary of Interior Stewart Udall, and HEW Secretary John Gardner—set out to answer that question. This prestigious triumvirate had been asked by President Lyndon Johnson to compile a list of Great Society achievements. Committee Chairman Ackley directed their staff economist, Jim Flannery, to "come back with the names of five places where the condition of the water after the Federal government did some work on it was better than it was before they got there." Flannery went over to water pollution headquarters late the same afternoon thinking he could simply pluck the information out of the files and have it ready for Ackley the next morning. "It seemed like an easy enough assignment," recalls Flannery, "but I waded through those file cabinets for hours and couldn't find a thing. By ten o'clock that night I had started to get panicky. But then I figured I'd just wait till Murray Stein came in the next morning and ask him."

Enforcement chief Murray Stein could provide little help. Several pollution control employees were assigned to work with Flannery, and after soliciting candidates from field personnel, they spent a week, as Flannery recalls, "running salinity counts and computer studies on data we collected just for this purpose." The net result of their dragnet was a report to the Ackley Committee that the North Platte River on the Nebraska-Wyoming border, serving such major urban centers as Scottsbluff, Nebraska (pop. 13,377), "no longer had any sugar beet waste pollution, or something like that." Ackley was dumfounded. "I lost all my credibility," Flannery says. "It wasn't just that I'd failed. We couldn't even find a river where they had shifted some pollution and swept it under the rug. I came up with a total zero— a nothing."

Three years later FWQA Commissioner Joe Moore decided to try again. He assigned his Regional Directors to search for "just one bucketful of water, anywhere, that we honestly could stand up and say we had improved." The report he got was no more encouraging. The Willamette River in Oregon, Moore learned, had "just had a trout run." That was about it.

The Task Force can report that if Flannery or Moore had searched again in 1971 they would have been just as disappointed as before. While FWQA still does not attach enough importance to the question of water-use improvement to have bothered to consolidate at headquarters the information needed to answer it, there are only a few places that Federal officials are willing even to name as possibilities.* The Task Force received several more votes of confidence on the North Platte River.

---

* Two noteworthy exceptions to the general pattern of failure of abatement efforts are Lake Washington in Seattle and San Diego [Cal.] Bay (*Environmental Quality,* First Annual Report, Council on Environmental Quality, August 1970, p. 51), both of which were severely polluted as late as 1963. Significantly, however, Federal participation in these two cases of dramatic improvement in water quality was minimal. In neither case was there any Federal enforcement action. Federal subsidies (for treatment plant construction) amounted to only $8.5 million of Seattle's $145 million cleanup program and $2.5 million of San Diego's $60 million effort. (See E. W. Kenworthy, "San Diego Cleans Up Once-Dirty Bay as Model for U. S.," *New York Times,* Sept. 25, 1970, p. 28; and E. W. Kenworthy, "A Study in Pollution Control: How Seattle Cleaned Up Its Water," *New York Times,* Sept. 18, 1970, p. 30.)

There are also rumors that the South Platte (in Nebraska) enjoys slightly better health. Murray Stein told the Task Force that "the sight and odor problem in the Potomac has been cleared up." And since 1969 the Potomac's bacteria level has been reduced to the point where much of the river would be swimmable were it not thick with silt and other solid matter. Finally, Stein claims, Lake Michigan "is better." *

The kind of statistics the enforcement office gathers would suggest that they see progress in much the same way other branches of FWQA do—namely, in terms of how busy they have been and how much money has been spent. The enforcement section keeps such records as "estimated total amount to be expended in compliance with enforcement conference recommendations, 1957–1972: $13,400,000,000"; and "municipalities involved in actions held to date: 1410." At first glance these figures are very impressive. No other subdivision of FWQA can point to such a substantial record of activity likely to result in actual pollution cleanup. But to people who remember a time past when water was not a dirty word, "money expended" and "tons of concrete poured" bring little consolation. One cannot swim or fish in a waste treatment plant.

This is an absolutely crucial point. It would not matter if a treatment plant were built for every inhabited acre if the best result we could hope for would be the same rotting lakes and turbid rivers that a century of reckless destruction has left us with today. No progress can be counted until our water has been restored to some of its original uses.

Assistant Commissioner Stein is the only FWQA official who has publicly acknowledged the need to look at what his agency is doing from this perspective. "Progress in most cases is measured by the number of municipal and industrial treatment plants constructed," he wrote in one of his many published articles on pollution control.

* When the Calumet River conference reconvened in 1970, the Technical Committee on Water Quality presented a detailed report that qualified Stein's assessment. The southern end of Lake Michigan, the report showed, had improved in some places with respect to some parameters of water quality. Several other parameters, however, showed no apparent change, and at least two—notably "oil and grease" and "threshold odor"—showed continuing deterioration over the area observed.

"While this is valid insofar as reduction of the pollution load ultimately reaching the waters is concerned, true progress might better be measured in terms of the quantities of polluted water restored to acceptable quality for the maximum number of legitimate uses." [1] These are inspiring words. But FWQA's performance in the 51 pollution spots that have been singled out for enforcement belies them.

To get an approximate notion of the FWQA approach to the enforcement of water pollution laws, imagine a convict named Louie who has been committing six murders, rapes, and armed robberies a week—and happens also to be a big spender in a certain neighborhood bar. The bartender-owner and a Federal enforcement officer get together to decide what to do with him. "I never did nothin'," Louie mutters. "Lock him up and throw away the key!" shout the frightened neighborhood residents. "What's the rush?" asks the Federal enforcement officer soothingly. "First let's give him a chance to straighten out on his own. We can put him in jail later on if it doesn't work out." "Yeah," the bartender agrees, never in a hurry to lose one of his better customers. "Let's give him seven years to see if he can stop the rapes and murders, and ten to stop the holdups—unless he runs short of cash." "Come on, Harry," rejoins the enforcement officer, giving him the elbow, "you want to make your other customers nervous? Let's give him two years —he could do it in two." They settle on five years for complete self-reform.

Every once in a while during the next five years the enforcement officer calls up Harry while he's tending bar to check on how Louie has been doing. "Oh, he's doin' fine," says Harry, "he came by the other day and he told me so himself." And there's actually some truth in this; Louie is tired of seeing the other patrons gravitate to the opposite end of the bar as soon as he walks in; moreover, he knows that if the Federal enforcement officer stops thinking he is trying, he will doubtless plaster Louie's old mug shots all over the Post Office, which could cut into the thriving mail order business Louie runs on the side. So Louie curtails the armed robberies. The enforcement officer goes on national television to commend him to the public and to convicts everywhere: "Let Louie be an example to you—he's going straight!"

At the end of five years Louie and the bartender and the Federal enforcement officer have a follow-up conference. An assistant shows up with some embarrassing snapshots of Louie with a gun in someone's ribs, Louie with his fist in someone's jaw, Louie with his fingers in somebody's pocket. "He may have had an altercation or two," admits the bartender a little sheepishly. "It was self-defense," Louie interrupts loudly, "and a couple a non-violent loans. And them only two, three times a week. Plus recently I cut out the rapes completely." "Way to go, Louie!" whoops the Federal enforcement officer, clapping him on the back. "We'll give you three more years to reform, and the next time we get together I want to hear you're a hundred percent clean, alright?" "Yeah," says Louie, shaking hands.

In the meantime a delegation of over-excited neighborhood residents starts complaining about this proven criminal on the loose (and all the other known criminals whom the enforcement officers don't have time to prosecute but promise will fall in line once they see Louie reform). Louie, the bartender, and the enforcement officer all chide the citizens for their impatience, pointing out that nobody ever proved that a knife in the ribs is harmful in the long run, and reassure them that the partnership approach is the most effective way to handle criminals anyway. "So how come every year we've had more crime as long as you've been doing things your way?" skeptics yell out from the crowd. "Don't worry, folks," says the enforcement officer brightly over his shoulder, "the only thing holding us back before was we needed to know you were behind us. Now we're all pulling in the same direction—you, me, Louie, everybody—we all want law and order. Where would we be if we hadn't stepped in and given Louie here a talking to? Why he'd be up to fifteen, twenty murders a week. You gotta remember, these things take time."

There isn't an element in this story that doesn't have its parallel in the way FWQA goes about enforcing water pollution laws. The temporizing starts even before the enforcement conference opens, when the Federal con-

ferees pack their bags and set out, cleanup proposals
in hand, for the Missouri River, Raritan Bay, or wherever. Rarely has the Federal government espoused a
bold and far-reaching pollution abatement program,
though that is usually what is needed (by the time the
Federal government intervenes) to get the water really
clean. Instead FWQA officials have asked pretty much
for what they think they can get state officials to agree
to—somewhat more ambitious programs where state
standards are relatively high anyway, and less exacting
ones where the state looks askance at tough requirements
for polluters. The focus is on relative improvements in
the existing situation, not absolute standards for clean
water.

Unassuming as the Federal recommendations are, they
frequently look too ambitious to the conferees from the
state agencies, who work with polluters on a day-to-day
basis and often pick up their point of view. So the state
conferees typically put forward still more limited proposals of their own. There is nothing in the Pollution
Control Act that says the state and Federal conferees
have to agree on abatement recommendations. But given
FWQA's "partnership" approach to enforcement, little
cleanup results unless they do. As long as the Federal
government restricts itself to the purely exhortatory
stages of enforcement, never hauling polluters before
a judge, the state pollution control agencies—not the
Federal government—are the only ones with any authority to make polluters comply with conference recommendations. And the state agencies will not push polluters to follow cleanup recommendations that the states
themselves never agreed to in the first place.

Probably the main reason that Federal enforcement
conferences have worked as well as they have is that
Murray Stein, FWQA's chief of enforcement and the
chairman of most of the conferences to date, happens
to have a rare talent for negotiating consensus. "In almost all our conferences we have unanimous agreement," Stein told the Senate Air and Water Pollution
Subcommittee in 1966. "One of my charms, I guess, is
that I keep the boys' noses to the grindstones and will
not let them out of the room until we do have that." [2]
Stein's insistence on unanimity at these public confer-

ences has often led the state pollution control agencies (which don't, after all, want to look as if they are coddling polluters) to endorse more stringent cleanup recommendations than they would have backed independently. This does represent a net gain in cleanup potential for a given conference area as long as state agencies are the only ones that do any actual enforcing. But the price the public pays for these little triumphs of compromise is the further dilution of the Federal government's initial cleanup proposals—most of which were too cautious to start with. The net result has been Federal enforcement conference recommendations that would not in most cases have cleaned up the water even if they had been carried out on the spot.

Recommendations for Perpetual Pollution

By the late 1950's and early 1960's it had already become clear that a minimum of secondary treatment for municipal and industrial wastes would be needed in most places just to keep pace with economic and population growth and keep polluted waterways from getting worse than they were already. "Even if, by 1980, we provide secondary treatment [with 80% BOD removal] by conventional methods for all persons served by sewers," wrote Murray Stein in *Natural Resources Journal* in 1962, "the amount of pollution reaching watercourses in 1970 and 1980 will be substantially the same as today." [3] Yet at the time Stein made this statement not a single enforcement conference (all of which were chaired by Stein himself) had issued a recommendation for secondary treatment, except the Potomac conference (1957 and 1958), where the District of Columbia had already made plans for secondary treatment on its own. With one other exception—the Coosa River (Alabama and Georgia) conference in 1963—the most the conferees could ever agree to ask for in the 26 areas where enforcement conferences were held between 1957 and 1963 was either "further study" or primary or "adequate" treatment (which was construed as primary). Primary treatment, a mechanical process (i.e., large settling tanks and filters) that screens solid objects or allows them to settle out of sewage, removes an average of 30–40% of BOD in organic wastes (50% at best) and none of the dissolved or non-settleable matter (e.g., fine sediment, bacteria, dissolved phosphates, oils, lead). Recommending primary treatment is like asking polluters who have been dumping 40 tons of raw organic sewage into a river to cut those 40 tons down to 20 or 25—but only temporarily, since the pollution output grows as the amount of waste going into the treatment plant grows. This may have looked like a big step forward to polluters who were accustomed to doing nothing at all, and to enforcers who viewed the slightest movement forward as a gratifying sign of "progress." But it was not enough to stop pollution from increasing.

By 1965 there was hardly a sanitary engineer left in the country who would have argued that primary treatment was adequate for seriously polluted water. At that point FWQA's enforcement conferences finally shifted gears and started coming out with recommendations for secondary treatment (at least for municipal sewage). Secondary treatment, a biological process in which bacteria consume the organic matter in wastewater which has already been given primary treatment, can now remove as much as 90% of BOD under optimal conditions. But the average BOD removal rate is more like 60%, especially in older or poorly operated treatment plants. And even a secondary system getting 80% of BOD out of sewage can remove only 60–70% of the phosphorus and nitrogen that lead to massive growths of algae. Thus, in many thickly populated or heavily industrialized areas, secondary treatment would not be sufficient to prevent continued deterioration. "If we don't go to tertiary [advanced] treatment * generally, we're going to choke to death on nutrient pollutants," said Assistant Secretary of Interior Carl Klein in an interview with *The New York Times* on May 25, 1969. "And with the population explosion, we're going to have to do this [provide tertiary treatment] or be short of

---

* Tertiary treatment can remove most pollutants, including poisons like arsenates, cyanide, and phenois. But tertiary processes are very expensive and easily fouled. And neither secondary nor tertiary can remove pesticides, radioisotopes, certain poisons, or bacteria. To keep these pollutants from contaminating the water, they either have to be kept out of untreated wastewater or, in the case of bacteria, disinfected (chlorination being the only widely accepted way to disinfect drinking water supplies in the U. S.).

water." [4] As of October 1970 only three conferences had ever recommended tertiary treatment (95+% BOD removal) or the equivalent for any of the polluters in a conference area.[5]

But to say that waste treatment recommendations (i.e., for primary, secondary, or tertiary treatment) have consistently lagged behind what is needed at the time to keep pace with pollution is to understate the case against the standards that have been set at Federal enforcement conferences. Most recommendations have also been so vague that state agencies and polluters have had wide latitude to interpret them as they see fit. Allowing polluters to choose their own method of reform is of course entirely in keeping with the easy-going conference approach to enforcement, and undoubtedly helps pave the way for consensus. The vaguer a recommendation, the less there is to disagree with. But the predictable result is that many polluters "comply" with vague conference requirements by doing whatever is cheapest and easiest, no matter how little it actually contributes to cleanup.

Even "secondary treatment" is a vague recommendation when it fails to specify the degree of BOD removal desired. "Secondary" is what pollution control officials like to refer to as a "rubber word." It can mean anything from 55% to 90% BOD removal, depending on the number of different treatment processes included and on treatment plant operating efficiencies. Yet in spite of this range only nine out of the 21 conferences that had recommended secondary treatment by October 1970 ever specified the BOD removal desired (80–90%).* The Hudson River conference in 1965 was one of those which failed to spell things out, and when the conference reconvened in 1967 Senator Robert F. Kennedy expressed his dismay at the way New York City was stretching its rubber recommendation:

---

* These tallies are based only on the most accessible records on conference recommendations: FWQA's conference summaries, and the Secretary of Interior's letters of transmittal in which he made the conferees' recommendations official (sometimes choosing among conflicting recommendations when the conferees didn't agree, or making other modifications of his own). It is possible that some of the recommendations which were not spelled out in detail in the conference summaries were fleshed out later in technical sessions or by the state agencies involved.

New York has designed a North River plant to be located on the East Shore of the Hudson at 145th Street. . . . But [this plant] will not provide secondary treatment—only intermediate treatment. The design now calls for the removal of only 60 to 65% of the wastes. The commonly accepted standard is 90 to 95%.

. . . New York City is redefining the secondary treatment level previously recommended by this conference as removal of 60 to 65% of the wastes it processes. If this standard were applied to all communities along the Hudson, we would have to deal with the equivalent of raw wastes of 5 million people rather than half a million people. This is not an acceptable solution.[6]

But even a recommendation as specific as "secondary treatment or the equivalent"—with or without the percentage of BOD removal—is the exception rather than the rule at Federal enforcement conferences, especially for industrial polluters. The industries at both the Hudson River conference and the Lake Erie conference in 1965, for example, were asked only to "improve practices for the segregation and treatment of waste to effect maximum reductions of the following." They were then given an unusually detailed list of 10 pollutants to watch out for in their effluents,* but the explicitness of the list was completely vitiated by the vagueness of the instructions; each industry was free to decide with its own state agency what "maximum reductions" meant to them.

* It is easy to see why the conferees were reluctant to tell the Lake Erie and Hudson River industries exactly what to do, since a strict interpretation of the conference recommendations would have implied costly tertiary treatment or changes in plant processes to achieve "maximum reductions" of most of the pollutants on the list; i.e., grouping the grab-bag of pollutants listed in the recommendation according to the kind of treatment implied:

a. acids and alkalis	
b. oil and tarry substances	very likely imply
g. toxic and highly colored wastes	changes in plant
j. foam-producing discharges	processes
c. phenolic compounds and organic chemicals that produce taste and odor problems	
d. ammonia and other nitrogenous compounds	imply tertiary treatment
i. oxygen-demanding substances	
e. phosphorus compounds	imply efficient
f. suspended material	secondary treatment
h. excessive heat	implies cooling towers

The recommendation for the industries along the tip of Lake Michigan at the 1965 Calumet conference was even more nebulous: to "institute housekeeping practices which will minimize the discharge of wastes from industrial sources." This time an extraordinarily detailed and comprehensive set of water quality criteria * was worked out by a technical committee in 1966 and incorporated into the conference recommendations, complete with instructions on where to take samples to see if the standards were being met. But once again the conferees shied away from translating the recommendations into specific instructions to polluters, just as they had with the effluent standards at Lake Erie.

There have been some noteworthy, if scattered, exceptions to the vague-recommendations rule—when the states involved in an enforcement conference have wanted it that way. Least exemplary, perhaps, among these gold-star recommendations are effluent requirements which specify that a certain percentage of a pollutant be removed from wastewater without considering the total volume of waste going into the river.[7] The problem with setting this kind of effluent standard—a simple percentage removal—was carefully explained to Carl Klein, the Assistant Secretary of Interior in charge of the water pollution program, at the 1969 Potomac River conference:

MR. [AL] PAESSLER [Virginia conferee]: I am wondering whether it wouldn't be better to tie requirements of the river into specific poundages of discharge of the nutrients rather than to tie them into percentage removals [e.g., 96–98% nutrient and BOD removal], because as the loads of population increase, the percentage of removal will remain the same and this will permit more nutrients and BOD to get into the river, and, as a result, the situation in the river is going to slip up on us as it has in the past.

MR. [GROVER] COOK [enforcement chief for FWQA's Middle Atlantic Region]: Yes, I suspect you are right. . . .

SECRETARY KLEIN: If I may, what you are talking about is a technical item between two biologists here. Which way is a better way of expressing what you want in quality,

---

* Examples of water quality criteria are maximum quantitative levels of different chemicals, organisms, and other materials permitted in the water; minimum levels of dissolved oxygen to be maintained in the waterway; maximum temperature of the water, etc.

whether you say 96–98 percent, or whether you tie it into poundage?

MR. COOK: Pounds per day of loading, yes.

SECRETARY KLEIN: Which is the better?

MR. PAESSLER: I would say expressing it in terms of pounds, which is translatable into concentration in the river, and on which presumably you would project what is going to happen in the river would be the preferable way of expressing this.

SECRETARY KLEIN: In other words, the 16,500 [pounds per day of BOD as a maximum total discharge] that he has in this would be better than the 96 or 98 [percent reduction in BOD]?

MR. PAESSLER: Yes, and that [16,500 pounds] represents 96 percent of what is being generated by how many millions [of people] there are now [i.e., around 3 million]; 96 percent of [the sewage generated by] 7.7 million [people] or even an intermediate figure is certainly going to be more than 16,500 [pounds].

SECRETARY KLEIN: And they think it is 3 million [people] today, and 6 million would give us 33,000 pounds [of BOD per day], which would be twice as much as we think is necessary.

In other words, you have said it as an absolute figure.

MR. PAESSLER: Yes. . . .[8]

Klein caught on fast. "No matter how good treatment facilities are," he informed the Task Force a few months later, in August 1969, "because of population growth, you really can't control pollution until you set a total maximum pollution load for a river and divide it up among the polluters. We're going to do this on the Potomac."

A few load limits had actually been set before the Potomac conference (though apparently they failed to make much of an impression on water pollution officials), but there had been something wrong with almost all of them. Some, for example, recommended a percentage reduction (80%) in the total load of phosphates going into Lake Erie (1967), Lake Michigan (1968), and Lake Superior (1969). The problem with this kind of load limit, enforcement personnel told the Task Force, is that polluters can get around it by designing facilities to remove 80% of the total amount of phosphate they had been dumping at the time of the conference—and then running more wastes through their treatment systems, thereby increasing total phosphate loadings once again. Three other conferences [9] set load limits for certain rivers in pounds of pollutant per day— a much better approach. But two of the three either neglected to split the load up and assign shares to individual polluters and/or failed to set effluent standards at the same time, an important precaution against polluters dumping their whole daily allotment of, say, BOD in one lethal slug of raw sewage.

There is no denying that the 1969 recommendations of the Potomac River conference far surpassed all of these precursors. Exact numerical limits were set on not one but three major pollutants—BOD, phosphorus, and nitrogen—in pounds per day of loading. Each load limit was apportioned out among the five major waste treatment facilities discharging into the relevant stretch of the river. Each treatment plant was told the percent of BOD (96%), phosphorus (96%), and nitrogen (85%) to remove from its wastes. And deadlines were set for each major polluter. Assistant Secretary Klein was justly proud of this model set of conference recommendations. The only question is why it took the Federal government three sessions of an enforcement conference on a grossly polluted river in the nation's capital, a conference that had started 12 years earlier, to make this basic mathematical calculation: that to multiply an ever increasing figure (the additional waste due to population growth) by the same percentage year after year will yield a product (the amount of waste going into the river) that must also inevitably increase.

After the Conference: Slippage

The issue here is no longer whether they are going to clean up and put in the remedial facilities. Everyone is committed to this. The issue is when.

—Assistant Commissioner
for Enforcement
Murray Stein, Calumet
Conference, December
1968

In *Bleak House,* Charles Dickens described the fictional case of Jarndyce and Jarndyce, which had been tied up in the British Court of Chancery for generations. The case was literally passed on from generation to generation of hapless plaintiffs. It went so long, Dickens tells us, that it reached the point where it could be said that "no man alive knows what it means." And when the most recent plaintiff in the case, Richard Carstone, inquires about its progress, he is told by his lawyer, Mr. Vohles—"a sallow man with pinched lips"—that there is no cause for concern, that "a great deal is doing." The Task Force was told much the same thing when it inquired about the fate of recommendations coming out of enforcement conferences, and we could not avoid the feeling that Jarndyce and Jarndyce is still on the docket at FWQA.

The history of the Potomac enforcement conference, for example, would strain the patience of even the pinched-lipped Mr. Vohles. The first session of the conference was convened on August 22, 1957. Coming so soon after the enactment of the Federal Water Pollution Control Act of 1956, the failure of this conference has had enormous practical and symbolic significance. What the conferees found in 1957 sounds very much like what Washington-area residents know is true of the river today. The conferees found that the river was unfit for recreation, commercial uses, and game habitation. In some places the Potomac was not even fit for viewing. The conferees found what they characterized as severe sight (and odor) problems. But nothing was decided in 1957 except to meet again soon and that "efforts be made to obtain Virginia's attendance at the next session of the conference."

In 1958 Virginia was convinced to join the District of Columbia, Maryland, and the Federal government at the next session of the conference held in February of that year. The goals agreed upon at that time must be considered significant in the light of the years of neglect that had preceded the conference. It was agreed that the District's Blue Plains plant should be brought up to 80% BOD removal in the very same year, with chlorination to kill bacteria in the effluent. The recommendation was important because Blue Plains was treating

83% of the sewage from the 1.7 million people in the Washington metropolitan area.[10] But since the District was already planning on finishing up in 1958 the secondary treatment program it had begun in 1949, this recommendation was nothing more than support for what was already being done; the more significant recommendation called for a major expansion of plant capacity by 1965 so that treatment efficiency would not be rendered hypothetical by overloading. And just in case District officials should be tempted to misconstrue the conferees' intention to clean up the Potomac, the District of Columbia was required to "make plans sufficiently in advance" for any additional construction needed after the 1965 project was completed so that the Blue Plains treatment plant would "not again become a significant source of pollution." Finally, in order to keep raw sewage from being bypassed directly into the Potomac every time it rained, the District was ordered to complete a storm sewer program, begun in 1937, by the end of 1966.

These were not earth-shaking proposals. They were not technologically unfeasible. They did not disturb established land uses or proposed new ones. The time limits were generous—it takes two to four years at most to build a large sewage treatment plant. And the recommendations did not threaten the profits of large industrial operations—there are none to speak of along the Potomac. Nevertheless, if they had been realized, they would probably have affected significantly the quality of the Potomac today. But the recommendations slipped away.

"Slippage" is an important phenomenon to recognize in order to understand why things everyone seems to agree should be done do not get done in government. When we asked Murray Stein and his colleagues what had happened to the recommendations from the Potomac and other conferences, the Task Force was routinely told, "Well, there's been some slippage there," implying slight deviations from the projected deadlines. What we found on closer inspection is that this small, deceptively mild word is used to mask enormous failures in surveillance and advocacy on the part of FWQA.

If anyone noticed the slippage on the Potomac in the

early years, he kept it to himself. At the Detroit River conference in 1962, four years after the recommendations of the second session of the Potomac conference had been formulated, Murray Stein pointed to the Potomac recommendations as an example of the kind of cleanup the Detroit River conferees could look forward to if they set their sights high:

We have a schedule established there and *by 1966* all parties agree and believe that the water in the Potomac River, and in the vicinity of the Metropolitan Washington area, from a bacterial view, *will be of swimming water quality*. You will be able to swim right up to above Roosevelt Island. . . . [Emphasis added.] [11]

In 1962, the year Stein made that statement, the average coliform (intestinal bacteria) count at Roosevelt Island during July, August, and September was 2620 MPN/100ml. By 1966 the average count at the same site was 44,000 MPN/100ml.[12] The usual coliform limit for swimming is 1000 MPN/100 ml.

Until 1969—12 years after the first conference session—the river that Lyndon Johnson had once called the "truly American river" was for all practical purposes forgotten by the Federal officials who live and work along its shores. The reason the Potomac was so thick with bacteria was of course that none of the major conference recommendations had been met. The chlorination called for nine years earlier was "inadequate during the summer months and . . . not provided during the winter," as one of the technical teams FWQA assigned periodically to study the situation noted blandly in a 1967 report.[13] By 1969, another report noted later on, "effective disinfection" had still not been accomplished.[14] The storm sewer plan intended to keep raw sewage out of the Potomac by 1966 was only 40% done in 1969.[15] And the Blue Plains secondary treatment plant was so overloaded as a result of the District's failure to meet the conference recommendation for advance planning that it was getting the shockingly low BOD removal rate of 57% in 1966. By 1969 Blue Plains had edged up to 71%. But it was 80% BOD removal that had been recommended over a decade earlier. The falling behind throughout the 1960's had occasionally been punctuated by optimistic statements like Murray Stein's quoted above. But after the final deadlines passed in 1966, FWQA just stopped talking about the Potomac. Delays were not met with expressions of concern or attempts to awaken Congress or local citizens. Delays were met with silence. This is the stuff slippage is made of.

By early 1969 the Federal government could not overlook the Potomac any longer. Public criticism of the inactivity found a political outlet. As one of his final acts before leaving office, Secretary Udall called for the conference to be reconvened in April of that year. When the conference rolled around, the public was deluged with *mea culpas*. The Department of Interior's report reiterated the conclusions of the conferees 12 years earlier, indicating generally that the river was not very good for anything. Secretary Hickel opened the session by declaring that the Potomac "is a shocking example of man's mistreatment of a natural resource. . . . We are here today . . . because the recommendations for corrective action approved by the conferences in 1957 and 1958 have not been fulfilled." The Secretary closed his statement by reminding his listeners that "delay is the friend of pollution." [16]

The conferees unanimously adopted the recommendations of the Interior Department. As would be expected, the most important recommendations concerned the Blue Plains plant. The call was for 96% BOD removal to meet Federal and state water quality standards by 1975–77 (in May 1970 Interior Secretary Hickel moved this deadline up to 1974). Of course there was a long way to go, since the 1958 goal of 80% removal had still not been achieved by 1969. The recommendations also called for 98% removal of bacteria and 96% removal of phosphorus. As of 1969 Blue Plains removed only 10% of the phosphorus in the wastewater. Load limits were also set for the first time on the total amount of BOD, phosphorus, and nitrogen that Blue Plains and other treatment plants could allow in the water; the plants were dumping seven times the desired BOD limit in 1969.

Since the 1969 conference the Potomac has been showered with Federal attention. In 1970 Secretary Hickel directed the conferees to meet an unprecedented four times annually to check on cleanup progress—at least twice as often as any other conference has ever been reconvened. The early returns indicate that the Potomac polluters are progressing right on schedule—though the District is is behind in one important area. The 1958 program to keep raw sewage from being by-passed into the Potomac during storms (scheduled for completion in 1966) had been only 40% finished when the conference reconvened in 1969. So the District of Columbia had been given a 1971 deadline for finishing another 38%. When the conferees last got together in December 1970, the District was again caught lagging. The conferees have threatened to sue the District if it doesn't bring in a satisfactory schedule the next time around. For the last leg of the project, however, the conferees have set no schedule. The District now plans to finish off this project begun in 1937 by the year 2000 —if there is no slippage, of course. And inquiries about the muck that pours into the Potomac every time it rains can be answered by pointing out that, in effect, "a great deal is doing."

The slipping that Potomac River polluters have now been doing for nearly 13 years is, unfortunately, not the least bit exceptional. Slippage on cleanup deadlines has occurred in virtually every Federal enforcement action to date; polluters are almost always several months late, frequently several years, and in some cases a decade or more. It is evident from its record in 51 enforcement conferences that the Federal government's strategy for dealing with slippage contains two major plans:

Plan 1. Do nothing at all. Simply let the deadlines slip by without saying a word. This is the plan that was followed on the Potomac from 1958 until 1969.

Plan 2. Reconvene the conference over and over again to review progress, or the lack of it. This is obviously preferable to Plan 1 because it focuses public attention on the recalcitrant polluters. The potential for citizen pressure is, unfortunately, muted somewhat by the two alternative ways in which Washington typically handles polluters who have slipped when they come before reconvened sessions. The government sometimes declares lagging polluters "in substantial compliance" (as opposed to full "compliance") with conference recommendations—a euphemism for abatement efforts that don't quite make it. At the 1969 progress evaluation meeting of the Lake Erie conference, 75% of the municipal polluters and 47% of the industries were rated "in substantial compliance" when they were in fact behind schedule; 53 municipalities and 28 industries found "in substantial compliance" were more than a year behind the recommended timetable.[17] Murray Stein told the Task Force he considered assigning polluters to this ingenious category preferable to what the Federal government usually sees as its only other alternative—extending their deadlines.

The one step the Federal pollution control program's strategy for dealing with polluters almost never includes is using the legal sanctions available to force cleanup—calling polluters before a Hearing Board and/or taking them to court.* Plans 1 and 2 described above—and the massive slippage that inevitably accompanies them—are the predictable outgrowth of the government's amicable approach to environmental crime. Like Dickens' Jarndyce and Jarndyce, the cases just go on and on. The result comes as no surprise. In those cases where conference recommendations are not obsolete on the day they are made, they are hopelessly out of date on the day they are met—if they ever are.

Following the Leader: U.S. Steel

> Having been involved in pollution control from what are supposed to be both sides of the fence—government and industry—I am not at all sure how big a fence there should be between the two and, if the fence could be removed, how much there really would be to divide the two.
>
> —Former Federal Water Pollution Control Administration Commissioner James M. Quigley at the annual meet-

* As noted earlier, the Hearing Board has been invoked only four times, all of them prior to 1960. Only one polluter has ever been taken to court under the Federal Water Pollution Control Act. FWQA has one other remedy available for use against industrial polluters, suit under the Refuse Act of 1899. As of February 1971, FWQA had asked Justice to sue industries for compliance with cleanup deadlines only four times. (See Chapter 15.)

ing of Miami University's Pulp & Paper
Foundation, Oxford, Ohio, May 20,
1969

The only discernible pressure the Federal government
puts directly on polluters between enforcement conference sessions consists of a "high-level" image game in
which the harshest penalty threatened has been damage
to polluters' reputations, and the strongest form of encouragement a compliment in public. The "leaders in
pollution control" that Federal officials single out for
praise in the follow-the-leader game they try to get industrial polluters to play are not corporations that do
not pollute the water, or even necessarily those that
pollute the least. They are the wealthy, powerful, and
highly visible giants that the Federal government presumably hopes will set the pace for their respective industries in pollution control as they do in other matters.
And if these companies cannot be flattered into setting
the pace in fact, a clean-water image will do.

U.S. Steel was a natural choice for one of the Federal
government's starring roles in its pollution control charade. An established leader in the business community,
the company has carried on a long and costly campaign
to create an image of concerned corporate citizenship.
Schoolchildren across the country grow up watching free
U.S. Steel movies of square-jawed steelworkers stoking
the nation's furnaces as deep offstage voices pay them
sober tribute and the brass orchestra swells. "We're involved," boast the corporation's ads, which describe,
as one public relations officer put it to the Task Force,
"what U.S. Steel is trying to do to help the problems of
the day." Its efforts in water pollution control are de-
picted in glossy pamphlets filled with pictures of gleam-
ing pools of polluted wastewater in all the hues of the
Caribbean; delicately referring to waste treatment processes as "water conservation systems" or "water quality
safeguards," the pamphlets describe how wastewater
"goes to the cleaners" in settling tanks and scale pits
"after performing its host of steel-making assignments"
at showcase plants like Fairless Works (on the Delaware
River) and Gary (Indiana) Sheet and Tin. For more
sophisticated readers, U.S. Steel runs its turtle ads in na-

tional magazines like *Saturday Review*. "I'm never coming out," says one turtle to another. "Pollution. Bad water. Bad air." "Look at U.S. Steel," his companion responds. "In the last 18 years they've spent nearly a quarter of a billion dollars on air and water quality control. That's a huge investment considering they don't get a red cent return on it. I'd call it faith in the future." * [18] Clearly, such a company would be in the forefront of industry's battle against pollution.

Yet airplanes surveying Chicago's Calumet Harbor and the southern end of Lake Michigan for the Calumet enforcement conference in 1965 found the water discolored by flue dust from the blast furnaces of U.S. Steel's South Works. Bathers at Calumet Park Beach would stub their toes on iron nuggets of popcorn slag that washed ashore from the U.S. Steel plant, and torpid pools of South Works' oily wastes lay on the harbor and the lake.[19] The Grand Calumet River downstream from U.S. Steel's Gary Steel Works, Gary Tube Works, and Gary Sheet and Tin in Indiana was reddish-brown and covered with oil.

Conference recommendations gave U.S. Steel a December 31, 1968, deadline for installing adequate treatment facilities at its Gary and Chicago plants. When the conference reconvened a few weeks before the deadline, on December 11–12, 1968, it was clear that none of U.S. Steel's major polluting plants (Gary Works, Gary Sheet and Tin, or South Works in Chicago) was going to make it. The special FWQA Surveillance Unit that had been monitoring the water in 1967 and 1968 reported that concentrations of iron, phenol, cyanide, and ammonia in the Grand Calumet River below the outfalls in U.S. Steel's Gary steel-making complex were as high as they had been three years earlier when the Calumet conference opened; the river was "still the same reddish-brown color," sludge beds were continually being formed, and oil had been found on the surface of the

---

* The company reported in the January/February 1969 issue of its house organ, *U.S. Steel News*, that its combined investment for *both* air and water pollution control devices over the past 18 years had been "over $235 million." This "huge investment" amounted to 0.3% of U.S. Steel's gross revenue (over $72 billion) and 5% of its profits (nearly $5 billion) for the same period, 1950–68 inclusive. (U.S. Steel, *Annual Report*, 1965 (p. 29) and 1969 (p. 29).)

water almost every time the team went out to look. Only one pollutant, the sulfate from pickling liquor wastes, had been brought under control.[20]

And in the mouth of the Calumet River in Chicago, where most of the pollution comes from U.S. Steel's South Works, the Surveillance Unit found nothing whatsoever to praise. The water there, "frequently discolored by iron from the [South Works] outfalls," consistently violated applicable criteria for "turbidity, surface oil, dissolved oxygen, ammonia nitrogen, phenol and phosphorus." [21]

Studiously ignoring this irrefutable evidence, at the January 1969 continuation of the conference U.S. Steel's Herbert Dunsmore made a valiant stab at maintaining that the South Works plant (which happened to be the farthest behind of the local U.S. Steel plants in its cleanup) was technically in compliance with conference recommendations. The conferees refused to go along:

[CHAIRMAN MURRAY] STEIN: In other words, would you think South Works is in substantial compliance with the Conference recommendations right now?

[HERBERT] DUNSMORE [of U.S. Steel]: Yes. Our position and that of the Greater Chicago Metropolitan Sanitary District is and has been that South Works is in compliance with the Calumet Conference, and we were pleased with the [conferees'] inspection team's analysis of our cleanup program at South Works which was made on January 16th and they, too, concurred that it did meet.

MR. [PERRY] MILLER [aide to the Indiana conferee]: Not me.

MR. DUNSMORE: Well, at least the question was asked [whether conferees agreed that there was compliance]—

MR. MILLER: Not me . . . (Laughter.)

MR. [H. W.] POSTON [Federal conferee]: Not me either.

MR. DUNSMORE: All right, the question was asked, then, if there was agreement and nobody said no, I will put it that way, and that is not a concurrence—

FROM THE FLOOR: Try it again. (Laughter.)

MR. DUNSMORE: No, I guess I had better not try again, because I was at the meeting and I thought—

MR. STEIN: Your hospitality must have overwhelmed them. (Laughter.) [22]

At the end of the session the conferees agreed to extend the deadlines for U.S. Steel's Gary Works an extra year, and Gary Sheet and Tin an extra half-year for two of its treatment processes and 15 extra months for a third. Only one industry's request for an extension of the 16 that were made at that session seemed so out of line that the conferees sent it up to the Secretary of Interior with the explicit comment "extension not recommended": U.S. Steel's South Works.

This was Secretary of Interior Walter Hickel's cue to take action against U.S. Steel to compel the South Works plant to comply with conference requirements, perhaps by calling the company before a Hearing Board (the prelude to going to court). Even a formal hearing would at that point have been a relatively mild step to take in Chicago. South Works had become environmental enemy number one in the eyes of the local populace and had already been sued by the Chicago Sanitary District for spilling oil into Lake Michigan.

But the Federal water pollution program does things differently. Nothing happened for three months after the Calumet conference recommended that U.S. Steel's request for a deadline extension be denied. Then on April 30, 1969, Secretary Hickel made his move. Calling in the press, he posed for pictures with U.S. Steel president Edgar B. Speer, along with Undersecretary of Interior Russell Train and Assistant Secretary Carl Klein. "What U.S. Steel has done is both good policy and good industrial practice," [23] Hickel announced, praising the company for "taking the lead" in abating water pollution at its Lake Superior, Lake Erie, and Lake Michigan plants. [24]

The next session of the Calumet conference was scheduled for August 26, 1969. On August 25 an estimated 70 citizens gathered in the streets of Chicago's financial district to picket the Midwest office of U.S. Steel, protesting, as one of them later wrote to President Nixon, "the fact that U.S. Steel has been dragging its heels on solving this pollution problem." But where exasperated citizens saw foot-dragging, Assistant Secretary of Interior Carl Klein, as usual, discerned remarkable strides forward. His glowing description of the company's progress was carried nationally on Huntley-Brinkley the next night.

Within three weeks Illinois Attorney General William J. Scott had filed a suit to speed up U.S. Steel's abatement program, seeking an injunction against further discharges. Jack Schmetterer, the first assistant U.S. Attorney in Chicago, and his boss, U.S. Attorney Thomas Foran, had also begun to prepare legal action against U.S. Steel and a number of other polluters, and in August had asked the Chicago regional office of FWQA for information on the polluters' discharges.

FWQA's regional office checked with Washington headquarters. But the spirit of cooperation so prized by Federal water pollution officials in their dealings with polluters and the states deserted them when it came to helping another Federal agency prosecute industrial criminals. On September 24, 1969, a telephone call came back from FWQA's Bryan LaPlante relaying Assistant Secretary Klein's simple plan for justice: give *no information whatsoever* to the U.S. Attorney's office— not even proceedings of meetings that had been widely distributed. When the suit was finally filed in the spring of 1970 against U.S. Steel's South Works and several other polluters, the Interior Department phoned Assistant U.S. Attorney Schmetterer. Why, they asked him, did he want to prosecute "those nice people"? [25]

In the meantime, however, while the U.S. Attorneys had been gathering evidence without assistance from FWQA, Assistant Secretary Klein had been busy arranging something more in the Federal water program's line: its first National Executives' Conference on Water Pollution Abatement. For two days in October 1969 some 700 industrialists, trade group representatives, waste consultants, and lobbyists who had paid $100 each for the privilege mingled with government officials and with each other at the Washington Hilton.[26] They had lunch together, cocktails, and dinner; they heard a chorus of burly steel-workers from U.S. Steel's Fairless Works sing a medley of show tunes, and they also listened to speeches. U.S. Steel president Edgar B. Speer was among the six presidents and chairmen of the board of major U.S. corporations who were invited to address the conference because of their "leadership" in pollution control.* (Carl Klein told the Task Force later, "We only

---

* The other so-called leaders were Consolidated Edison of New York, DuPont, International Harvester, Scott Paper, and Standard Oil of Indiana.

put the good ones up front.") As far as the Task Force could tell, the most substantial signs of commitment anyone carried out of the conference ballroom were the "Clean Water" tie tacks FWQA handed out as souvenirs.

On January 28, 1970, Big Steel received yet another favorable plug from the Nixon Administration. Washington press correspondent Joseph Kraft wrote in his syndicated column: "There obviously are white hats all over the private sector [in relation to pollution control]. The United States Steel Company, for instance, has made a noble effort to reduce the pollution caused by its activities." [27] Kraft told the Task Force he got his information from Russell Train, then Undersecretary of Interior (and now the Nixon Administration's official environmental spokesman as chairman of the President's Council on Environmental Quality), and John Whitaker of the White House pollution staff. Train's tip was based on the fact that U.S. Steel has been more cooperative than Republic Steel—which is to say that U.S. Steel's cooperation is a little better than none at all.

The continuing gross pollution violations at U.S. Steel's South Works are not the only ones that both the corporation and the Nixon Administration neglect to mention in their promotional material. The Chicago plant is only one of the more visible polluters in U.S. Steel's vast empire. To mention just a few of the others:

1. After the 1965 Mahoning River conference, U.S. Steel's Youngstown (Ohio) Works made minimal improvements and then flatly refused to comply with the conference recommendation that it eliminate all its oil, acid, and iron discharges. When its December 1969 final cleanup deadline passed over four years later, Youngstown Works had not even *begun* to construct the necessary facilities.

2. While attention remained focused on South Works at Chicago, U.S. Steel's Gary (Indiana) Works topped the Chicago plant as the Calumet River–Lake Michigan area's worst polluter. After receiving an extension in its original December 1968 abatement deadline until December 1969, the Gary plant has failed to meet the extended timetable. By December 1970 Gary Works was demanding new extensions and making no pretense of intending to cooperate with the original requirements.

The Justice Department sued the plant in February 1971.

3. U.S. Steel's Lorain Works at Lorain, Ohio, told Federal authorities it would meet the 1965 Lake Erie conference recommendations by hooking up to the local municipal treatment system, but the city refused to go along. So Lorain Works is still pouring its coke effluent —phenol, cyanides, and ammonia—into the Black River in defiance of a December 1969 final cleanup date.

4. By December 1970 the Waukegan Works at Waukegan, Illinois, had still not complied with a Federal cleanup deadline of September 1969.

There are, in addition to this small sample, scores of other polluters among U.S. Steel's steel and alloy mills, iron and coal mines (U.S. Steel is the fifth largest producer of coal in the nation), limestone quarries, processing plants, cement manufacturers, real estate developments, and railroads extending throughout most of the 50 states. Many are in actual violation of state or Federal anti-pollution laws; many push behind the scenes to get pollution control requirements lowered.

In view of U.S. Steel's disgraceful record in pollution control across the country, why, the Task Force asked Assistant Secretary Klein at an August 1969 interview, did Administration water pollution officials continually praise the company so highly? Klein acknowledged U.S. Steel's violations readily, but went on to explain:

ASSISTANT SECRETARY KLEIN: We patted U.S. Steel on the back deliberately, of course.

TASK FORCE: But why?

ASSISTANT SECRETARY KLEIN: So we can say to other polluters, "If U.S. Steel can do it, so can you."

The moral of the U.S. Steel story does not appear to have been lost on other large polluters. In the words of former Assistant Secretary Klein, "If U.S. Steel can do it, so can you."

It Pays to Cheat

It would be misleading to suggest that polluters never comply with enforcement conference recommendations. Many eventually do—at least on paper. Going back through the box scores the Federal government keeps on old enforcement actions, one will find a number of former violators logged into the "in compliance" column. But to assume that what the scorecard reads and what the polluter does are identical would be a serious mistake. It is one thing to get polluters to install visible control equipment. It is quite another thing to make sure they operate it conscientiously when no one is watching —which is most of the time. With minimal surveillance and minimal penalties for pollution infractions, many "reformed" polluters find it pays to cheat. As a consequence, if the average citizen were to travel up any one of our urban rivers on almost any day, his experience would in all likelihood be similar to *Chicago Tribune* environmental editor Casey Bukro's when he made a random check of Detroit's River Rouge in October 1970.

On his trip down the river Bukro first encountered "a torrent of blood" pouring out of a sewer, then "great clouds of white wastes" coming from a Wyandotte Chemical plant. The state officials who had assured the reporter that all the river industries except Great Lakes Steel were in compliance with pollution control requirements were puzzled by Bukro's discoveries. The floodgate that should have been directing the blood from the slaughterhouse to the Detroit treatment plant must have gotten stuck, they said. And "Wyandotte has been in compliance for a year," chief engineer Jack Frost told Bukro; "the white plume was a horrible example for years, but we thought it was gone. Something must have gone to hell." [28]

Federal enforcement officials take a more cynical view. "It turns out something goes to hell at least once every three weeks," one Federal enforcer told the Task Force, "when they pull the plug."

Vigilantes

> The fox went out on a chilly night,
> > He prayed to the moon to give him light,
> He had many a mile to go that night,
> > Before he reached the town. . . .
> > > —English Folk Song

Had anyone been looking closely one Sunday night in late 1970, he might have seen a dark figure crawling slowly along a railroad bridge that overpasses the In-

diana toll road near Chicago. Closer inspection would
have revealed that the shadowy figure was a very ordi-
nary-looking man busily engaged in hanging a 60-foot-
long homemade sign along the side of the bridge where
it would be seen by people driving on the toll road
below. The sign read: "We're involved—in killing Lake
Michigan. U.S. Steel." The corner of the sign bore the
inscription "The Fox," with the "o" drawn as a smiling
fox face. The Fox had struck again.[29]

The man who calls himself The Fox has become well
known in the Chicago area for his imaginative exploits
in defense of the environment. But even *Chicago Daily
News* columnist Mike Royko, who has interviewed The
Fox several times and given his deeds wide publicity in
the press, claims not to know The Fox's real name or
what he does when he is not out prowling. From his
conversation with the self-styled pollution fighter, Royko
tells us only that The Fox is a normal everyday middle-
class citizen who several years ago became so troubled
about the destruction of his environment that he decided to do something about it. Since he began, The Fox
has pulled off a number of commando raids on industrial
polluters, all calculated to bring environmental violence
back home to the corporate executives who cause it. He
has jammed chimneys and plugged effluent pipes so the
companies' noxious wastes backed up inside the factory
instead of pouring into the air and water. He has strolled
into executive offices, dumped mounds of stinking fish
on the carpet, and vanished before he could be identified.
He has been shot at in the night by industrial plant
guards and tracked by the sheriff's office in Kane County,
Illinois, near his home. But he has never been caught.
And he has remained anonymous.

The Fox's latest target is U.S. Steel and he has been,
as columnist Royko puts it, "warming up" for more
dramatic tactics with a home-grown advertising campaign. After The Fox hung his sign from the railroad
bridge, Royko tells us:

He slapped up a couple more big signs in the nearby Gary
(Ind.) area, all of them directed at U.S. Steel, which has a
big plant there.

On Tuesday he showed up in Chicago's Loop with a stack
of cardboard signs under his arm.

He strolled in a Woolworth store and put one in their window.

"Nobody paid any attention to me. I just put it in the window and walked out." Then he watched with pleasure as people stopped, their eyes caught by the wording: "I can't stop killing your environment. I need the profits. U.S. Steel."

He walked over to a woman's wear store and slipped a sign into their display. It read: "Please stop me before I kill more wildlife. U.S. Steel." [30]

He pasted one up on an elevated ramp leading into a department store. While he was on the "el" platform, he stuck one on the side of a train that was pulling in. It said: "I will go on killing your environment as long as the courts let me," again signed "U.S. Steel."

"After that, I took a break for a cup of coffee," he said. When he finished his coffee, he paid his check and slid a sign in the window of the coffee shop. "Making steel is my business, murdering your environment is my sideline," it read.

He went into a men's store, looked at some sweaters and popped a sign into their window. A couple of empty newspaper stands were handy and he left signs in them.

In the corner of each sign was the familiar inscription —"The Fox." Columnist Royko explains that The Fox took his nickname from the Fox River Valley in Illinois, where he lives and where he first began his one-man campaign against environmental contamination several years ago. According to Royko, The Fox concluded the account of his latest venture this way:

"It's funny," he said, when he was done, "people never notice me leaving the signs. But they notice the signs. And that's the idea. People have to be reminded over and over again. That's what I'm trying to do." [31]

The Fox is only one man. He does not have the millions to spend on advertising that his wealthy adversary does. He does not have the authority to convene an enforcement conference nor the power to hand down a legal abatement order. But he has ample reason to be proud of his accomplishments. While the Federal government has been lulling the people of his city into a false sense of security by praising corporate criminals, The Fox's hand-printed signs have been alerting Chicagoans to the environmental violence taking place every day in their midst. While government officials only talk,

The Fox exercises his common-law right of self-defense. By so doing, The Fox challenges the government to act. Until state and Federal governments respond to the challenge with action appropriate to the crimes polluters are committing, The Fox and other concerned citizens like him throughout the country will continue to serve the public well by using the humble tools at their disposal.

11

A Uniquely Local Problem

Why has the Federal government stubbornly clung to the cooperative approach that has proved so futile? Why won't Washington move beyond the "consensus" stage of enforcement to the adversary stages—Hearing Board and courtroom—when polluters balk at the small changes they are asked to make? Have Federal enforcers grown so fond of the *kaffeeklatsches* they have been holding at various scenic spots around the country that they don't want to risk rupturing these friendly relationships? Or is this, in some vague way, simply "the way we do things in America" (as Murray Stein once told the Task Force) when dealing with the crimes of corporate and political officialdom?

The Task Force began asking questions like these at the outset of our investigation, and in sorting through the answers we received, one conclusion was absolutely inescapable: the forces that shield polluters are effective primarily because of the weakness of the basic Federal laws regulating pollution.

Because high-level discretion invites political pressure, the broadly discretionary Federal Water Pollution Control Act is essentially a political instrument. It has been used—or not used—primarily as politics has dictated. But even if enforcement action were mandatory and Federal officials were required to move from discussion to legal confrontation, the Pollution Control Act would still place insuperable obstacles in the way of effective enforcement.

Among persons familiar with the Washington pollution control scene, there is no disagreement whatsoever over the source of those obstacles. Thanks to persistent and well-financed industrial lobbyists on Capitol Hill, the Federal Water Pollution Control Act was for the most part designed by industrial polluters for industrial polluters. Polluting municipalities have also reaped the benefit of industry's efforts to keep pollution control laws as weak as possible, but it is American industry that has always done the work it takes to make sure they stay weak.

Although the Act has been amended several times, the provisions that set forth the details of the government's authority to hold enforcement conferences and to follow up with court action have remained essentially unchanged since 1956, when the basic legislation was enacted. At that time the rallying cry of spokesmen for industry and states' rights, both of whom wanted to hold down Federal power to abate pollution, was that water pollution is a "uniquely local blight." Their premise was that the Federal government should act only when the states for some reason were not handling pollution problems effectively on their own. Consistent with this notion, or so it seemed, a number of limitations were placed on the Federal government's enforcement authority. It has since become obvious to all who did not know it then that most of these restrictions were calculated to insure not so much that Washington wouldn't jump in where it wasn't needed, but that Federal enforcement would be weak and ineffective where it *was* needed.

The first of the crippling restrictions in the Act followed from the description of pollution as a "uniquely local" problem. The Federal government was empowered to initiate an enforcement action only when pollution from one state injured persons in another state. When the pollution was merely *intra*-state (i.e., confined to a single state), the Federal government could not intervene on its own. It needed the request of the state's governor. This limitation on Federal authority has never been altered, and it remains a tremendous barrier to effective enforcement. The indictment against it is impressive. All by itself this jurisdictional limitation on Federal enforcement has held back water cleanup at the state level, caused inequities in the cleanup burdens given different polluters, and made Federal abatement efforts expensive, time-consuming, and sometimes utterly useless.

One fact accounts for all these problems: there are

innumerable sites where an industry can locate (or a polluting municipality may be located) outside the purview of Federal enforcement. A polluter is safe as long as none of his wastes flows into a different state. That gives him an automatic haven in over a third of the states—the coastal states—where many rivers flow straight to the sea. And even when a river does flow into another state, a polluter may be so far from the state line that he can be fairly confident of immunity from Federal enforcement, since his wastes are unlikely to make it over the border with his trademark still intact.

The absence of a nationwide enforcement standard handicaps state efforts by putting states in competition with one another on the pollution deals they can offer. When a large corporation implies that it will take its jobs and tax money to a more lenient state (or a new company threatens not to settle there at all) unless waste treatment requirements are relaxed, it may not be serious; but few states feel they can risk calling the bluff or asking for Federal help.* This is why critics of the fragmented Federal jurisdiction have always pointed out that state enforcement will remain weak until Washington has the power to act on its own against all big polluters—interstate or intrastate—where the states can't or won't.

Quite apart from damage done to state enforcement, the "interstate pollution" requirement has handicapped the Federal abatement program in its own right. Its obvious potential for creating inequities between polluters who are subject to Federal regulation versus those who are not has forced the government into roundabout procedures to bring intrastate polluters up to interstate standards. These sidelong approaches are excessively time-consuming, as Murray Stein conceded when he explained to the Senate Air and Water Pollution Subcommittee in 1963 how ticklish jurisdictional problems are handled:

---

* Out of the eleven Federal enforcement conferences on intrastate pollution which have ever been held at a Governor's request, eight, it turns out, were initiated in the Governor's final year of office. The Task Force was told by state officials of instances in which a governor intentionally held off making his official request until just before his term of office expired, when the enmity his request would create could no longer jeopardize the Governor's other programs.

SENATOR MILLER: I would like to ask you one more question relating to your policy on enforcement. To illustrate the problem, let us take State A that is upstream, State B right downstream next to it.

State A has a packing plant which is polluting the water and State B also has a packing plant which is endangering its own people but it is not endangering the people of State A.

Would it be your policy to require State A to take care of their problem if State B was not taking care of its problem at the same time?

MR. STEIN: No, sir. That is a key question and I am happy to have an opportunity to talk about that. That is putting your finger on one of our most vexing problems, Senator. . . . where you get a river running across a State line, it is very, very difficult from an equitable point of view to ask the upstream State to clean up just to send clean water down to a State line where it will again be grossly polluted.

We have never done that and I hope we never will.

Now, what we have to do in a case like that is try to set up a schedule for the upstream State where we clearly have jurisdiction and depend upon the pressure, the moral suasion of the conference technique to get the downstream State to undertake a voluntary program also to clean up and get their river clean.

We have had this problem in several States. As you can appreciate, it is a rather delicate one to solve; the way you generally put it is that we are asking the polluter in State A to clean up by X date, but only if State B, the downstream State, has a commensurate program. If this is done, then we can have a clean stream.

So far we have had a fair amount of success. Of course, this takes a little more time for us to get a corrective action in such cases. But we do not have jurisdiction in the lower State.[1]

Still more preposterous are delays that stem from the prohibition on starting Federal enforcement until pollution crosses state lines—even in emergencies. At the same 1963 hearings Stein described to an incredulous Senator Gaylord Nelson how the "interstate" technicality often prevented the government from acting until damage had already been done.

MR. STEIN: . . . we have a much more dramatic case in that oil spill which finally got to Wisconsin and caused a duck kill. We were waiting while that oil was frozen in, so to speak, during the winter in Minnesota and just had to wait until it got down to Wisconsin and Lake Pepin before we could take action.

SENATOR NELSON: Do you mean to say this 2 million gallons of oil which poured into Lake Pepin which may leave that damage on that lake for years to come, that you would not have the power to move at the time in the winter when you could have stopped it from destroying—

MR. STEIN: That is correct, without the request of the Governor. It wasn't interstate in effect and we couldn't anticipate—

SENATOR NELSON: Did you know—

MR. STEIN: Yes, we did; we watched it. . . .

SENATOR NELSON: . . . [So] you don't have the power, even though the threat is imminent, and the pollution is going to occur.

MR. STEIN: That is correct.[2]

Even when the Pollution Control Act is applied most aggressively, there may be no way to get around the intolerable waiting game Stein described. But the jurisdictional limitation also provides a convenient out when the Federal government is reluctant to initiate an enforcement action for political reasons—as it was at Lake Erie and Lake Superior, for example. At Superior the government insisted that pollution was only intrastate from the time Senator Nelson first asked Murray Stein about it at the 1963 hearings, until public pressure forced an enforcement conference in 1969.

The interstate requirement has not only been the government's favorite excuse for evading cleanup responsibilities, but has also sapped the enforcement program of much of its potential for receiving high-level support. Politicians like to spend money where it shows. Back in 1965 a Presidential advisory committee chaired by Gardner Ackley was prepared to recommend an expanded enforcement program as the best way for LBJ to produce some measurable water cleanup accomplishments. The plan, as it emerged in initial discussions, was to conduct a wide-ranging Federal dragnet—frequent and vigorous enforcement actions backed up by a heavy increase in the enforcement budget. But after Ackley saw a list of places where the government had the legal authority to embark on such a crusade, he quickly scrapped the idea. The only waterways that the Federal government had jurisdiction to cleanse in their entire length, it appeared, were a number of highly polluted creeks. Nowhere were the glamour projects he had envisioned, certainly nothing impressive enough to merit the boldness of his original scheme. On most larger rivers only tiny segments straddling state borders were included, and there was nary a lake to be found.

While some of the omissions from Ackley's list of candidates for Federal enforcement derived as much from political considerations as from jurisdictional ones, the sparseness of the list also reflected the difficulty of proving interstate pollution—at least to a court's satisfaction. In order to obtain an abatement order from a court, it is not enough to prove pollution is present. The government must also show *to whom it belongs*. Because of the jurisdictional constraints the government operates under, this is by no means easy. Federal investigators must trace the offending substance from the scene of the crime—where the pollutant crossed a state line—all the way upstream to a single source. This is a complicated business involving fluid dynamics, chemistry, biology, and often a good deal of inventiveness.

The FWQA teams that do this detective work have taken up the challenge and become truly excellent at it. But modern science is simply not good enough to cope with the Pollution Control Act. Federal investigators quickly concede that they still can't pin down all polluters, especially where several contribute the same pollutant to a common pool. It is easy to measure phenols, say, next to a company outfall and conclude that a steel plant is responsible for phenol pollution. But to prove that some of the phenols 50 miles downstream at the state line come from that same steel plant is harder—especially when five other steel plants are close by. Unpredictable currents, particularly in lakes but also in some of the larger rivers, have also blurred many a polluter's trail.* And sometimes polluters are just too far away from the state line—even when too far is not really

* Conducting these investigations is expensive, too. The consensus of Federal investigators we interviewed was that somewhere around half the government's investigatory resources are spent simply to demonstrate that polluters who are causing obvious and serious damage immediately downstream from their outfalls are also causing some slight but still measurable damage many miles away down at the state line.

very far at all. As Ken MacKenthun of FWQA's National Field Investigation Center explained to the Task Force:

The waste from the big paper mill in Rhinelander, Wisconsin [St. Regis Paper Company, Rhinelander Division] travels about 40 miles down the [Wisconsin] river in the summer and about 70 miles in the winter. Since they're further up [from the Iowa state line] than that, there's no way we could get them. [They are approximately 350 miles upstream from the Iowa border.]

Distances like these—40 miles, 70 miles—marked off the short stretches of river that Gardner Ackley was told were within Federal jurisdiction in 1965. FWQA has extended these limits since then, suggesting that some of the difficulties in tracing pollution can be surmounted. And some can—by improving measuring techniques, and by sending enough scientists to collect evidence. But another reason FWQA has later held enforcement conferences on several lakes and sections of rivers that were considered out of bounds in 1965 is simply that it is no longer always gathering the kind of evidence it would need to go to court. The Act's jurisdictional requirements are not nearly so restrictive at the conference stage (where the government need only show which *state* the pollution belongs to) as they are if the government tries to take a polluter to court. But those who would rejoice over FWQA's expanding horizons (the Task Force is among them) must nevertheless remember that as the government takes on more territory, it dilutes what little legal power to abate pollution it now has.

The Missouri River is a good example. The first conferences held on the river were limited to the usual small sections upstream from state borders. But the government later went back and included the stretch of the river that cuts through the middle of the state of Missouri. As FWQA investigator MacKenthun explained it, the conference "took in polluters over a stretch of 630 miles" and "lumped them all together." He continued:

However, we couldn't trace a single polluter's organic wastes that far. So that means that most of those we've called into the conference we couldn't go against individually.

Or as Chicago regional enforcement officer Frank Hall said:

We've never *proved* that phosphates from Green Bay [Wisconsin] cross the [Michigan] state line in Lake Michigan. There's no way to put a little tag on them and follow them across. We were lucky the conferees just *accepted* this.

Another Federal investigator confided to the Task Force that "we've gotten pretty sloppy in the last few years. We're really just trying them in the court of public opinion now." If that is the only court available, as it sometimes is, then that is the best one to use.* But knowledgeable polluters undoubtedly appreciate how "difficult it is to prove this stuff if a judge ever starts asking the questions," as FWQA regional enforcer Grover Cook put it; if those polluters ever bother to comply with Federal recommendations, it will be at their own elephantine pace.

The original justification for limiting Federal jurisdiction to interstate pollution was that pollution was primarily a state and local problem. Local effort is obviously essential for pollution control. But these efforts have for some time received Federal boosts. State pollution agencies get Federal funds to use in planning, and municipalities get Federal grants to assist in constructing waste treatment plants. Even an *industry*, on a tiny intrastate stream, may have its waste treated by a plant built with 55% Federal funds. The Federal tax bill passed in December 1969 gives industry a direct subsidy for pollution control efforts. It does not ask whether their pollution is intrastate or interstate. To put these other Federal activities beyond the reach of enforcement deprives them of the backup needed to insure that government's efforts—and taxpayers' money—will not be wasted.

Ever since the Federal Water Pollution Control Act was passed in 1956, the "interstate pollution" requirement has outraged conservationists, burdened Federal investigators, dragged out abatement proceedings,

---

* This is not to suggest that there are not also a large number of polluters whose wastes the Federal government *can* follow all the way across the state line, and to whom this justification for avoiding the use of the court stage does not apply.

masked Federal inaction, discouraged going to court, penalized states that enforce their laws and rewarded those who don't, and given great comfort to all manner of polluters. And by keeping alive the chance to escape pollution control standards, it has diverted industries' attention from the absolutely essential task of developing better methods of waste treatment—a diversion neither they nor the public can afford. That the Federal law still makes a distinction between interstate and intrastate pollution is testimony to the power of the pollution bloc at both state and national levels.

12

Private Property, Keep Out: Information on Pollution

"In arriving at proper solutions to water-control problems, it is essential that decisions be based on facts, not on suppositions or suspicion," said John E. Swearingen, Chairman of the Board of Standard Oil of Indiana, at an Executives' Conference on Water Pollution Abatement held by the Interior Department in October 1969. "Industry needs reasonable assurance that a proposed solution is not motivated solely by a desire to solve the problem, but is backed up by enough facts to insure that it is the best available solution." What industrial spokesmen always neglect to mention when they sound this familiar refrain is that industrial lobbyists have permitted the government only token authority to collect "facts" on industrial wastes and treatment processes under the Federal Water Pollution Control Act.

There are four questions enforcement officials have to answer in order to see that a stream gets cleaned up. Who are the polluters? How much and what kind of pollution is each dumping? What cleanup measures should each polluter be required to take to improve the water? And finally, once requirements have been set, are they being followed? Under the Pollution Control Act, Federal enforcement officials are free to look for answers to these questions anywhere but at the source. They can ask state pollution control agencies for information, or take water samples by the polluter's outfall. But unless the polluter gives permission, Federal water pollution investigators have no legal right to inspect either industrial plants and their waste treatment processes or municipal treatment plants. "We can knock on the door of the worst polluter," one FWQA regional enforcement official lamented, "and if he doesn't feel like letting us into his factory, he can just tell us to go to hell."

The Pollution Control Act does allow the Administrator of EPA, once an enforcement conference has been held, to demand that polluting industries or cities file reports listing the composition of their discharges and any actions they have taken to purify them. But even then the polluter may omit any information he claims might reveal "trade secrets or secret processes." And enforcement officials have no legal right to check the accuracy of his reports by inspecting in person. (This authority to subpoena information that cannot be verified has never been used to date.)

If EPA's Water Quality Office could find out all it needed to know from state pollution control agencies, the first place it normally turns to, the cost of relying on roundabout methods of collecting information might not be too great. State laws generally require an aspiring polluter to get a permit to discharge wastes, and the information on the nature and volume of discharge on the application form goes into state files. Some polluters, however, prefer to remain anonymous. One of the Water Quality Office's representatives told the Task Force he had "no idea how many unidentified polluters have pipes that run nine miles underground to the river that nobody knows about." Neither state nor Federal officials were aware that the U.S. Steel plant in Joliet, Illinois, had any waste disposal problems until a local citizen reported in 1969 that the plant was discharging its wastes to the Des Plaines River through a storm sewer, avoiding both the cost of treating it and the city's sewer-user charges. One Scott Paper plant in Washington State, whose noxious wastes are too well known to be slipped into Puget Sound without anyone being the wiser, has simply refused to file for a discharge permit. The case is still in the courts.

Even polluters that do give the states information on their effluents and disposal practices often omit crucial data. In 1969 the U.S. General Accounting Office investigated the state permit files on 80 industrial plants

discharging wastes into a 170-mile stretch of the Mississippi River. The GAO found that the Louisiana state
files included information on the volume of waste for
only 52 out of the 80 plants, and BOD figures for only
30 of them.[1] The GAO had no way of telling, of course,
whether or not even these sketchy figures were up to
date.

The Water Quality Office's right to rummage through
state records has been a mixed blessing in other respects.
Even when a state and its industrial playmate start out
with a show of accessibility toward Federal investigators,
they sometimes change the game to "keep-away"—each
one claiming that the other will supply any additional
information needed. The information eventually comes
—after the Water Quality Office has wasted a great deal
of time running from one source to the other, or too
late to serve the purpose for which it was intended.
When the Interior Department called a hearing in Cleveland on pollution violations in the Cuyahoga River in
October 1969, for example, Republic Steel avoided
answering questions on its treatment program by refusing to supply information except through the State of
Ohio:

MR. MILLS [counsel for Republic Steel]: . . . I regret to say
that we have come to the conclusion that having hearings
with the Federal Government really only results in Republic being branded as a polluter. . . . We will [not] go
into any details which, in effect, would be construed as
bypassing the State of Ohio.

MR. HARLOW [representing FWQA]: Are you saying, then,
that you won't answer my questions?

MR. MILLS: I am saying: submit your questions to the State
of Ohio, which is my interpretation of what the law requires, that you go to the State of Ohio. . . .[2]

The state endorsed Republic's position, promising to
provide the requested information within a week—but
presumably after the public hearing was over.

Federal investigators are on even weaker ground when
they need information the state agency cannot give. They
can always ask to inspect the polluter's premises, but
the polluter is free to say no. "The only way we got into
the big steel plants [U.S. Steel and Republic Steel] be-

fore the Calumet conference," one Federal investigator recalled, "was that one of our men told them, unknowingly, that the information was only going to be used for a comprehensive plan and not for enforcement. If they had known it would be used for enforcement, they wouldn't have let us in."

If polluters do not deny the government access altogether, they can at least postpone the visit for as long as they need to put their best foot forward. "At the U.S. Steel plant on the Mahoning River," one investigator recalled, "it took us four days to get permission and by that time, of course, they had everything all tidied up." Presumably very few Federal investigators ever see plants at anything but their best, since none the Task Force interviewed could recall getting into a plant less than three or four days after requesting permission. Sometimes the negotiations take longer, as Murray Stein testified before the Senate Subcommittee on Air and Water Pollution in 1966:

We are having a meeting next week with one of the pesticide manufacturing plants [Velsicol Chemical's Memphis, Tennessee plant]. I think it was a year or two before we could make arrangements to go into the plant. The plant is under new management now; it was bought by another corporation. It was not until a month ago, or maybe it is as much as 3 or 4 months ago, that we consummated the agreement for our people to go through the complete process. This was several years in coming.[3]

When enforcement officials have lost all hope of getting into the plant to inspect, the last resort is to gather data at the company outfall. Murray Stein described one of these expeditions at the 1966 hearings of the Senate Subcommittee on Air and Water Pollution:

. . . where we have not been able to get the information [we] had to resort, as we always have to in those cases, to posting a man outside the effluent discharges, outfall pipes, on a 24-hour basis. We use boats. After the industry refused to let us into plants to get the information, they finally asked us how the Federal government had the nerve to go around picking up samples in the boats and not inviting the industry people on the boat.

Our man in the field, who is a fast thinker, indicated that

it would violate the safety regulations to overload the boat with all those industry people. At that time, they decided that we would have joint investigations and split the samples, and also they would allow us in the plant. As you know, this takes months of operating back and forth to get information of this kind.[4]

But outfall measurements by themselves may not detect all the pollutants going into a stream. Dangerous levels of mercury, say, may be so diluted by other wastes that the amount at the outfall at any one time will be undetectable. And even when pollutants can be discovered by laboratory analysis, and the investigators also have an idea of what they should test for, outfall measurements without plant inspections are still grossly inefficient. As one Federal investigator pointed out to the Task Force:

Most industries run all their wastes from many different processes together into a single stream sometime before the discharge point. If you can get into the plant, you can analyze the wastes from different processes to find out their content. But if you can only measure at the effluent outfall, you have to separate out the different constituents. For example, GM [General Motors] has combined wastes in their discharge point. Knowing that plating wastes contain cyanide and chromium, we would run chemical tests for these elements. But if we could get in the plant, and discover that they don't have that particular process, or they run those wastes out somewhere else, we could eliminate that test. That would cut your analysis by more than half in most cases.

Even at current low levels of enforcement activity, FWQA personnel have estimated that in the Chicago region alone, 50 or 60 man-years were spent in 1969 doing investigations that would not have been necessary if FWQA had the right to inspect polluters' premises. In the Middle Atlantic region (which has less industry under Federal jurisdiction than most other regions, and thus has lost a minimum number of man-hours investigating recalcitrant industrial polluters) the estimate of time and effort lost was 8 to 10 man-years. Figuring conservatively, this means that between $700,000 and $1 million out of a total enforcement budget of just over

$3 million is currently spent gathering information that industry already has in its files and in its plants.* Our nation's industrial polluters have not only destroyed billions of dollars' worth of the public's most valuable natural resources. They have used the dividends they have reaped from their stolen subsidy for an even more profitable investment. For the comparatively minor cost of lobbying, industry has purchased a Pollution Control Act that allows them to charge the American people a million-dollar admission fee just to see how the crime is committed. A criminal statute that permitted a search at the scene of a robbery only when the thief "agreed to it" would be unthinkable; one can only be thankful that professional holdup men do not have an organized lobby as powerful as their industrial counterparts.

But the most serious handicap that the absence of a right to inspect imposes on the Federal enforcement effort is neither lack of evidence, delay, nor needless expense; it is the pattern of day-to-day activity that the Act encourages in enforcement officials. Only by cultivating relationships of trust and confidence with the polluters they regulate can Federal authorities both get the information they need and avoid the frustration of continued confrontation and dragged-out negotiations. While this usually is not the only way to do the job, it is by far the easier way, and the Federal government has taken it far too often. When pollution control officials become dependent on a polluter's good will for information, they are afraid to push hard. If they do, the trickle of information may run dry. Their lack of authority debilitates the entire enforcement effort.

Whenever an expansion of Federal investigatory authority has been suggested, industry trots out its stock objection: that full disclosure of the contents and volume of industrial effluents would reveal trade secrets. In order to see how well this argument stands up, apart from any considerations of the public welfare, the Task Force asked several industrial representatives and trade

* This assumes that half the enforcement team in each region is paid at the GS-8 salary level and half at the GS-4 level—both low estimates —and uses the estimated number of man-years lost in the less industrial Middle Atlantic region as the figure for each of those regions for which the Task Force received no estimates.

organizations two questions: (1) "Is there any danger of members of your industry revealing trade secrets and thereby losing a competitive advantage by disclosing the contents and volume of their effluents? If so, give specific examples"; (2) "Can you cite any examples where a governmental body at any level has dealt improperly with information they were given in confidence, thereby causing the industry which had provided them the information competitive damage?"

Over and over the Task Force got the same answer to the first question: that disclosure would present no particular problem for the industry in question, but that it surely would for someone else. After admitting he didn't "know of any problems," John Blake of the American Paper Institute's National Council of the Paper Industry for Air and Stream Improvement thought up a putative example that, it turned out, had more to do with trade secrets of pollution control than with closely guarded secrets on paper manufacturing per se:

But somebody *may* have discovered a way—in terms of treating water—we use a lot of water you know, somebody may have discovered a way to, say, recycle or reuse their water so that less of the wood particles go out in the stream but are captured in the process. This means both less pollution and also a greater efficiency in putting out more of the product. I can conceive of the possibility of a company discovering a way to do this. Of course, you might hope that it would become public information because it would reduce pollution everywhere. But we have to deal with people as they are, and unfortunately they are more interested in their own profits naturally. So if someone could find a way to do this successfully, they naturally don't want their competitors to find out about it.

The notion that a new and effective method for controlling paper waste pollution should be kept secret so the company that discovers it can reap a competitive advantage while surrounding paper mills destroy the water is highly dubious, to say the least. But even in this example Mr. Blake conceded, "I'm not sure that revealing what was in their effluent would give this pollution control process away anyway."

Douglas Trussell of the National Association of Manufacturers also was asked if he knew of any case in all the industries his organization represents in which trade secrets would be revealed by information on the industry's effluents. He couldn't name one. George Best, technical director of the Manufacturing Chemists' Association, Sam Thomas of Owens-Corning, and William B. Halliday, Air and Water Conservation Coordinator of the Atlantic-Richfield Company, all gave the same answer: "No problem," they said.

P. Nick Gammelgard of the American Petroleum Institute, which represents the oil and petroleum industries, said the trade-secrets argument was "B.S.—that's right, I said B.S. That's a bunch of crap. If you want to know what industries are making, you can get most of it right out of catalogues. . . . All those people know what processes are available. There just plain aren't any secrets in the refining business anymore." He also made a particularly good point: "If you've got anything good for a process, it's covered by a patent anyway." In closing, he confirmed the Task Force's growing suspicion: "The only thing they want to cover up when they give that excuse," he said, "is how much pollution they're putting in."

Nick Gammelgard was the only respondent who gave an affirmative answer to the Task Force's second question, whether to his knowledge a government body had ever caused a company competitive damage by dealing improperly with confidential information.

MR. GAMMELGARD: Yes, I can think of only one instance that I've heard of. Bob Jones' Committee [the Natural Resources and Power Subcommittee of the House Committee on Government Operations] got some information at a hearing one time and later used it to embarrass the company in the press.

TASK FORCE: Was it a trade secret?

MR. GAMMELGARD: No, just embarrassing information about how much pollution they were putting out or something like that.

If there are any genuine instances in which manufacturers would be put to a competitive disadvantage by revealing effluent information, they have a clear-cut option available: they can stop polluting and the government will leave them alone. As long as polluters persist in unleashing dangerous contaminants in public waterways, the Federal government must have access to whatever facts it needs to protect the public from danger. Private profit from trade secrets can never take precedence over public health and welfare.

When FWQA was confronted with a full-blown mercury-contamination crisis in the spring of 1970, it found out who the mercury polluters were by going to the Manufacturing Chemists' Association, a private trade group, and the U.S. Bureau of Mines for their lists of mercury users. Emergency investigation teams fanned out to check for high mercury levels, and which of the 600-odd possible users were located in those areas. FWQA's regional offices (working with the states, which in most cases had no better idea than FWQA did where the mercury was coming from) started calling industrial users, one by one, and inspecting plants when possible, to see whose mercury wastes were ending up in rivers. The responses varied: the Goodrich Chemical Company in Calvert City, Kentucky, let the FWQA investigators in, but the Pennwalt Chemical Company next door refused (despite the urgency) and a scuba diver had to plunge into the fetid Tennessee River to take water samples next to Pennwalt's submerged effluent pipe. FWQA asked the states to ask the cities to check their municipal treatment systems for detectable mercury dumped via the city sewer systems. (Many states balked at the request.) Given the extraordinary momentum a well-publicized scare provides, FWQA was able by dint of special effort—pulling staff off other jobs and working them overtime, making countless phone calls and trips—to track down many of the most obvious sources of a single pollutant, mercury, in less than half a year. By then the fishing industry in many parts of the country had already suffered devastating and irremediable losses. Unknown stocks of tainted fish had been eaten or were piled on grocery shelves. Sport fishermen in more than 20 states had been urged to take up safer hobbies. And FWQA's search for mercury polluters is still going on, though Thaddeus Rajda of the Office of Enforcement assured the Task Force in October 1970 that "any big ones we didn't find have cut down on their own, since we're finding much less mercury in the water now."

The mercury is less, but beryllium, cadmium, lead, and other toxic metals may still be increasing. FWQA doesn't know exactly where these or any other contaminants are being displaced into public waterways, or by whom, or in what quantities. The agency is no better prepared to head off an epidemic from these poisonous wastes than it was with mercury.

Comprehensive national inventories of municipal and Federal wastes have been taken for almost a decade. But the nation's industrial organic pollution wasteload, known to be more than three times greater than the sewage load and growing rapidly, has never been accurately measured. (See Chapter 2.) The same goes for industry's lethal inorganic waste output—toxic metals, acids, cyanides, arsenic, etc. In 1970 the Federal government required many private households to supply detailed information on subjects ranging from the amount paid for rent to whether or not showers are available in the home and whether privies are indoors or outdoors. But Washington has never asked all the nation's private industries to tell how many pounds of which dangerous contaminants they are spewing into streams across the country.

A modest first step toward remedying the government's abysmal ignorance of these facts on industrial pollution was proposed to the Secretary of HEW by Congressman Robert E. Jones' Natural Resources and Power Subcommittee * of the House Committee on Government Operations as early as 1963. The mild proposal, for a National Inventory of Industrial Wastes, with industries participating on a wholly voluntary basis, aroused enough opposition from industry to hold the plan up for seven years.

The business organization that carried the ball against the proposed inventory was the little-known quasi-official

* Now the Conservation and Natural Resources Subcommittee, chaired by Congressman Henry Reuss.

Advisory Council on Federal Reports,* a group set up
in 1942 by the National Association of Manufacturers,
the U.S. Chamber of Commerce, and three other business sponsors to advise the Office of Management and
Budget (OMB) on Federal questionnaires. Under the
Federal Reports Act of 1942, all questionnaires to be
sent from Federal agencies to 10 or more businesses or
persons have to be cleared by OMB first. The law's
ostensible purpose was to protect small firms from being
deluged with overlapping requests for information. But
in practice it has given OMB a powerful policy-making
role in deciding what kind of information shall be collected by government agencies. The only formal non-
governmental group assisting OMB is the Advisory
Council, which describes itself as "appointed by and
responsible only to the business community." Until
1969, when under Congressional pressure OMB and the
Advisory Council opened their meetings to outside observers, all the council's guidance was proffered behind
closed doors—without labor, consumer, or environmental representation, and without public transcripts of the
proceedings (except in summary form).

The Advisory Council and other interested businessmen are notified of proposed Federal questionnaires by
way of OMB's "Daily List of Reporting Forms and
Plans Received for Approval"—unofficially, the "yellow
sheet." The businessmen on the yellow-sheet mailing
list † are known collectively as Birdwatchers: if one of
them spots something of interest to his industry among
the thousands of forms received for approval each year,
he asks to see the questionnaire involved, and sometimes
will suggest an Advisory Council meeting. The 1964 and
1968 versions of the proposed National Industrial Waste
Inventory aroused sufficient interest to prompt the Advisory Council to assemble *ad hoc* 27-man panels of
steel, coal, paper, power, petroleum, and other repre-

* Now the Business Advisory Council on Federal Reports.
† Other non-governmental groups which have an interest in the kind
of questionnaires the Federal government sends out—environmentalists, for example—have been free since 1964 to request copies of the
yellow sheet as well. But distribution has been limited, as in so many
other cases, by the fact that few private groups apart from business
organizations are aware of the yellow sheet's existence, or can devote
comparable resources to vigilant monitoring.

sentatives of polluting industries.* While the voluntary nature of the proposed questionnaire left companies the option of omitting data they were eager to keep secret, the industry representatives knew that companies discovered covering up their crimes would look even worse than those reporting them. The best way out of this dilemma was clearly to have no waste inventory at all. Barring that happy solution, the next best thing was to pledge the government to secrecy. So, despite repeated badgerings by Jones' Natural Resources and Power Subcommittee, OMB effectively quashed four successive versions of the inventory (in 1964, 1965, 1967, and 1968), either by failing to approve it outright or by attaching such stringent confidentiality restrictions that the water pollution agency could not accept them.

The latest inventory form would probably have stayed buried indefinitely if the Reuss Subcommittee had not held public hearings on the inventory in September 1970. On the first day of testimony, FWQA Commissioner Dominick and an OMB spokesman announced with a flourish that the two agencies had finally reached an agreement on the questionnaire. In October 1970, 250 industrial firms, a small test sample, were finally asked to report the quantity and composition of their wastes, where their discharge points were located, and what treatment, if any, they provided.[5]

---

* The industry representatives at the 1968 meeting were: CHEMICALS: George Best, Manufacturing Chemists' Association; Harry Robbins, Manufacturing Chemists' Association; J. H. Rook, Manufacturing Chemists' Association; Charles Welch, Manufacturing Chemists' Association; Everett Call, National Paint, Varnish & Lacquer Association; Robert Balmer, E. I. du Pont de Nemours & Co.; W. H. Garman, National Plant Food Institute; FOOD: A. Dewey Bond, American Meat Institute; William DeWitt, Corn Refiners Association; Fred Greiner, Milk Industry Foundation; Harry Korab, National Soft Drink Association; Fred Mewhinney, Millers' National Foundation; Stephen Palmer, National Association of Frozen Food Packers; Austin Rhoads, National Canners Association; Robert Shields, U.S. Beet Sugar Association; OIL AND PETROLEUM: S. O. Brady, American Petroleum Institute; P. Nick Gammelgard, American Petroleum Institute; METALS: Jules A. Coelos, Jr., U.S. Steel; H. J. Dunsmore, U.S. Steel; G. Don Sullivan, American Mining Congress; DeYarman Wallace, Youngstown Sheet & Tube; PULP AND PAPER: James Clabault, American Paper Institute; Floyd O. Flom, American Paper Institute; AUTOMOTIVE TRANSPORTATION: A. R. Balden, Chrysler Corporation; Robert Smith, Ford Motor Company; Daniel Cannon, NATIONAL ASSOCIATION OF MANUFACTURERS; Jack Coffee, U.S. CHAMBER OF COMMERCE.

Even if this trial run is moderately successful and the inventory is eventually extended to all industrial polluters, it is doubtful that the information will be as complete or as useful as it should be. Participation is still voluntary. This virtually insures that some polluters will refuse to respond, or will give incomplete information, thereby distorting the overall picture of industrial pollution that the inventory is supposed to provide, and exacting an unfair penalty on the polluter who sends in an honest report.

Furthermore, another industry advisory council—in the Commerce Department this time—is, with FWQA Commissioner Dominick's encouragement, taking a strong interest in the inventory. The National Industrial Pollution Control Council (NIPCC), which President Nixon created by executive order in April 1970 to give the government advice on environmental programs affecting industry, is composed entirely of polluters—61 top executives of major oil, automobile, electric utility, mining, timber, coal, airline, and manufacturing companies along with chief executives of the U.S. Chamber of Commerce, the National Association of Manufacturers, and the National Industrial Conference Board.* NIPCC's chairman is Bert Cross, chairman of the Board of Minnesota Mining and Manufacturing (3M Company, makers of Scotch Tape), which in July 1970 had still not complied with a 1966 Wisconsin order to stop dumping sulfurous wastes into municipal sewers.[6] The Commerce Department provided this discreet intragovernmental lobby with office space and staff. Following an unpublicized October 1970 meeting from which 10 environmental and consumer groups were barred,[7] NIPCC sent out letters to each of the industries that had received a waste inventory questionnaire, urging them to cooperate—and assuring them that NIPCC had "been advised by Commissioner Dominick that . . . proprietary information will not be divulged." [8] NIPCC's assurance appeared to create a vague and potentially broad exception to FWQA's statement on the questionnaire itself that the answers would *not* be confidential.

Keeping confidential the details on the wastes individ-

* The NIPCC members and their affiliations have been listed in Appendix D.

ual polluters are discharging would destroy any value the inventory may have for the public at large. Aggregate figures on industrial pollution are not much good to private citizens who want their local lakes and rivers cleaned up and kept clean. What people do need to know is what is going into their water, and who the polluters are—so they can check to make sure their water pollution laws are being enforced, and defend themselves against waterborne poisons when they find that law enforcement agencies have been looking the other way. It should also be emphasized that the waste inventory itself will be worthless as an enforcement device, since FWQA still will not have the authority for a personal tour of inspection to verify whatever information is collected. But there is some chance that the inventory may have a salutary, if indirect, effect if the names and the crimes of polluters are not kept secret.

One large group of industrial polluters will soon be *required* to disclose to the government the information that the National Industrial Waste Inventory asks polluters to supply on a voluntary basis. Under the Refuse Act of 1899, it is illegal for industries to discharge wastes directly into navigable waters without a permit from the Army Corps of Engineers. President Nixon announced in December 1970 that the Corps would start using its permit-granting authority to require mandatory disclosure of effluent data by all industrial polluters who fall under the purview of the Act. A mandatory national permit system of this sort will be a vast improvement over the voluntary waste inventory. Some major information holes will still remain after it is instituted, however. Although the great bulk of industry's waste is now discharged directly into lakes, streams, or bays, industries that discharge into municipal sewer systems are responsible for a rapidly growing but unknown proportion (believed to be at least half and possibly much more) of the overall *municipal* wasteload; what that proportion actually is will continue to be a mystery since only polluters discharging into navigable waterways fall under the Corps' reporting requirements. (See Chapter 15.) The industrial burden on municipal systems can in fact now be expected to increase even faster as corporate polluters begin to seek out this convenient avenue of escape from the new Federal scrutiny. In addition, as pollution control officials found out during the mercury crisis, even the most thoroughgoing permit system cannot provide data on contaminants that polluters "don't know" they are discharging.* What is really needed is an accurate accounting of all raw materials an industry uses † and where they ultimately go—especially those raw materials, like mercury, that somehow get "lost" in the manufacturing process. No inventory permit system by itself, moreover, would provide Federal water pollution officials with what they need most where information-gathering for enforcement is concerned: the legal right to enter and inspect polluters' premises to see for themselves exactly how and where the pollution originates.

For too long the instruments of environmental crime have been afforded a privileged status they do not deserve. While the weapons of individual violence have always been subject to public confiscation for use as evidence in court, the weapons of industry's abuse of the environment may still be secreted on the private property of their corporate owners, where they are protected by the law from public scrutiny or control. Industry has continued, successfully so far, to press upon Congress its claim that the dangerous pollutants it produces and information about them—the amount deposited in the water, their harmful effects, and especially their sources —are all properly in the private preserve. It is sufficient, industry has contended, that this information be released to the public only as an occasional act of beneficence on the part of the companies involved. This notion of private property is, in 1971, an intolerable anachronism. There can be no private property in information where lethal contaminants are concerned.

---

* Some companies that were found to be discharging mercury profess not to have been aware, before it was found in fish, that so much of the element was being washed away in industrial effluent. All they knew, they say, was that a great deal of the mercury being used as a raw material was somehow "disappearing."
† The State of Michigan passed a law, in response to the mercury crisis, which requires at minimum a listing of all the raw materials used in any manufacturing process. The law, which became effective in April 1971, also requires a detailed listing in loadings of all the constituents in Michigan dischargers' effluents.

13

Time Is Money

If polluters typically waste their own time and the government's trying to escape what seem but minor inconveniences—opening the door to a Federal inspector, for instance, or releasing information already in their files—it is not because they are fools. They recognize that whatever delay they can introduce at the beginning of their confrontation with the public will stave off a greater inconvenience—spending money on pollution abatement. If time can be purchased more cheaply than a waste treatment system, it's worth it.

So far the cost of buying time has been low—whatever it takes to compile a list of confusing technical terms to "explain" one's engineering difficulties, and to rehearse the stock excuses for delaying abatement (labor troubles, unsettled relocation plans, and mechanical difficulties if the polluter is an industry; stingy citizens and jurisdictional conflict for a city). Legal counsel (in case worse comes to worst and someone threatens court action) is a must. Optional extras are a small staff of scientists to question the government's pollution data and a team of consultants to kill time evaluating alternative abatement plans. The aim is to create an illusion of movement, of bona fide effort, while going as slowly as possible, and polluters—both public and private—are very good at it.

They have done so well, in fact, that the snail's pace at which they have complied with enforcement recommendations—when they have—is the most outstanding characteristic of the Federal record on pollution abatement to date. This halting progress reflects in part the time it takes to build treatment plants, and, more important, the inexcusable credulity of Federal officials who content themselves with polluters' assurances of good intentions.

But for the most part the colossal delay characteristic of Federal enforcement is thoroughly predictable—part and parcel of the conference strategy embodied in the Pollution Control Act. Delay has been consciously built into each of the Act's enforcement provisions, and it permeates every step of the process. The process is so interminable that it has been played out in full only once in the entire history of the Federal enforcement program: in the Missouri River city of St. Joseph, Missouri.

The saga started, really, when Dr. Samuel J. Crumbine, a Kansas state health officer living downstream in Atchison, Kansas, could no longer in good conscience remain silent about the pollution coming down the Missouri from St. Joseph and points north. When he expressed his concern to the Public Health Service in 1912,[1] they investigated the danger promptly, notifying the public in a Hygienic Laboratory Bulletin in 1913.* By 1949 impressive body counts of durable Missouri River catfish had been reported for several years. And when a Federal enforcement conference was held at St. Joseph in June 1957, the sandbars downstream were coated with blood, grease, and paunch manure from meatpackers Armour & Co. and Swift & Co. and other industries in the city's adjoining stockyard district; the water itself, said one witness, was "a combination of city dump and open sewer," and industries were said to be shying away from locating on the shore because the water was too polluted to use.[2]

The conference gave the city and her stockyard industries a January 1, 1959, deadline for awarding contracts for waste treatment facilities. The next time the government looked, the date had rolled by and nothing had been done. So in July 1959 the Surgeon General called both St. Joseph and her stockyard industries before a Hearing Board. Though the board found "continuous danger in St. Joseph of introduction into the public

* In the Public Health Service archives, pollution buffs will still find this classic in its original form—Hygienic Laboratory Bulletin No. 89, "Sewage Pollution of Interstate and International Waters—the Missouri River from Sioux City to Its Mouth."

[drinking] water supply of disease-producing organisms," [3] they moved back the deadline for awarding contracts two and one-half years more, to June 1961. [4]

At this stage the businessmen in the South St. Joseph stockyard district got themselves off the hook by making stirrings toward cleanup.* But the citizens of St. Joseph voted down a second bond issue in May 1960. At this point the long-suffering Kansas State Board of Health had had enough. Two weeks after the bond issue was defeated, they requested immediate legal action against St. Joseph. Their demand was the quickest move of the epic, and HEW Secretary Fleming was clearly impressed. On August 1, 1960, he made his historic request to the Attorney General to bring suit. [5]

At a preliminary hearing on November 18, 1960, Federal District Court Judge Richard M. Duncan agreed to let the case be settled out of court if the electorate would approve a bond issue. With (as one local pre-election advertisement put it) "the Federal government breathing down our necks," the voters authorized a $6 million bond issue for a treatment plant on April 11, 1961, and later approved a $3.6 million issue for extending the sewer system. The consent decree Judge Duncan handed down on October 31, 1961, set a June 1, 1963, deadline for all facilities to be in operation. [6]

The city reacted to its court order with all the zeal of a cranky Missouri mule on a hot summer day. Although St. Joseph's bond issue had been enough to cover the entire estimated cost of the project (as was required by the court), the city balked at spending the money until the Federal government would give them a respectable subsidy to go with it. [7] The Federal government knuckled under, sternly ordering cleanup with the right hand and slipping the city funds to supplement its own ample monies with the left. Even so, the city wanted to wait until Congress made it eligible for *larger* subsidies;

* The barest hint that the South St. Joseph industries that had been before the Hearing Board were making plans to treat some of their wastes exempted them from further enforcement. Unfortunately, the primary treatment plant they proposed remained in the planning stage for several years thereafter. In spite of a $250,000 Federal subsidy to help the project along, the industries' wastes continued to pour into the Missouri without even primary treatment until five years after the Hearing Board meeting.

until property easements could be obtained; until a new state highway was built. The 1963 deadline came and went before the ground had even been broken.

The plant was finally finished at the end of 1965. But when Judge Duncan held a second hearing on the case in 1967, it was treating only half the city's sewage. Five million tons of raw sewage a day still went straight to the river. As the U.S. attorney explained to the Judge, the treatment plant "is in operation and is a satisfactory, good plant, but the difficulty is getting the sewage into it." Judge Duncan was unimpressed: "That was the purpose of it, to get the sewage in." As he cogently observed,

it seems to me progress has been pretty slow, I don't know, but this case has been on the docket now for six years. I realize there are problems, but I wonder if there has been any dragging of feet along the line. . . . it has been a long, long time since they [the city] started this thing.[8]

Since that hearing the Federal government has handed out over $1 million more in subsidies, and 25% of St. Joseph's raw sewage still pours into the Missouri untreated. It is anticipated that the work necessary to provide primary treatment for 98% of the city's sewage will be completed by 1972.* Primary of course is no longer considered an acceptable level of treatment, as FWQA well knew when it announced its latest grant for St. Joseph's primary system in October 1969. In return for an additional $484,700 (with a pending increase that would bring the grant up over $1 million), the city promised not only to finish up the old treatment system, but to move its tentative deadline for secondary treatment up to 1974—"pending the arrival of [further] State and Federal funds," of course. Not much of a promise, but enough to bury the news of the grant in a release hailing this major breakthrough in the Federal effort to get secondary treatment installed.[9]

Does compliance with a Federal abatement requirement have to take 15 years or more, as it will in St. Joseph? Clearly not. The consensus of pollution control engineers and equipment manufacturers interviewed

* The Task Force was assured that the construction necessary to get the last 2% of St. Joseph's sewage into the treatment system was "in the mill."

by the Task Force was that installing primary treatment (including laying sewer connections) for a city of 100,000 (St. Joseph's population was 79,673 in 1965) should take eight months to two years. Converting to secondary treatment should take no longer than 15 months. And the time required for installing an industrial treatment system (once the polluter decides to), according to industrial spokesmen, ranges from 5 to 27 months.

Then what's to blame for the glacial pace of pollution abatement? Lack of interest, first—not only on the part of polluters but from the courts and the government as well. In the St. Joseph case Judge Duncan apparently considered it sufficient to follow up his court order with a polite reminder every six years or so. And judging from the government attorney's performance at Judge Duncan's first follow-up hearing in 1967, the Justice Department had all but forgotten the case.

THE COURT: Mr. Depping, what do you have to say about this matter?

MR. [HENRY] DEPPING [the government attorney]: Well, I didn't come here for this purpose, but I hadn't been here for some time, but we had a hearing here I think it was in 1961, and since then the City voted bonds and semi-annual reports have been filed. I have been informed there are some problems. . . . Well, I didn't know exactly what would come up this morning.[10]

Follow-through by FWQA was also lackadaisical, judging from another of Depping's statements:

And they [St. Joseph] have been discussed with me from time to time by the people in Washington [the FWQA's Office of Enforcement], and I have always advised them [FWQA] to get in touch with the officials at St. Joseph and see if they couldn't work out whatever it was they were talking about and they [FWQA] have been unable to tell though where the trouble is. It may be that if the St. Joseph people will just sit down and talk with the Regional people in Kansas City [FWQA's Missouri Basin Regional Office], they might be able to get a little help.[11]

FWQA and the Justice Department still debate who should have done what. But it is incontrovertible that a major factor in the St. Joseph debacle was inadequate—

indeed, almost nonexistent—Federal follow-up supervision.*

Yet the history of abatement litigation at the state level suggests that something more than bureaucratic buck-passing is to blame for the missed deadlines at St. Joseph and elsewhere. Though the record is as sparse in many states as it is at the Federal level, there are far too many chapters as inexhaustible as the St. Joseph epic to conclude that inadequate supervision has always been the culprit. At first glance it seems as if litigation itself may be to blame. Going to court is such a time-consuming process, in fact, that in many places it has been discredited as a viable way to abate water pollution.

Industry representatives have seized on this record of failure, citing crowded dockets, multiple continuances (which industrial clients have asked for and obtained), and lenient decrees (often arrived at out of court) as proof that "cooperation" is a better approach. The legal system, they argue, just takes too long. State and Federal officials have accepted industry's contention and repeat it as a commonplace of pollution control. What more could industry ask for than the attitude of North Carolina State Stream Sanitation Committee chairman J. V. Whitfield, expressed in his testimony before the Senate Subcommittee on Air and Water Pollution in 1966?

We in North Carolina have a slogan: The longest way to clean up the streams of any Nation is through a courthouse door. We have never had a case in court. We go the first mile, the second mile, and then sometimes the third mile.[12]

Former FWQA Commissioner Joe Moore, who came to the Federal program from a career with the Texas Water Quality Board, passed along to the Task Force the Lone Star State's version of Whitfield's maxim: "You can't get water out of the courthouse."

* Since 1968 FWQA has sent an enforcement official to the city on a weekly basis. The city has also gotten a new and more energetic Public Works Director, Captain Frank Endebrock. The net result has been much quicker movement toward completing the primary treatment system. For example, the city had argued that the only place it could build an important sewer line was along an as yet unfinished state highway; since the State Highway Department would not grant an easement, this project had languished for more than 11 years. Endebrock decided to put the interceptor underground and proceeded with construction plans. The decision to use tunneling rather than open-cut construction not only broke an 11-year deadlock, but lowered the cost of construction, perhaps by as much as $750,000.

When the assumptions of pollution control officials and of industry coincide so perfectly with what actually happens when a polluter is taken to court, they begin to sound like an immutable law of nature. It would be easy to forget that most of the delay and frustration spring from state and Federal pollution control laws. The delaying devices written into these laws all reflect the powerful influence—direct or indirect—of private interests on the legislatures responsible for them. The crowning achievement is the Federal Water Pollution Control Act, with more delaying devices packed into it than in any other—delays that plague every step of the Federal enforcement process.

The rationale for all the time allowed at the initial stages of Federal enforcement (the only stages most polluters ever have to contend with) is the old industry chestnut: polluters should be given time for "voluntary compliance." Until the final court stage, polluters are subject to nothing more binding than "recommendations." What this means is that a decision with far-reaching public consequences is left up to the polluter; he may make plans to abate his pollution or not, as he chooses. If he doesn't, he may eventually be given a court order to clean up, but in the meantime he is free to continue polluting. To the polluter who willingly goes ahead with abatement plans, it would make little difference were his compliance required and not simply optional. But for the polluter who doesn't plan to clean up at all, the period for "voluntary compliance" is simply a reprieve.

Not content with a basic process designed to stave off abatement, industry saw to it that several more guaranteed delays were written into the Act. The most gratuitous of these requires the Administrator to wait at least six months before going from the conference to the Hearing Board stage, and another six months before taking a polluter to court. To begin with, the idea that the bureaucratic procedures involved could take any less than six months is far-fetched. In St. Joseph (typical of the few cases that have gone beyond the conference stage), 25 months elapsed between the conference and the Hearing Board, and 27 months more before the case arrived on the court docket.

Even if the six-month waiting periods were eliminated

from the Act—as critics have recommended—and a superhuman Administrator found some way to crash through the bureaucracy, the polluter would still be shielded from unseemly haste. The Administrator can call a hearing or seek a court order only if "*remedial action . . . or action* which in the judgment of the [Administrator] is *reasonably calculated to secure abatement* of such pollution *has not been taken* [emphasis ours]." And the "remedial action" spelled out in conference and Hearing Board recommendations has never prescribed instant cleanup. The recommendations attempt to set a "reasonable" timetable for construction of treatment facilities, which includes target dates for such intermediate steps as engineering design, bidding by construction firms, and commencement of construction. The patience with which these preparatory activities are usually carried out is described in the following response to a Task Force questionnaire sent to manufacturers of pollution control equipment:

If an industrial company violates the water pollution law . . . , the first thing they do is to hire a professional engineer or an engineering firm. . . .

In many locations we have found that it has taken these engineers *as long as two years before they can come up with* the proper equipment or *any solution at all* to correct the problem. In my opinion it is evident that some of these consulting engineers are *retained only to buy some more time for the companies,* or else they have little or no knowledge in the water pollution field. [Emphasis added.]

If the polluter is a municipality, the schedule will also set aside time for passage of a municipal bond issue to pay for the treatment facilities and the campaign for public support that must precede it. (St. Joseph was given three chances to get its bond issue passed over a period of almost four years.) The *least* sympathetic recommendations promulgated thus far for either municipal or industrial polluters have allowed two years for these activities.

If a polluter fails to take any tentative steps toward abatement after all this time, there is an outside chance that the government will sue him. This final stage is of course the first time the polluter can be required to clean up under the current Act, and, late though this comes in the process, the fact that a *court* must issue the abatement order insures that the polluter will get even more time—a year perhaps until his case comes before the court, and another year or so of postponements if the polluter or his lawyer can make a minimal showing of being unprepared for trial.*

By giving the authority to issue abatement orders to a judge rather than an "expert" administrator, the Act does more than enlist the well-known sluggishness of the legal system to buy time for the polluter. If the case ever comes to trial, a judge (or jury) unfamiliar with the technical aspects of pollution control will have to wade through long, complex testimony from experts on both sides—rather than simply reviewing the Hearing Board's findings (the usual role of the courts in an administrative process and one for which they are better suited). And if in the end the judge issues an abatement order, the press of courtroom business (and the unlikelihood that any of his law clerks will be trained sanitary engineers) makes it improbable that he will provide any more supervision than Judge Duncan did at St. Joseph.

Other provisions in the Act make it questionable whether an abatement order will even ensue. For one thing, the Act instructs the court to consider both the "practicability" and the "physical and economic feasibility" of abatement.† In combination with other phrases in the same section of the Act,‡ this language might sug-

---

* These delays not only impede cleanup. Since legal rules of evidence are strict, the freshest evidence is required. The Act makes it hard enough for Federal investigators to collect good evidence against polluters in the first place; with every subsequent delay, that evidence grows less and less reliable. If enough time elapses after the conference to make it necessary for enforcement officials to conduct their entire investigation all over again, the taxpayers would have to pay twice or even three times for information on effluents and waste treatment processes that polluters should be required to hand over free in the first place.

† Taken together, these provisions in Section 10(h) make the deficiencies in other sections of the Pollution Control Act even more critical than they would be otherwise. If the government is to prove in court that it is "practicable" and "physically and economically feasible" for a polluter to clean his wastes, then it needs legal authority to find out what goes on inside the polluter's plant. The Act withholds this authority from enforcement officials.

‡ The clause that follows in Section 10(h) of the Act instructs the court "to enter such judgment, and order enforcing such judgment, as the public interest and the equities of the case may require." This looks suspiciously like a reference to the old common law on water rights, which has come to define both "the public interest" and

gest to a judge that no abatement order should be issued
if the polluter could convince the court that he would
have to close down his plant—or even cut back on production—to curtail his pollution. And even if the
polluter were unable to conjure up such drastic consequences, the wording of these provisions virtually guarantees that any pollution abatement decree would be
generous in its time allowance. As a member of one State
Attorney General's anti-pollution staff said to the Task
Force, "Look at that —— language. It's worthless."

In 1966 a proposal came before the Senate Air and
Water Pollution Subcommittee to eliminate just one of
all the delaying provisions from the Act, the six-month
waiting periods, requiring instead that the government
wait a "reasonable" time before proceeding against polluters. Industrial spokesmen invited to comment on the
proposal naturally urged, for all the usual reasons, that
the waiting periods be retained. But it was P. Nick
Gammelgard, testifying on behalf of several national oil
interests including the American Petroleum Institute,
where he is Director of the Committee on Air and Water
Conservation, who got to the heart of what industry likes
about these six-month periods, although he expressed it
obliquely:

The record shows that protracted proceedings have been
avoided under the law as presently written, perhaps at least
partly because time is left for compliance at each step along
the way.[13]

"Under the law as presently written"—that is, according to the specifications of Gammelgard and other trade
group representatives—protracted court proceedings
have indeed been avoided. Not because compliance with
cleanup recommendations makes them unnecessary, but
because all the "time left for compliance at each step
along the way" has given weary enforcement officials a

"equities" in a way particularly favorable to industry. There was normally a strict prohibition in the common law against spoiling the water
for the next user downstream. To spur economic growth, however, it
was considered to be in "the public interest" to grant exceptions to
industry. Opinions which took this point of view suggested that "the
equities," the jobs and products which industry was generating, justify
a special dispensation.

convenient excuse for avoiding every step but the first—the initial conference stage.*

When pressed to defend the delays written into the Act on grounds other than their own preferences, industrialists who spoke with the Task Force eventually fell back on the argument that polluters should have the same right to due process as anyone else. The Bill of Rights stipulates that "life, liberty, or property" not be taken away without "due process of law." To the degree that legal precedents and the Pollution Control Act now extend these protections to the polluter at all, both the courts and the Congress have operated on a mistaken set of assumptions. The polluter stands to lose no life, nor will he be incarcerated. He will be asked only to give the people back the freshness of their water, "property" that was never rightfully his to take. He will be charged no back rent for what he has used so long, nor will reparations be assessed for the damage he has caused. The ultimate penalty the Act provides is "abatement of the pollution." The most the court may do is tell the polluter that since he has already ignored the findings of Federal and state conferees, the recommendations of a Hearing Board, and two sets of recommendations by the Administrator of EPA, he must now begin preparing for the day when he will have to stop committing his crime. How much "due process" is due process?

---

* For example, FWQA's replies to citizens who write in demanding that legal action be taken against their favorite recalcitrant polluter receive a standard form-letter reply that could just as easily have been composed by industry. The reply mentions the Federal government's concern and involvement in the situation and then goes on to lament that "the Federal Water Pollution Control Act has built-in delays, and consequently, direct State action under their laws can be applied more speedily to resolve a particular pollution situation."

14

The Rise and Fall of a Mirage: Federal Water Quality Standards

By 1963 it had become clear to enforcement chief Murray Stein and the staff of Senator Muskie's Subcommittee on Air and Water Pollution that the spasmodic approach which had been passing under the guise of water pollution "enforcement" had to be altered. Stein assisted the Subcommittee staff in developing what was billed as a more systematic approach—water quality standards. The idea was to set standards that would indicate the maximum permissible levels of pollution in a given waterway. This would provide a clear-cut yardstick against which to measure stream improvement or degradation. FWQA Commissioner James Quigley was later to describe the scheme this way:

The laborious, costly, time-consuming process of providing water pollution damage as the basis for corrective action will become a thing of the past. In the future, the states—or if necessary the Federal government—determine what standards should apply to a given river or stream. Then, anyone who persists in dumping wastes into that lake or stream which reduce or threaten to reduce the quality of water below those standards is subject to legal action.[1]

The concept seemed appealingly simple, although it was hardly revolutionary. Many states had been setting stream standards for years—though without much apparent success. Now the idea was to get the Federal government involved in the process and to approach the problem of water pollution through comprehensive standard-setting and planning for pollution control. But things were not quite as simple as they appeared in 1963.

While Senator Muskie got the Senate to go along with the new approach right away, Congressman Blatnik couldn't get his more conservative Rivers and Harbors Subcommittee to agree on a similar bill until two years later. Finally in 1965 each chamber passed a bill that embodied the idea of water quality standards, but important differences between the Senate and House versions remained to be hammered out in conference committee sessions.

Both versions had contained some valuable provisions. But when they were finished, the conferees had expunged the strongest points from each and strapped the remaining sections of the joint product with debilitating qualifications. For instance, the House version had contained a provision empowering the Secretary to subpoena witnesses and documentary evidence for use in public hearings. The Senate version had contained a section that would have prevented private parties from profiting from patents on pollution control equipment developed with Federal grant money. When the dealing was completed and the dust had cleared, the final act contained neither of these provisions.

On the all-important question of who should set stream standards—the Federal government or the states—the strongest features of both bills again got lost in the shuffle. The Senate bill had provided for the establishment of stream standards by the Federal government, while the House had left that power with the states. The House, however, had included a sanction against states that did not indicate an intention to cooperate—cutting off their Federal grant money. The cut-and-paste job that emerged from the Senate-House conference left standards-setting with the states, but removed the threat of the loss of pollution control grants. If the standards initially set by a state were not high enough to meet with Federal approval, the Federal government was given the final crack at setting the standards for that state itself—though only after a tortuous conference-and-hearing procedure. Instead of sitting down to a tough poker game and strengthening their respective hands, the conferees on the Water Quality Standards of 1965—Senator Muskie, Congressman Blatnik, and Congressman William Cramer (R.-Fla.) the leaders among them—built a house of cards.

Briefly, the scheme set forth in the Water Quality Act is supposed to work this way: after conducting public hearings, the states draw up and submit to the Federal government a set of "water quality standards" defining the outer limits of permissible pollution for each interstate waterway within their jurisdictions. The standards are to contain two elements: "criteria" and an "implementation plan." The "criteria" portion indicates the projected use or uses for each waterway—e.g., recreation, public water supply, fishing, agricultural or industrial use—and the maximum allowable concentration of each pollutant consistent with these projected uses.

It is noteworthy that the standards approach assumes that the waterways will continue to be used as receptacles for waste. The "implementation plan" is thus essentially a plan for allocating the privilege of dumping wastes. It is supposed to indicate the methods by which each polluter will reduce his discharge in order to keep total pollution levels on a given waterway within the bounds set by the criteria. Once a state's standards (in other words, its criteria and implementation plan) have been approved by the Federal government, they are, in theory, enforceable by either the Federal government or the state.

It should be emphasized that the new approach did not replace the older conference procedure. The 1965 Water Quality Act's provisions were simply amendments to the 1956 Pollution Control Act, and they took their place in the enforcement section of the old Act alongside the previously existing enforcement conference provisions. The idea was to supplement the older enforcement scheme and expedite water cleanup by giving the Federal government an additional tool. Actually, the new procedure expedites cleanup in only the most modest way. Whereas under the old approach the government has to take the polluter through two preliminary stages—conference and then Hearing Board—before a court injunction can be sought, water quality standards can be enforced directly by court injunction. But the government still must notify the polluter first and then give him a minimum of six months—half the old waiting period of one year—to rectify his violation before the court resort is available.

In all other respects, this bold new addition to the Federal government's cleanup kit is actually weaker than the conference procedure it was intended to supplement. The 1965 amendments to the Federal Water Pollution Control Act not only incorporated all the other defects of the old law intact—its lack of investigatory powers, *de novo* court review of administrative determinations, crippling jurisdictional restrictions, the broad discretion allowed the Secretary to stretch out and delay enforcement or forget about it altogether—but also came complete with additional obstacles to enforcement that were all its own.

Whereas the government's authority to conduct enforcement conferences extends to all navigable waters, for example, the Water Quality Act applies to interstate waters only. This is more than a minor technical point. While most of the nation's estimated 26,000 waterbodies are legally defined as navigable,* only about 4000 are interstate.

By omitting this simple word, the 1965 amendments raised serious questions about the applicability of water quality standards to some of the Great Lakes and their tributaries; large intrastate waterways such as the Detroit River; all rivers, streams, and lakes in Alaska, Hawaii, the Virgin Islands, and Puerto Rico; international boundaries like the St. Lawrence, Niagara, and Lower Colorado Rivers; and waters flowing across the borders of the U.S., such as the Red River in Minnesota and Lake Champlain in New York. The pollution of any of these waterways is clearly as much a matter of national concern as the pollution of any interstate stream, yet they are ignored by the water quality standards scheme. On the other hand, a national landmark such as the Corney Creek Drainage Ditch on the Louisiana-Arkansas border is embraced by the amendments. Taking these jurisdictional limitations into account, the "comprehensive na-

---

* Murray Stein summarized the importance of the term for the Muskie Subcommittee in 1963: "According to the *Appalachian Coal* decision and other decisions, a stream is navigable when either it is navigable in fact or has once been navigable or by reasonable expenditure of funds can be made navigable and being navigable means carrying some kind of commercial traffic which possibly would mean a loaded canoe."

tional approach" begins to look more like a scatter-gun approach, with pellets falling without rhyme or reason.*

Even on that small percentage of the nation's waters where the standards are applicable, the "laborious, costly, time-consuming process of proving water pollution damage" has not become what former Commissioner Quigley optimistically characterized as "a thing of the past." If anything, the water quality standards approach thrusts additional burdens of proof on the government. Under the older approach the government has only to show that a discharge presents a danger to the health and welfare of persons in a state other than the one in which it originates (loss of a fish species, property damage, or restrictions on swimming, for example). But under the newer scheme, in addition to showing a threat to the health and welfare of persons in downstream State B, the Federal government must show that a polluter in State A is discharging matter which is reducing the quality of water in State A or B below the acceptable water quality standards, and also that the polluter is in violation of his implementation plan. Only then—if the Federal government is reasonably certain that it can make all of these showings—*may* the government proceed to take the next step: give the polluter six months to stop before the EPA Administrator *may* ask the Justice Department to seek an injunction restraining further excessive discharges. The Justice Department *may or may not* honor the request, as it chooses. It all sounds very familiar, only worse.

FWQA was handed the prodigious task of turning the convoluted new water quality standards legislation into

---

* Some of the least effective pellets of all in the water quality standards scheme landed on interstate waters that flow from, but not into, another state (e.g., a stream in coastal State A that has flowed in from State B—and is thus interstate—but empties, say, directly into the ocean). Because the 1965 amendments were simply grafted onto the old Pollution Control Act's requirement that pollution injure persons in another state before it can be stopped, the "Federal standards" that downstream states were required to set on these interstate streams are, for all practical purposes, unenforceable at the Federal level. There are, according to the calculations of former FWQA Commissioner Joe Moore, some 22 states with interstate waters in this category. Many (or in some cases all) of their so-called Federal water quality standards are nothing but meaningless numbers as far as Federal enforcement is concerned.

a practical enforcement tool at the beginning of 1966. Having just been liberated from the Public Health Service, the agency was in the throes of organizational upheaval at the time. Its new water quality standards responsibilities were caught in the middle of a conflict between Murray Stein, who had been instrumental in formulating the standards idea several years earlier, and newly appointed FWQA Commissioner James Quigley. Several conservation groups and organized labor had just waged an intensive but unsuccessful campaign on behalf of Stein for the Commissionership, and the enforcement chief had been anything but unenthusiastic about their efforts. But Quigley won out instead, and the immense job of supervising the standards-setting was too choice a bureaucratic plum to go to anyone but a Quigley faithful in the organization. It meant control over what promised to be the agency's most significant activity for several years to come. So the assignment went to Dr. Allan Hirsch, then the Assistant Commissioner for Program Plans and Development.

Hirsch is a brilliant scientist, and his staff was extremely competent in the scientific area. The plan was for the standards to be set by the scientists first, and then at some time in the future, when all the numbers had been refined and polished, they would be handed over to the enforcers to enforce. Some outside observers questioned the wisdom of the plan, since Hirsch's office had neither the legal expertise to make sure the standards would be enforceable, nor the experience in dealing with the states that Stein's Office of Enforcement possessed. But even the most concerned skeptic could have had no idea just how far in the future lay the date when the first enforcement of water quality standards would actually take place.

Hopes that FWQA might somehow salvage some potential for pollution control out of the wreckage Congress had deposited on the agency's doorstep vanished in late 1967 and early 1968 when the agency finally began to approve the standards states had submitted. FWQA, it appeared, had simply picked right up where the legislators left off. In the first place, the agency emasculated the Water Quality Act's "enhancement" policy, one of the few tattered remnants of the original

Muskie bill that had made it through the two years of House and Senate debate completely unscathed. The preamble to the Water Quality Act says that its purpose is to "enhance the quality and value of our water resources. . . ." FWQA endorsed the principle on paper, specifying in its Policy Guideline Number 1 to the states in 1966 that:

Water Quality standards should be designed to "enhance the quality of water." . . . In no case will standards providing for less than existing water quality be acceptable.[2]

But the standards FWQA was actually approving showed the Guideline had no meaning in practice. Take the State of Illinois' standards, for example, which FWQA approved in January 1968. Lake Michigan's actual measured level of dissolved solids is about 155 milligrams per liter (mg/1). The Illinois standard permits the annual average of dissolved solids to go to 165 mg/1—10 mg/1 higher than it is already in this polluted lake.[3] The present level of cyanide in the lake is about 0.01 mg/1; the Illinois standard is 2½ times as high, 0.025 mg/1.[4] In some Illinois streams there is presently no oil to be found. Yet the standards say only that oil shall not be permitted in such amounts as will create a fire hazard, coat boat hulls, or injure fish,[5] thus implying that at least some oil is acceptable where none existed before.

Soon after FWQA started approving standards like these, the National Wildlife Federation and other conservation groups began to complain that such standards were an effective license to pollute. Assistant Secretary of Interior Frank DiLuzio agreed with them. He informed Interior Secretary Udall in early 1968 that the standards which FWQA Commissioner Quigley was approving and sending up to Udall for official signature contravened the Department's own Guideline No. 1.

What was needed, of course, was a forceful commitment on the part of FWQA to apply its own Guideline when reviewing specific state standards. If the present level of cyanide in Lake Michigan is 0.01 mg/1, for instance, no more permissive standard should ever be accepted. Instead of sparking a review in 1968 of specific state standards, however, the controversy burned itself out by focusing on what turned out to be an essentially irrelevant issue. This was whether or not state standards should contain what came to be called a "non-degradation clause"—a promise not to lower the quality of the waters.

The lines were drawn publicly, with Assistant Secretary DiLuzio and the conservationists on one side and FWQA Commissioner Quigley and a coalition of Western political and economic interests on the other. The Western interests saw the non-degradation clause as a potential threat to their attempts to increase the exploitation of natural resources in their states. The Interior Department, always susceptible to pressure from the groups that form its main constituency—timber, mining, oil, coal, and ranching interests—sought to strike a balance between the two opposing camps. Secretary Udall adroitly engineered a compromise that the press described as a victory for clean water but that was in fact a disguised victory for the Westerners over the conservationist camp. Udall decided that each state would henceforth have to include a non-degradation clause as a part of its enforceable water quality standards. But there was one giant kicker. Degradation would be permitted where "such change is justifiable as a result of necessary economic and social development." [6] This meant that a state bent on "necessary" economic development needed to require only that water quality be consistent with its economic plans.* What was to have been a revocation of the license to pollute turned out to be a renewal instead.

As far as FWQA was concerned, however, even Secretary Udall's half-a-loaf was more than the conservationists were due. The water pollution agency has still never demanded that states describe the present condi-

---

* Udall's non-degradation statement did specify that no "assigned uses" are to be impaired. But to get around that restriction, a state need only make sure that it hasn't "assigned" any water uses which it wishes to impair. Even if the Federal government is careful to plug that obvious loophole by requiring that a state's "assigned uses" always include all uses presently possible on a stream, the non-degradation clause still provides only the most limited restriction on water quality impairment. A crystal-pure lake, for example, would not then be legally allowed to become so grossly polluted that a fish species would be wiped out or that physical contact would be indisputably unhealthy. But the lake *could* be legally degraded *just up to* the point where either of these uses would be lost.

tion of the waters when their standards are submitted. Thus there is no baseline against which to measure the value of a state's promise not to degrade its waters. Signing a non-degradation clause is like taking an oath of abstention from alcohol—the value of the pledge depends entirely on the maker's will power. Francis T. Mayo, FWQA's Regional Director in Chicago, admitted as much in a September 1970 statement to the Illinois Pollution Control Board: "to our knowledge, in no areas has a baseline for non-degradation . . . been formally defined. *This lack of a baseline makes the non-degradation concept impossible to enforce*" (emphasis added). The Department of Interior's non-degradation compromise gave conservationists paper promises while its execution at FWQA has given polluters the nation's waterways.

As part of the same generous package to polluters, FWQA also undermined another essential part of the water quality standards—the implementation plans. Without an implementation plan that translates general stream criteria into specific discharge requirements for each polluter, the standards are useless as a cleanup tool. Because the implementation plans are so crucial to the standards' enforceability and because they ultimately determine the amount of money each polluter must spend on control, the position the government should take on this matter also became the subject of a heated controversy late in 1967.

On the surface, the debate centered on the intent of the Water Quality Act. The Act's meaning on the question of implementation plans is particularly difficult to fathom, mainly because of its tumultuous legislative history. When Senator Muskie pushed his original water quality standards bill through the Senate in 1963, it contained a number of items that he was later forced to expunge because of strong resistance in the House. The most critical of the elements that went by the wayside was a provision that would have permitted the government to set effluent standards for polluters along with water quality standards for streams.

The failure to include an effluent standards provision is the major weakness of the 1965 Act. Effluent standards—as they were described in the 1963 Muskie bill—

are controls over the "type, volume or strength of matter permitted to be discharged." If the FWQA had been given the authority to enforce effluent requirements directly, it would now be able to move against a polluter whenever the contaminants in his discharge exceeded permissible levels—instead of having to show in each case, as FWQA does now, that those discharges cause a violation in water quality criteria out in the stream.

Greater ease of enforcement is not the only thing Congress gave up by opting exclusively for limiting stream pollution. Because so little is known about what are called synergistic effects—the various ways individual pollutants operate in combination to produce more serious damage than any of them would produce separately —a stream standard plan would be ecologically unsound even if it were not so difficult to enforce. Under the stream criteria approach, any pollutant is presumed to be permitted in unlimited amount unless it is specifically restricted or unless it can be shown to be dangerous. Only by setting effluent requirements which *forbid* all discharges that are not *expressly allowed* can the government insure that unpredicted or understudied pollutants will not be continually creating unpredictable hazards.

The resistance of the pro-industry House Public Works Committee to the effluent standards provision in Senator Muskie's initial 1963 proposal stemmed precisely from the fact that it promised to make the law so much more effective against polluters. When Muskie tried to get the bill passed again in 1965 his task was clear: make an end run around the most basic objections of the House Committee—including effluent standards.* This he did with some success. But the resulting law can only be described as an abominable piece of legislative draftsmanship.

When FWQA was faced with the necessity of passing along guidelines for the states to follow in making up their implementation plans, the Water Quality Act posed what appeared to be a curious dilemma. Congress had clearly meant to deny the Federal government the power to set "effluent standards." But the only way to imple-

* The more inclusive term "navigable waterways" in the 1963 version of the Muskie bill was also eliminated in the 1965 version, which restricted the scope of water quality standards to interstate waters.

ment the criteria in a manner reasonably calculated to achieve water quality goals was for someone to set specific limits on the type and amount of waste matter that each polluter discharged. If FWQA went ahead and demanded what water cleanup required, would the courts ultimately uphold the agency's decision?

While a cynical reading of the legislation might have indicated that Congress really hadn't wanted the law to work at all, this would hardly have been an acceptable premise on which to base administrative action. The courts normally give an administrative agency some latitude in applying an ambiguous statutory mandate to serve its basic purpose—in this case, presumably, pollution control. It was obvious, of course, from the Water Quality Act's legislative history that the law does not permit the *Federal government* to set "effluent standards" that would be *enforceable independent of violations of water quality criteria.* But the 1965 amendments do require that the state standards include a "plan for implementation and enforcement of the water quality criteria adopted." It would not have been inconsistent with the prohibition on Federally-set effluent standards for FWQA to demand that the *states* set effluent requirements for polluters in their implementation plans, *specifically tied to the need to attain water quality criteria* and *enforceable only as they related to water criteria violations.** The only thing it seemed clear that the Federal government could *not* legally do was promulgate a flat national across-the-board discharge requirement unrelated to the need to attain specific water quality criteria.

This is just about what FWQA went ahead and did anyway. Mindful of industry's historic opposition to effluent controls, the agency chose not to require that the states set effluent requirements directly related to stream criteria. Instead, FWQA took what it perceived to be the easy way out. It instituted a flat requirement

---

* As Senator Muskie stated at the June 1970 hearings before his Subcommittee during a discussion of the "effluent standards" question, "The Federal government has the responsibility under the 1965 Act for Water Quality Standards. To enforce them, the Government must have a handle on the effluent of particular industries. *That was our intent in 1965 but we didn't spell it out clearly.*" (Emphasis added.)

that secondary treatment be required in all states' implementation plans, unless a polluter could show that a lesser degree of treatment would suffice to meet the stream standard.

Industry wasted little time demonstrating that FWQA's favor had been wasted. In 1968 the Chamber of Commerce commissioned the high-powered Washington law firm of Covington and Burling to make the business community's views official in a widely circulated legal opinion. The Covington and Burling opinion was, predictably, just what its client wanted to hear. It said that the Water Quality Act permitted neither a flat secondary treatment requirement nor any effluent limits (and it took a shot at the non-degradation clause requirements as well). What the Chamber was saying, through its obliging counsel, was that it believed Congress had enacted a law which required the setting of standards but which provided no way to implement them, no way to measure progress, and—most important for the Chamber—no way to enforce the entire law.

By sacrificing effluent requirements for what it apparently hoped would be a change in industry's obstructionist posture, FWQA wound up with neither. All it has left is a stack of state implementation plans whose relationship, if any, to water quality goals is purely coincidental. If the government ever takes a polluter to court, it will of course have to demonstrate to the judge's satisfaction that such a relationship exists, and for that reason some of the treatment requirements which polluters have been given are almost certainly unenforceable.

A much more serious problem, however, is that even if the implementation plans were to be carried out, they would simply not bring the water cleanup for which they were supposedly designed. It is openly acknowledged that secondary treatment for waste is not enough in the nation's most highly polluted areas. And a decade of futile Federal enforcement conference experience has shown that in those places where secondary might be enough right now, treatment requirements which specify only the percentage of pollutant to be removed can never do more than make a temporary cutback in water pollution as long as population, production, and absolute wasteloads continue to grow. Only pollution-strength limits combined with load limits on discharge—in other words, effluent controls—will do the job.

FWQA's decision to avoid effluent controls was, unfortunately, only the beginning of a long downhill slide on the subject of implementation plans. The agency required no implementation plans whatsoever—even long-range plans—for eliminating pollution from non-point sources (urban stormwater runoff, pollution from farms, feedlots, etc.). Setting water quality criteria for substances like dissolved oxygen, phosphorus, oil, or bacteria without developing implementation plans for non-point sources is like asking a river to clean *itself* up. Yet those few state standards which acknowledged problems like agricultural or urban runoff at all simply defined pollution so narrowly that non-point polluters were exonerated. Illinois' dissolved oxygen (DO) standard for the Mississippi River, for instance, is 5 mg/1 "due to effluent discharges" (thus excluding DO reduction caused by farm runoff).[7] To take another example from the Illinois standards, on the Calumet and Grand Calumet Rivers the coliform and fecal streptococci bacteria criteria do not apply "during the periods of storm runoff." [8] In other words, pollution is against the law except during periods when there is pollution.

For municipal and industrial polluters—already free of effluent controls—FWQA's permissive policy on implementation plans has also provided another bonanza: mixing zones. Each polluter's Federally approved implementation requirement identifies a stretch of river in which he is permitted to "mix" his wastes. These areas are privileged sanctuaries from pollution control; no water quality measurements taken inside the "zone" may be used against the polluter in a legal action. The Federal government has in the most literal sense given these sections of our nation's waterways away to the polluters.

Most of the Congressmen who worked on the Water Quality Act have probably never heard of mixing zones. The Act states simply that any reduction of water quality below the applicable criteria in the area where they apply is a violation of the law, and anyone unfamiliar with FWQA's courtship of industrial polluters might be tempted to assume that the Federal government's pollution control agency would read it that way. Not FWQA. When the agency was making up its guidelines, industry and the states argued that the old traditional practice of allowing waste dilution areas ought to be continued. FWQA agreed. The Federal guidelines specify that a polluter may be given up to three-quarters of a mile in any direction from the point of discharge for mixing.* [9] That means a large polluter might have up to a mile and a half of lake or stream which is legally his to pollute.

The water quality standards are little more than a fraudulent joke in those areas where one legally pollutable area overlaps another. Take the Buffalo River, for instance, in Buffalo, New York; there are 18 industrial outfalls located in a single 2.5-mile stretch. Or the Cuyahoga River in Cleveland, Ohio; in one lethal four-mile stretch of river there are 32 distinct industrial discharges, 4 raw sewage overflow points, and about 100 stormwater outfalls. Just upstream is a huge treatment plant, presently overloaded, for Cleveland's municipal wastes. The Federal mixing-zone rules specify that polluters must always leave a "corridor" for fish, but the likelihood of large numbers of aquatic creatures ever successfully threading their way through these chemical minefields is believed by many to be slim indeed.

The problem is even greater in areas ostensibly set aside for human recreation. In one 30-mile stretch of the Fox River near Green Bay, Wisconsin, there are 34 industries and 20 cities. While the pollution zones on the Fox are interrupted periodically by stretches of river where the water quality standards apply in full, huge portions are nevertheless wiped out. The Federal guidelines limit mixing areas to no more than one-fourth of any stream's cross-sectional area so that waters will still be "suitable" for their designated uses. [10] But how does the unsuspecting swimmer know as he wades out from shore where his "recreation corridor" ends and a polluted "mixing zone" begins? Proceed at your own risk!

By the middle of 1969, however the most obvious diffi-

---

* States have latitude to set their own mixing-zone policies within this maximum. Mixing zones are negotiated on a case-by-case basis for each polluter.

culty with the standards program was not that the standards themselves were practically worthless but that they
were still not complete, despite the June 30, 1967 deadline which the Water Quality Act had set forth. Responsibility for what was supposed to have been the
Federal government's powerful new enforcement device
still lay with Dr. Allan Hirsch in FWQA's Office of
Operations *—it had never passed over to the Office of
Enforcement. FWQA officials had been quoted from
time to time as saying that standards for all 54 jurisdictions had been "approved." But that statement was always qualified by another: there were still a number of
small "exceptions" (e.g., certain criteria set too low in
some places, implementation plans for some polluters
not acceptable, the absence in some states' standards of
a non-degradation clause, etc.) remaining to be cleared
up.

The standard-setting took so long because FWQA had
chosen, as usual, to follow the path of least resistance
with the states and their polluters instead of using the
authority it had been granted under the Water Quality
Act. As noted earlier, when the Act was undergoing
final tailoring in the House-Senate Conference Committee prior to passage in 1965, the biggest battle between the two chambers was over who should set the
standards—the Federal government or the states. Senator
Muskie was not able to save much else from his original
1963 standards scheme, but he did wrench from the
House conferees one important concession: after each
state set its standards, the Federal government would
review them; if Washington didn't consider the state
standards high enough, the Secretary of Interior could
hold a standard-setting conference and, on the basis of
the conference discussion, promulgate new standards
that would take effect within six months. The process
was slow, but it wouldn't have taken nearly as long as
FWQA took without it. Had the agency used this authority, it could have had a complete set of approved standards in every state by the middle of 1968 at the latest.

---

* In 1968 Hirsch was shifted from head of the Office of Programs,
Plans and Development to the head of FWQA's Office of Operations,
and he took the standards-setting program with him when he made
the move.

As early as the beginning of 1968 it was apparent that many of the states' criteria were too low to pass Federal muster. Instead of following the simple but untested standard-setting procedures outlined in the Water Quality Act, FWQA preferred to stick with a technique it had used without much success on so many previous occasions: friendly persuasion. Standards were bounced back and forth between Washington and the states; issues dragged on unresolved for months and then years; many of the standards were in such a state of limbo that no one knew precisely what their status was. Not until May 1969 was the first Federal standard-setting conference held (with the State of Iowa), and none has been held since.*

In October 1969 the standards program was finally shifted to FWQA's Enforcement Office. A systematic inventory of the "approvals" and the "exceptions" was taken. The exceptions turned out to be the rule. FWQA had approved a complete set of standards for a grand total of only four states and two territories. And even this total is misleading. The standards in the two territories (Guam and the Virgin Islands) will always be completely unenforceable unless the law is changed, since polluted waters in these territories cannot possibly cause interstate damage. Allan Hirsch and his people had been sleepwalking through a maze of exceptions, promises, and numbers for almost four years. When they woke up, they found that they had only four complete state standards which were acceptable under the Water Quality Act of 1965 and at the same time had any chance of ever being used in an enforcement attempt.

Since that time FWQA has stepped up its efforts to resolve the disagreements with the states—though still by friendly persuasion, naturally. In March 1971, however, one year and five months after the shift of the standards program to Murray Stein's office, there were still only 28 states whose water quality standards had

---

* The agency threatened to hold a conference on Virginia's standards in December 1969, but dropped it when the state agreed to accept the Federal guidelines. As of February 1971, one other conference had been scheduled—for the State of Alabama—to take place in April 1971.

been fully approved.* And of course none of those 28 "fully approved" standards has a baseline from which to measure non-degradation; none has imposed effluent requirements; none has done away with mixing zones or given implementation schedules to non-point source polluters.

These defects alone are enough to make FWQA's belatedly "approved" standards nearly useless as enforcement tools. Unfortunately, however, they all bear another less visible—though no less incapacitating—scar from their several-year bout with Allan Hirsch's technical troops. Competent though Hirsch's people were on technical matters, they did not have the legal background necessary to understand the intricacies of this unusually abstruse piece of legislation. Becky Neurenberg of Hirsch's office told the Task Force, for example, that her group had been working on water quality standards for an entire year before they realized that their enforcement was tied to the same old delays and difficulties as the older enforcement conference procedure. "Naturally, we were just sick about it," she said.

Sick or not, the record shows that Hirsch's people lacked even the remotest notion of how the water quality criteria and implementation plans would have had to be written in order to preserve what hope existed for their eventual enforceability. FWQA's "fully approved" standards are riddled with language that lawyers could wrangle over endlessly: streams must be *"substantially free of visible floating oil"* [11]; bacteria criteria on some rivers apply only "when there is swimming," yet there is no designation of what areas of the river are to be used for swimming [12]; waters must be free from specified pollutants "in amounts sufficient to be *unsightly* or *deleterious"* [13] or "free from discharge materials that produce color, odor, or other conditions in such degree as to create a *nuisance"* [14]; a pH (acidity) criterion of 6.5–8.5 is to be "preferably" [15] met; and so on.

In the final analysis, it may make very little difference that so many of the Federal government's water quality

* The District of Columbia, Puerto Rico, the Virgin Islands, and Guam also had fully approved standards, making the overall count 32 out of 54 jurisdictions in March 1971. See Appendix E for a complete status report on state standard approvals.

standards are virtually unenforceable, because the evidence suggests that FWQA has never had any serious intention of enforcing them. Polluters' intermediate deadlines (for presenting cleanup plans, starting construction, etc.) under their implementation schedules began falling due early in 1968. Reports filed with FWQA by the states indicate that slippage in compliance schedules occurred almost immediately. New York's survey in the summer of 1968 showed that 21% of her industries and 19% of her municipalities were making unsatisfactory progress. Massachusetts reported 75% of her industries and 70% of her municipalities "not on schedule." FWQA was not interested enough in these widespread early violations to try to find out how much worse the picture was in the states that didn't even bother to report on compliance. Georgia, whose standards had been approved in 1967, indicated only that there had "been insufficient time to complete facilities and make specific evaluations." Indiana merely listed the problems that certain industries were working on, without specifying particular compliance dates or violations. As more detailed reports began to trickle in to FWQA in 1969, however, they showed that slippage was the rule in virtually every state. Still no action was taken.

Skeptics were temporarily taken aback in September 1969 when notices of violation (referred to at FWQA as "180-day notices" because of the 1965 Act's requirement that six months' notice be given prior to taking a violator to court) were served on six polluters. Eagle-Pitcher Industries (a mining company) was charged with violating the water quality standards for the Spring River in Kansas and Oklahoma. The City of Toledo (Ohio), Toledo's Interlake Steel, and three other steel plants located in Cleveland, Republic Steel, U.S. Steel, and Jones & Laughlin—all polluters of Lake Erie's tributaries—were also served with notices that could have been a prelude to court action after the mandatory six-month waiting period expired. Court action was never actually taken against the six polluters because—according to FWQA—the 180-day notices had a salutary effect on all of them. Eagle-Pitcher agreed to cease its violation of the water quality standards. And the Erie polluters, who had fallen behind on implementation dates, promised to step up the installation of their controls. In addition, FWQA won an agreement with the four steel companies permitting Federal personnel to visit the installations in order to determine whether compliance was proceeding effectively and whether further controls were needed. Only time will tell, of course, whether these promises will be kept. But if they are not, the government now at least has the option of taking the polluters to court immediately, rather than waiting six months.

To those who hoped this first round of action under the Water Quality Act might be the prelude to a broad new government offensive against polluters, it became clear as time wore on that repeat performances were going to be few and far between. FWQA waited nearly eight months until May 1970 before giving out its next set of 180-day notices to five more standards violators.*

If the pace of the standards enforcement program has not been such as to inspire confidence, neither has been the government's selection of targets. In the Cleveland area, for example, the steel companies received 180-day notices in 1969, but the city itself was passed over, though it is several years behind its cleanup schedule.† Cleveland, the second largest municipal polluter in the Erie basin, discharges more waste than the next three largest municipal polluters combined. Far behind Cleveland in cleanup progress, and also overlooked when the 1969 notices were handed out, is the city of Detroit.‡ As Toledo City Councilwoman Carol Pietrykowski wrote in an angry letter to Secretary Hickel:

* The five were: Penn Central Railroad, at Harmon, New York; General Aniline and Film Corporation, Linden, New Jersey; Midwestern Feeding Company, Manley, Nebraska; and two municipal jurisdictions: Fairfax Drainage District, Kansas City, Kansas; and Fargo, North Dakota.

† On the largest of Cleveland's three municipal treatment plants, secondary treatment extensions were scheduled to have been under construction by September 1968. They were not begun until the summer of 1970, and their estimated time of completion is 1973—more than two years after the September 1970 final deadline. Final construction plans on another of Cleveland's three plants were due in June 1969. By December 1970 they still had not been drawn up.

‡ Detroit's detailed specifications were due in November 1968 and her construction was to have been entirely completed by November 1970. By December 1970 the city had neither *begun* construction nor even submitted specifications for several parts of her facility.

Your department's decision to attack Toledo as its first municipal target is most paradoxical when one realizes that Detroit, Michigan, presently only has primary sewage treatment. [Toledo has secondary.] As you know, primary treatment consists of merely removing solids from sewage. Detroit via the Detroit River injects more municipal pollution into Lake Erie than all other cities combined [64% of the overall pollution in the lake from all U.S. municipal sources, compared to only 3.5% from Toledo]. . . .

The decision to single out Toledo was, it appears, an example of the Federal government's time-honored pattern in pollution control: pick on the pushovers and forget about the rest. It is difficult to avoid concluding that the temporary flurries of enforcement notices in the fall of 1969 and the spring of 1970 were designed primarily to fill their respective paragraphs in FWQA's annual reports rather than bring about any real cleanup.

After the May 1970 notices, all was silent again in the standards program until December 1970, when the Federal government took its 180-day notices out of the closet for the third and, to date, the last time. Within one week of the formation of the new EPA, Administrator William Ruckelshaus was capturing headlines with his announcement that another flurry of standards notices would be handed out. This time the government elevated its sights to draw a bead on some of the nation's large cities. Atlanta, Georgia, whose 32 million gallons of untreated daily discharge turn the Chattahoochee River into what Ruckelshaus called "virtually an open sewer," was awarded a 180-day citation. EPA also went back to Lake Erie to give the government's most conspicuous municipal oversights, Cleveland and Detroit, their due.

Counting these final three notices, the government has now initiated enforcement proceedings against 14 polluters (none of whom has yet been taken to court) under the Water Quality Act of 1965. To put this figure in its proper perspective, consider Lake Erie, whose polluters have received well-deserved special attention under the standards enforcement program with seven out of the total 14 notices of violation. Lake Erie has approximately 200 municipal and 200 industrial polluters. By June 1970, 78 of the 110 cities with water quality standards deadlines were behind schedule; 49 of them were more than a year behind. In the case of municipal pollution, however, the government's coverage under the enforcement program has been impressively efficient: the three municipal targets so far—Detroit, Cleveland, and Toledo—together contribute about 75% of the pollution in the lake from all municipal sources. Even if these cities should stick to their cleanup vows, however, the improvement in the lake that their efforts might otherwise produce is likely to be largely obliterated by unabated contamination from Erie's many industrial sources. Forty-four of the lake's largest industrial polluters had fallen behind on their cleanup schedules by June 1970. Thirty-eight of these companies were more than one year behind. Mobil Oil in Buffalo was 32 months behind. Multiply the number of Erie's municipal and industrial standards violators by several hundred and you get some idea of the vast extent of FWQA's non-enforcement of the Water Quality Act throughout the entire United States.

The last time anyone at FWQA ever mentioned the possibility of someday actively enforcing the water quality standards may well have been back in January 1967 when Commissioner Quigley addressed a memorandum to Assistant Commissioner Murray Stein (with whom he was not on speaking terms at the time) asking for some suggestions on the standards program. Quigley noted that "the ultimate test of the standards is not in their setting, but whether they will do the job of effectively abating pollution in this country." The Commissioner concluded his memorandum with an unequivocal declaration that EPA should take to heart. "The job is to enforce the water quality standards," Quigley wrote. "If we can't enforce them, we may as well quit."

15

Hope Springs Eternal: The Refuse Act of 1899

Late in 1969 environmentalists who had despaired of seeing any water pollution cleaned up with dispatch by the Federal government "rediscovered" an old law that held out the promise of speedier action: the Refuse Act of 1899.* Congressman Henry S. Reuss of Wisconsin generated tremendous interest in the law with public hearings on its use, by turning the names of polluters in his home state over to U.S. Attorneys (and in at least one case collecting his half of the fine levied on the polluter, as the Refuse Act provides), and by making information kits available to the thousands of interested citizens who inquired how private citizens could file *qui tam* † actions against polluters under the Refuse Act if U.S. Attorneys refused to sue. Ralph Nader, the Environmental Defense Fund's Victor Yannocone, and Environmental Action's Denis Hayes were mentioning the law in speeches and on television. By July 1970 the Bass Anglers Sportsmen Society had started legal proceedings in three Federal courts against 214 Alabama polluters under the Refuse Act. To clean-water advocates, the resurrected law looked like a godsend.

Enacted in a simpler age, before legislators had mastered the art of convoluted legislative draftsmanship, the act says flatly, "It shall not be lawful to throw, discharge, or deposit . . . any refuse matter of any kind or description whatever . . . into any navigable water

* The Refuse Act (33 U.S.C. 407) is Section 13 of the Rivers and Harbors Act of 1899.

† *Qui tam* actions are common-law suits that private citizens are authorized to file in order to collect their share of a fine when a statute provides for a reward to informers but the government does not prosecute the violator.

of the United States . . ." without a permit from the
Army Corps of Engineers. "Refuse matters," according
to a 1966 Supreme Court ruling, includes industrial pollutants; the term covers all "foreign substances and pollutants" [1] except for municipal sewage, which is exempted by the statute.

There are some 40,000 industrial plants in the United
States that discharge into navigable waters. Out of those
40,000 polluters only 4 to 415 * have bothered to obtain
a Corps permit since 1899. In 1970 the Engineers let it
be known that they planned to issue no more permits.
This means that somewhere between 99% and 99.99%
of the industries in the United States are committing a
crime when they dump anything but pure water into our
navigable waterways.

There are criminal sanctions for violating the act that
can amount to as much as $2500 for each violation and
a prison sentence of up to one year. One possible problem with using the criminal penalties is that the law does
not explicitly make each day's pollution a separate violation (this point has not been tested in court); and the
fines are of course too small to deter industrial polluters
with much larger daily grosses. The Supreme Court has
ruled, however, that in addition to the criminal sanctions, the government may also seek an injunction
against a polluter who has violated the Refuse Act to
require him to prevent all future discharges and to clean
up those substances already discharged.[2] Thus, while far
from perfect, the Refuse Act represents the high point
in effective national pollution control legislation. As far
as industrial pollution is concerned, all of the 20th-cen-
tury control legislation has been a retreat from this law
enacted in 1899.

But the Federal government, not accustomed to moving expeditiously against polluters, has welcomed this
exciting "new" law with considerably less enthusiasm

---

* Only 4 of the 415 permits granted by the Corps since 1899 were
issued under Section 13 of the Rivers and Harbors Act, which pertains
to the discharge of waste. The rest were issued under Section 10,
which relates only to the building of physical intrusions into navigable
waters. Although we did not examine the individual Section 10 permits, some of them may contain language granting the industry permission to dump. While such permission is not technically legal, it
might nevertheless operate as a bar to prosecution of the discharger
under Section 13.

than the environmentalists. The history of the law since its 1969 rediscovery has been largely one of attempts by the Federal government to confine its scope. The first attempt came early in 1970 when the Department of Justice, after consulting with the Department of Interior (where FWQA was housed at the time), began formulating a set of policy guidelines to instruct United States Attorneys on the appropriate use of the 1899 Act. The basic concept of the guidelines was to be a policy of "deference" in pollution control matters to the weaker Federal and state laws enacted since 1899 and to the Federal and state agencies that administered them—i.e., to the state pollution control agencies and to FWQA.

Upon hearing that Justice was planning to restrict enforcement of the Refuse Act, Congressman Reuss immediately became concerned and dashed off a letter to Attorney General Mitchell to find out what was happening. Reuss received a reply on June 2, 1970, from Assistant Attorney General Shiro Kashiwa, whose letter explained that there had indeed been a policy decision at Justice to "defer" to FWQA and state cleanup efforts. Kashiwa told Reuss that the Administration felt that it was necessary to fit the Refuse Act into the "regulatory scheme devised by Congress to combat pollution. . . ." The Act would not be applied where it "would have a disruptive or devitalizing effect upon programs [already] designed or approved" or where a polluter was spending "significant amounts of money" to abate pollution under an FWQA program.

Reuss got to the heart of the matter when he responded to Kashiwa that the Refuse Act "doesn't exempt polluters who spend money to clean up their mess. . . . The Justice Department should obey the law."

But Justice was not about to obey the law. The guidelines went out to all U.S. Attorneys on July 10, 1970. Whereas the Refuse Act explicitly commands the Justice Department to "vigorously prosecute" offenders, the net effect of the guidelines was to insure that prosecution would in fact not be very vigorous. Justice told its U.S. Attorneys around the country that without clearance from Washington they could prosecute violations under the Act only "to punish or prevent significant discharges, which are either *accidental or infrequent,* but which are *not of a continuing nature resulting from the ordinary operations of a manufacturing plant"* (emphasis added). Cases of on-going pollution (which, the guidelines conceded, constitute "the greatest threat to the environment") could be prosecuted only with clearance from Washington.

Clearance has not been easy to come by under the "deference" policy. Despite unprecedented public support for water pollution control, the government has used the 1899 Refuse Act to stop continuing pollution against a grand total of 28 industrial polluters from the beginning of 1970 through the middle of March 1971. Only 13 of those 28 suits were initiated at the request of FWQA. Ten of the 13 cases requested by FWQA came in July 1970, when the frightening prospect of large-scale poisoning from mercury pollution was widely publicized in the press. Good environmental politics required that the Federal government move quickly, so the 10 suits were filed against large mercury dischargers. By FWQA's account, the mercury suits were eminently successful. Under the threat of court action, the 10 companies quickly agreed by consent decree to reduce their discharges immediately and to discontinue them entirely (except for "negligible" unmeasurable amounts) within six months. But despite the unusual effectiveness of these actions, FWQA has shown no disposition to put the law into more general use. Except for this handful of injunctive cases against continuing polluters, the rest of some 40,000 corporate criminals who are now openly and flagrantly breaking the law on a routine basis have been permitted to dump to their hearts' content. This is the policy of "deference."

The thrust of the Justice Department guidelines has been to infuse enforcement of the Refuse Act with the same timidity that has always characterized enforcement of the other water pollution laws. Congressman Reuss got a first-hand picture of what the new guidelines meant in practice. Between March and June of 1970 he forwarded a list of 139 polluters in his home state of Wisconsin to the Department of Justice, requesting investigation and prosecution under the Refuse Act. The Department reported back in February 1971 that 116 of the polluters were "in compliance" or in "substantial

compliance" (i.e., not in compliance, but ostensibly trying) with state standards and timetables. Only two of
the 139 had been prosecuted under the Act; the rest were
in various stages of investigation or subject to current
state proceedings. The Justice Department did not say
that the 139 companies were not polluting. They were.
The Department maintained that, under its guidelines,
state standards—no matter how weak or how feebly
enforced—provided a protective shield against prosecution under the Refuse Act.

When Justice's Kashiwa found himself before Reuss
at a joint Senate-House hearing on the Refuse Act in
February 1971,* the Congressman asked him about one
of the plants supposedly "in compliance" or "in substantial compliance"—the DuPont Chemical Company plant
in Barksdale, Wisconsin. Reuss held up, for all who
were in attendance at the hearing to see, a picture book
on water pollution published by the Sierra Club. The
book contains color photos of DuPont's dark red chemical discharges gushing into a thoroughly mutilated
stream, appropriately named Bloody Creek, that flows
into Lake Superior. "The Attorney General," said Reuss,

and the Department of Justice are the exemplars of law and
order. They are supposed to enforce the law. What kind of
impression does it give the young people of this country if
the chief law enforcement man, Attorney General Mitchell,
lets DuPont go free where the pollution is so open, so destructive, so long-term and so flagrant that books on everybody's coffee table show how it is going on as we sit here
in Washington talking about it.

Is that a good way to run a Department of Justice? [3]

That is, of course, no way to run a Justice Department,
and the entire approach to enforcement of the law il-
lustrates the almost limitless capacity for sophistry that
can be tapped whenever the law places Federal officials
charged with pollution abatement in the uncomfortable
position of having to prosecute "respectable" businessmen.

The irony of this deliberate non-use of the law is that
the failures of the past have come full circle and now

* The hearings were sponsored by Senator Philip Hart's Subcommittee
on the Environment of the Senate Commerce Committee.

serve double duty. Endless enforcement conferences, culminating in meaningless recommendations, uneven surveillance, and continuing pollution—all this can now serve to shield polluters against Federal action under the 1899 law. Among the protected are the National Lead Company on the Mississippi River at St. Louis, which has been ignoring Federal conference cleanup recommendations since 1958, and Youngstown (Ohio) Sheet and Tube Company on the Mahoning River, which has flatly refused to accept the 1965 recommendations of the Mahoning River conference. These two, as well as the more sophisticated industrial polluters who shrewdly promised action but have never delivered, are, according to the Justice Department policy, immune from prosecution under the Refuse Act. Why? These polluters have been, in the words of the letter from Justice's Kashiwa to Congressman Reuss, "subjected to an enforcement proceeding conducted by the Federal Water Quality Administration or a State."

Of course if FWQA could ever summon up the will to call a polluter "recalcitrant"—a non-exempted category under the Justice guidelines—it could still ask the Justice Department to prosecute. And well-timed suits under the Refuse Act could short-circuit many of the delays that have crippled the enforcement conference approach. When asked about the use of the Refuse Act to speed up the enforcement process, Thaddeus Rajda, an associate of Murray Stein's, told the Task Force, "We're thinking about that with every polluter."

While FWQA thought about polluters, in December 1970 President Nixon announced, with full fanfare, a "new" approach to the Refuse Act—the so-called "national permit system." The plan, basically, is to require every industrial polluter to come forward prior to July 1, 1971, and make application to the Corps of Engineers for a discharge permit. What this means is that industrial polluters who have been dumping in violation of the law up to now will henceforth be dumping with the government's official permission, a questionable improvement in itself. From a practical standpoint, however, a permit plan under the Refuse Act could, if properly administered and vigorously enforced, bring about a dramatic improvement in water quality on all the country's navigable waters. The reason is that the Refuse Act gives the government sweeping authority to impose whatever terms it chooses on permits it grants to each polluter, including specific discharge limits at any level, all the way down to zero (i.e., refusing to grant a permit). The first news of the plan called forth words of praise from editorialists around the country. Since then, however, the details of the "permit system" have gradually begun to surface, and it now appears that the new scheme will prove to be only the latest in a series of moves by the Nixon Administration designed to whittle down the scope of the Refuse Act.

In essence, the permit plan is a package deal with two parts. The Administration will be giving polluters permission to dump up to a given level, but it hopefully plans at the same time to end its self-imposed moratorium on enforcing the law. (As another compensating benefit, the government will have available, for the first time, accurate information on industrial discharges, except for those into municipal sewer systems.*) New guidelines for Justice Department suits under the Refuse Act were made public in February 1971 in response to criticism of the earlier version.[4] The revised guidelines, scheduled to go into effect along with the permit system in the summer of 1971, meet many of the objections raised against the old guidelines. They state that the U.S. Attorneys around the country need not apply for clearance from the Justice Department to bring an action under the Refuse Act. Instead they are required only to seek approval directly from the Corps of Engineers and from EPA (presumably at the regional level). There is still the possibility of Administration interference with the enforcement program exerted through EPA Water Quality Office officials in the regional offices, of course. (There is, for example, nothing to prevent the Water Quality Office's instructing its field personnel that they must get clearance from EPA headquarters in Washington before giving the go-ahead, or any assistance, to local U.S. Attorneys on suits against polluters.) But even if enforcement under the new Justice

---

* The penalty for false statements on permits, under 18 U.S.C. 1001, is five years in jail and $10,000 in fines.

guidelines should remain free of high-level constraints, it can safely be predicted that the program will still not open the floodgates to thousands of belated prosecutions. This is because of an additional legal snarl presented by the Administration's proposed scheme for granting the permits.

It is important to recognize in this regard that polluters who live up to the terms of their permits will be insulated from prosecution under the Refuse Act. Thus the permit system can be no stronger than the permits themselves are. The Nixon Administration is apparently committed to a plan for issuing permits which will virtually guarantee that most of them will be little more than "licenses to pollute."

The plan began to unfold on December 31, 1970, when the Corps of Engineers published its proposed regulations for implementing the permit system.[5] The regulations specified that the mechanical task of actually handing out the pieces of paper will be performed by the Corps. EPA (i.e., the Water Quality Office) will make the final decision on each permit's terms regarding water quality matters. But there is no mention of any public participation in the permit-granting process, no clue to whether or not specific effluent limits will be set on each source, and no indication of what guidelines the Water Quality Office will follow in assessing what the regulations refer to as "water quality considerations." Given the Water Quality Office's penchant for expanding loopholes to the point where all but the grossest of polluters can squeeze through, the regulations' vagueness on these critical questions was hardly calculated to inspire confidence about the success of the permit system.

Doubts were confirmed in February 1971 at the joint Senate-House hearing on the permit program. The question the legislators were most interested in having answered was how EPA planned to handle applications for permits from industries on waters not now covered by Federal water quality standards (which apply to interstate waters only). One of the Refuse Act's major advantages is that its framers felt no compunction about placing the full responsibility for its enforcement with the Federal government. The Administration thus has authority under the Refuse Act to demand compliance with Federally-set or Federally-approved discharge standards on *all navigable waters,* be they interstate or intrastate. What the Senators and Congressman at the February hearings wanted to find out was whether this authority would be fully used. They were, to say the least, disappointed to discover that, under the Nixon permit plan, it will barely be used at all. The Administration's plan is to surrender this one trump card by tying the permit-granting scheme to the moribund pollution control systems in the states.

The plan was spelled out at the hearings by EPA General Counsel John R. Quarles, who explained to the Hart committee members that EPA will normally review only the permit applications from polluters on *interstate* waters. On all other waters, Quarles made it quite clear, a state's determination that a discharge will not violate the state's own water quality standards will be given enormous—perhaps dispositive—weight in deciding whether a particular permit should be issued. EPA's procedure will be, with only minor exceptions,* to rubber-stamp whatever requirements the states have imposed on those industries which are not on interstate waters. Since interstate waters make up only an estimated 14% of the nation's stream mileage, this means that the Administration's vaunted "national permit system" may in fact be little more than a license to pollute with impunity for industrial polluters on as much as 86% of the country's rivers and streams. The "deference" policy has returned through the back door.

The Nixon Administration's position on this new form of "deference" is ironic. Quarles reminded the Senate-House committee that the Administration had

---

* Administration officials have indicated that there may be only three possible exceptions to their general rule of subordinating existing Federal law to state regulation, or non-regulation, as the case may be. EPA will be permitted to review state requirements when the industry is discharging "hazardous substances not covered by water quality standards." (Presumably if the industry's "hazardous substances" are covered in the state's water quality standards, the state requirement will stand, no matter how low it may be.) EPA may also, the Administration has indicated, make an independent study of state requirements for industrial polluters "where there are conflicting fish and wildlife values." Finally, EPA may alter state requirements where it finds that a state has applied its own low standards incorrectly in any given case.

recently proposed legislation expanding the Federal jurisdiction under the *newer* pollution control laws to cover all navigable waters. As Quarles explained it, the Administration would be happy to set standards on intrastate waters and enforce them, but not until its proposed legislative package has passed. By taking this position the Administration is saying, in effect, that it refuses to use the authority it has under the Refuse Act unless Congress grants it a second time. Until that happens, the Federal government will continue to "defer."

Upon hearing Quarles' presentation at the February hearings, Congressman Reuss was incredulous. "You testified," Reuss asked Quarles,

that unless Congress passed another Refuse Act of 1899, you were going to dish out these permits first come, first serve. Isn't that what you testified?

Quarles responded with astonishing candor:

MR. QUARLES: I would not agree with your expression of the thought; basically I am in agreement of [sic] the way you understand our testimony.

CONGRESSMAN REUSS: That is why I am appalled at your testimony this afternoon. It is like a big insurance policy. The big print [gives] it to you and the little print takes it away. . . . Now that I hear what you have to say, there isn't any permit program. . . . It is just a device for giving immunity to polluters.[6]

What about *interstate* waters, which do fall within the Administration's tortured interpretation of the Refuse Act? The Administration's vagueness on this score still leaves a shred of hope that the permit system may become a guarantor of clean rivers and streams—at least the 14% of the nation's rivers and streams that are classified as "interstate." However, there is not yet any assurance that here also the permit system will not become just a license to pollute.

An effective permit system must establish, first of all, a high minimum level of treatment for each industrial group which all plants in the industry are required to achieve. Beyond that, the government must give each industrial source of pollution a precise effluent standard, set so as to positively insure that the water in the rivers and the lakes will be as good as the water quality criteria say it should be. Where the best available or projected technology is not advanced enough for a polluter to be able to meet his effluent standard, he should be given a maximum of two to three years to do his own research and development on a better method. When his time is up, he should be required to meet his effluent standard —one way or another. Water Quality Office staffers buried deep in the EPA bureaucracy are now busy devising, with all good intentions, a permit-granting scheme that has a number of these elements. At this point, the public can only wait and see how much of the scheme falls to the cutting-room floor as it ascends the Administration ladder on its way to becoming official policy.

Given the way the Administration's permit system has been handled so far, the public may be lucky if it gets to see anything at all. The most critical unanswered question about the permit program remains the amount of citizen participation there will be in the process of forming effluent guidelines for different industries and of awarding permits. All that Water Quality Office staffers could tell the Task Force in March 1971 was that industrial representatives would undoubtedly be given an opportunity to take part in approving the standards they will be operating under. What about the rest of us? The Nixon Administration maintains an ominous silence on this crucial point. Citizens must demand a voice now —before their birthright is bargained away behind closed doors.

It now appears likely that the new permit system— scheduled to begin in July 1971—will go the way of earlier "bold new" approaches to water pollution control. Not only does the Administration's unnecessary surrender of the Refuse Act to the states promise to result in increased pollution in the short run. It may also produce unfortunate side-effects for some time to come. If Congress should, sometime soon, pass a new pollution control law that extends the Federal jurisdiction to set water quality standards to all navigable waters, EPA may have placed itself in the unenviable position of having to demand of some industrial polluters that they upgrade their treatment (to meet the new law's standards) just shortly after having ordered them to meet a much more modest requirement under its ill-conceived permit system. The greater danger, of course, is that the permit system may effectively prevent needed legislative change. Congress should give the new permit program a chance to work, opponents of control legislation will argue, before legislating yet another "bold new" approach. Should those opponents prevail, we will be stuck for a long time with the pollution that EPA plans to license under the permit system.

The Refuse Act of 1899 has tremendous potential for achieving water cleanup on all our navigable waters. This potential is now about to be thrown away. The policy of "deference" is about to be stamped on most of the permits outside of interstate waters. What the Federal government plans to do is, as the cliché goes, snatch defeat from the jaws of victory. So much for the Refuse Act of 1899.

Federal officials routinely trip over each other in their frenzied retreat from any dealings with polluters that have even a faint air of confrontation. This is especially true of lawsuits or criminal prosecutions against corporate polluters. In describing his enforcement experience, FWQA's Murray Stein says, "Our success is measured by the number of times we *don't* go to court." By this standard the Federal government has been eminently successful. Since 1956 over 3000 polluters have been involved in Federal enforcement actions. The number of polluters violating their schedules for implementing Federal water quality standards may be even larger than that. Only one of these polluters has ever been taken to court under the Federal Water Pollution Control Act. Although nearly half of the 3000 or so polluters given conference cleanup requirements (and around half the standards violators) have been industries, court action has never been taken against an industrial polluter under the Act. Around 40,000 industrial polluters are routinely violating the 1899 Refuse Act. Yet FWQA has asked Justice to initiate suits against only 14 of these industries under the 1899 law as of March 1971.

When the Task Force tried, in the summer of 1969, to fathom the thinking behind this avoid-court-at-all-costs policy, we drew an inordinate number of blanks at first. Alan Kirk, the Associate Solicitor in the Interior Department who had been assigned to answer FWQA's legal questions, referred us for details to Shiro Kashiwa in the Attorney General's office. "He and a small group of lawyers under him in Justice would know much more about the enforcement problem than anyone in my department," he said. When the Task Force asked Kashiwa about FWQA, however, he was puzzled by the initials. "Is that the Forest Service?" he asked. Walter Kiechel, the attorney in Kashiwa's office with special responsibility for pollution matters, knew what the initials stood for. But when asked why FWQA had retired its court weapon, Kiechel replied that the agency didn't communicate very much with his office. "I don't have the basis for second-guessing," he demurred. "I hope they get their thing in gear."

A handful of suits under the 1899 Refuse Act have given even Shiro Kashiwa a nodding acquaintance with the Federal water pollution agency since the summer of 1969. But Kashiwa's Land and Natural Resources Division in the Justice Department, the unit responsible for Federal water pollution suits (along with such related matters as Land Acquisition, Condemnation and Titles, General Litigation, and Indian Claims), is still not geared to preparing water pollution cases on any routine basis. Only nine of Kashiwa's 100-odd attorneys work primarily on water pollution. The Illinois State Attorney General, by way of contrast, has a special staff of 11 full-time lawyers to handle water pollution cases for his state alone. Legal action against polluters is clearly not a high-priority affair at the Federal level.

That the Justice Department lacks experience in water pollution suits was, curiously enough, one of the first explanations the Task Force received when we began asking FWQA employees why they had allowed their legal weapons to atrophy. The U.S. Attorneys had let polluters off too easily with consent decrees in the past (i.e., at St. Joseph and on some of the 1899 lawsuits), we were told, and a weak court order might preempt continued FWQA pressure. Valid though these observations are, the Task Force sensed in them a wish to preserve what had always been the water pollution agency's exclusive bailiwick. If U.S. Attorneys have fumbled their few water pollution suits to date, one reason is surely that they haven't had much practice.

The most familiar argument water pollution officials make for staying out of court is the old saw that legal action takes longer than persuasion, particularly in view of the compounded delays in the Pollution Control Act. But there need be no delay whatsoever in bringing the 1899 Refuse Act to bear on *industrial* polluters. If the government wanted to, it could demand a jail sentence and fines ($2500 per violation) for past pollution and an immediate injunction against all future industrial discharges. It might also have tried to expand the scope of the Refuse Act by suing the lake's municipal polluters too. Though the 1899 law exempts municipal "sewage," the courts might find that cities which host industrial effluent no longer qualify for the exemption. So far, however, FWQA has been slow to stretch a point for the environment and quick to expand a loophole for polluters.

But what if the 1899 law were held not to apply to cities, leaving only the Federal Water Pollution Control Act to halt their pollution? Federal officials typically take no longer to dismiss the possibility of a suit under the Pollution Control Act than it takes to say "St. Joseph, Missouri." They usually then attempt to change the subject, an understandable reaction. The details of the St. Joseph case are painful—10 years after the Federal government's lone court decree, one-quarter of the city's sewage is still pouring raw into the Missouri River without even primary treatment. But the St. Joseph fiasco hardly gave the law a fair test. For the first six years after the court order no one—neither FWQA, the court, nor the Justice Department—troubled to check up on how St. Joseph was doing. It is almost as though the only reason the government bothered to use the court stage at all was so Federal officials could shrug and say, "Well, we tried it once and look what happened. . . ."

While it is hard to say exactly what would happen if FWQA were to sue another city, it is important to remember that the climate for anti-pollution enforcement has radically improved since 1960. Assistant Secretary of Interior Carl Klein told the Task Force in the summer of 1969, "No judge would dare find for a polluter today." This may overstate the case. But well-chosen cases against the most unbending municipal polluters, coupled with vigorous follow-up supervision, could hardly fail to get faster cleanup than "voluntary compliance" has. As many state pollution agencies have discovered, a court injunction against new commercial and residential construction until the municipality provides adequate sewer and waste treatment can produce heavy pressure from real estate interests and building trade unions in favor of the necessary bond issues.

In any event, whether the Federal government has forfeited faster cleanup in a given case by leaving its court weapon in retirement is not the only important question. Individual pollution cases are not entirely unrelated. What the government does in one case is watched closely by other potential defendants. Water polluters have monitored FWQA for years and know that the probability of court action under the Pollution Control Act is near zero. Whatever deterrent effect the court remedy had when it was written into the Act has been blunted by years of non-use.

While officials usually point to the obvious deficiencies in Federal pollution control laws as an excuse for not using them, the greatest barriers to enforcement are really not legal but political. Both the Federal Water Pollution Control Act and the 1899 Refuse Act permit the government to choose whether or not the public will be protected with a hearing, a 180-day notice of violation, or a suit. Having been given the power to choose, Administration officials are subjected to powerful political pressure to choose for the polluter. The serious legal deficiencies in water pollution laws, particularly the newer ones, have made it easy for the government to rationalize bending to this pressure—by claiming, and probably believing, that vigorous enforcement of the law could not accomplish much anyway. As a result, law enforcement has never been given a real test.

Only by using the tools available can the need for new ones be fully exposed. If the government's court remedies were used, they might succeed. If so, their success would not only constitute a heretofore nonexistent deterrent to pollution. It would also suggest the greater gains that might be realized if more adequate legislation were passed. And to the extent that a court try failed, it would highlight the need for change. If the new EPA follows the sorry example of its predecessors at the Federal level by not using the laws it has been given, there can be little remaining doubt about where the government's loyalties truly lie.

Despite a few brief forays into enforcement of water quality standards and a limited fling with the Refuse Act of 1899, the overwhelming emphasis in the Federal anti-pollution effort has been on the purely exhortatory enforcement conferences outlined in the 1956 Water Pollution Control Act. The clearest purpose of the 1956 law was to encourage state performance, and in a superficial way FWQA has done just that—prodding, cajoling and harassing state agencies into enforcing anti-pollution laws. The fundamental defect in this approach, however, is that it fails to deal realistically with the major problem in pollution control: the power of large polluters. In dealing with their worst polluters—big cities crying bankruptcy and major industries threatening to take their jobs and tax money elsewhere—the states do not need encouragement; they need help.

Because of the Federal government's inability to provide concrete assistance, many state personnel working on pollution problems believe that what the Pollution Control Act refers to as Federal-state "cooperation" is often a mutual waste of time. James Coulter, head of Maryland's highly regarded state agency, said that all the enforcement conference does is make "recommendations which in many cases never materialize." He thought that because state anti-pollution laws were generally quicker and more efficient, FWQA should leave the "small fry" to the states and come in primarily when big polluters had too much political clout for the state to tackle them. He suggested that the Federal Water Pollution Control Act be changed to allow the Federal government to deal swiftly and more directly with the powerful polluters the state couldn't handle.

If the Federal government both could and would provide effective enforcement support, the cheerleading services it now offers to state agencies would no longer be necessary. Freed of the pressure from their most powerful polluters to keep state pollution control laws weak or not to enforce them, the states could concentrate on the multitude of smaller polluters. The large cities and industries that the Federal government had forced to install expensive waste treatment equipment might even reverse their lobbying efforts, urging more vigorous state enforcement so they would have some clean water to show for their investments. The "consensus" approach embodied in the old Pollution Control Act wastes this potential for real cooperation between state and Federal governments. Instead, to establish what it calls a "partnership" between the Federal and state participants in water cleanup, the conference approach has simply tied one state and one Federal leg together and set the two of them off on a three-legged race. Not surprisingly, pollution, which had a head start, has steadily increased its lead.

Part Five

THE NEW FEDERALISM

16

Talk Is Cheap

The whole effort is lagging now for a number of reasons, one of which is that the Federal government hasn't put money on the line.
—Stewart Udall, 1969

Public officials rarely speak on environmental topics without adding that more money will be needed to restore ecological sanity. But even the few officials who follow that warning by expressing their determination to find the funds somewhere have usually found it convenient to leave their pledges at home when they sit down to the practical work of making up a budget.

President Johnson was one of the all-time masters of the forgotten environmental promise. He was at his very best on February 23, 1966, when he gave Congress his stirring message on "Preserving Our National Heritage." "We see that we can corrupt and destroy our lands, our rivers, our forests, and the atmosphere itself—all in the name of progress and necessity," the Chief Executive said. He continued:

Such a course leads to a barren America, bereft of its beauty and shorn of its sustenance. We see there is another course —more expensive today, more demanding. Down this course lies a natural America restored to her people. The promise is clean rivers, tall forests, and clean air—a sane environment for man.[1]

The President then hailed the recent passage of the Water Quality Act of 1965 and pledged: "We mean to make full use of these new instruments. They will require increased expenditures, in a year of few increases for urgent domestic programs. We shall make them."

Congress took the President at his word and passed

a new authorization for Federal matching grants to municipalities for water pollution control totaling $3.4 billion for the next four years. When the President's next budget request came out, however, it was clear that his enthusiasm for pollution control had not survived the financial squeeze he had predicted in his message the preceding February. He asked for—and got—only $203 million of the $450 million in grants that had been scheduled for fiscal 1968. The year after that, he demonstrated that this memory lapse had been no aberration. Out of an authorized $700 million, LBJ wanted only $225 million. Congress had caught on by this time and went him one better, appropriating a puny $214 million for fiscal 1969. Johnson's parting gift to the incoming Republicans was a 1970 budget proposal that requested another $214 million for sewage treatment plants—this time out of an authorization that had risen to $1 billion.

By then the cumulative gap between authorizations and appropriations under the 1966 legislation was $733 million, and conservation groups were restive. The changing of the White House guard brought no improvement in pollution control financing. President Nixon labored a few months over the Johnson budget and made it known in April 1969 that he could find no more money for fiscal 1970's municipal grants than his parsimonious predecessor had—$214 million.

This time there were stirrings in the environmental camp that spelled trouble for the penny-pinching Executive branch. A coalition of conservation groups and municipal and county government associations calling itself the Citizens' Crusade for Clean Water had formed to pressure Congress to buck the President and appropriate the entire $1 billion. Up to this time the Nixon Administration had remained largely silent on environmental issues—apparently unconcerned that its subsidy-slashing budget decisions would tell the public how little it cared. But once the rumblings below got the new Administration worried enough to break out of its shell, it demonstrated a facility for underfinanced rhetoric rivaling LBJ's.

The Nixon Administration's first opportunity for a display of verbal footwork arose when the President sent Congress his alternative scheme for filling the ap-

propriations gap. Essentially similar to one half-heart-
edly offered by the Johnson Administration in 1968, the
plan would have let the Secretary of Interior make long-
term contracts with state or local governments to pay in
installments the Federal share of building their treatment plants. Interior Secretary Walter Hickel touted the
Administration's plan around the country with remarks
like those he made before the Executives' Club in Chicago on September 20, 1969. "The technology for
cleanup is here," said Hickel, "but without money, we
cannot do the job." The Secretary reminded the executives that the 1970 authorization was $1 billion. "This
may sound like big money," he continued, "but we now
find that one billion dollars will be merely a token sum
compared to what we really need. . . . We have come
to the realization that we must spend a much larger
amount of money, and that we *should* spend it, regardless of what that amount is." [2]

Unfortunately, few shared Hickel's excitement over
the new plan except the President and the Budget
Bureau. The plan, in effect, asked states and cities to
make interest-free, long-term loans to the Federal government. When we asked one Senate aide what would
happen to the proposal, he pointed to his wastebasket.

True to the time-honored pattern, the substitute financing plan had no sooner been tabled by a skeptical Congress than Secretary Hickel swallowed his billion-dollar
speeches and defected to the Budget Bureau's side.
When the House Appropriations Committee upped the
President's $214 million ante to $600 million, Hickel
quickly agreed that the higher sum was necessary. But
when Representative John Dingell (D.-Mich.) began
preparing for a floor fight to boost the $600 million
recommendation up to the $1 billion authorization, the
Secretary dug in his heels.

Hickel first sent FWQA into the fray armed with a
packet of unsubstantiated statistics reportedly sent from
the Budget Bureau. Administration loyalists in the House
quoted the water pollution agency as believing there
were not enough qualified engineers and contractors in
the U.S. to design and build the waste treatment plants
that would be generated by $1 billion in Federal aid.
A nationwide survey of the pollution control engineer-

ing profession conducted on short notice by the Consulting Engineers Council (a professional organization of consulting firms) laid that claim to rest: on October 6, 1969, the Council reported to Congressman Dingell that the Administration's estimates were "totally without foundation." Their data showed that more than $2 billion would have to be spent before the profession would even begin to feel the strain.

Two days later, however, when the spending measure finally came to a vote in the House, Hickel brought out his heavy artillery. A motion had been introduced from the floor, as expected, to raise the appropriation to the full $1 billion, and heated debate had begun. Just before the vote, opponents of the full appropriation relayed an eleventh-hour message from Hickel claiming that his department would be unable to spend the full amount. Whereas less than a month earlier Hickel had argued publicly that $1 billion was "merely a token sum compared to what we really need," he now contended that no matter how much money was voted, the Federal government could not find enough people willing to take subsidies to be able to spend more than $600 million. It worked. Minutes later the $1 billion proposal was defeated—by two votes. The compromise $600 million figure was approved shortly thereafter.

After the Administration's last-minute shelling had set back the conservationist cause in the House, the Citizens' Crusade for Clean Water shifted its lobbying to the Senate. The Senate took longer to allay its suspicions about the Administration's arithmetic than the House had. In December 1969 the Senate Public Works Appropriation Subcommittee asked FWQA to substantiate the $600 million figure that Secretary Hickel had invoked to defeat the House amendment. FWQA admitted, in a written response, that enough requests for construction grant funds were already on hand to use up an appropriation of over $1 billion, even if no new applications were accepted. Said FWQA:

On August 1, 1969, applications were being processed in the [FWQA] regional offices and in State agencies that, if fully supported, would require $841 million in Federal funds. There was also outstanding, as of that date, an additional

$514 million that States or local communities could be entitled to because of their pre-financing some portion of the Federal share.

The final weapon a President has in a battle to keep down an appropriation is an announcement (or a "leak") that the Executive branch will spend no more than it asked for, no matter how much Congress makes available. Rumors like this were rampant in December 1969, both before and after the Senate voted the full $1 billion and the House compromised with them at a final $800 million figure. Only after a month of Presidential silence did the Budget Bureau announce, early in February, that it would actually spend the money appropriated—an astounding 80% of the amount that Congress had originally said the cities should have.

About this time pollution was erupting into a full-blown national issue and Administration speech-writers were busy creating a new conservationist image for President Nixon. But the new mood of national militance on ecology could not prevent the pollution program from being short-changed again. The 1966 Clean Waters Restoration Act provided for $1.25 billion in fiscal 1971; the Administration asked that $250 million be held back, and Congress honored the request, passing a $1 billion appropriation in late September 1970.

The whole history of authorization/appropriation gaps focuses attention on the only significant question for the long run: how much money is actually needed to do the job? The only indisputable fact about the cost estimates made so far is that, over time, they all go up. In 1968 FWQA estimated that $8 billion in capital outlays would suffice to finance treatment plant construction during 1969–73.[3] In *The Economics of Clean Water 1970*, FWQA economists calculated that in the four-year period commencing in fiscal 1971 a $10 billion investment would be required to meet national needs for municipal treatment works; [4] this happily coincided with the four-year $10 billion "Federal program" (all but $4 billion of which was to be paid for out of state and local revenues) announced by President Nixon in his 1970 State of the Union Address.

FWQA CONSTRUCTION GRANTS:
AUTHORIZATIONS
vs.
APPROPRIATIONS

DOLLARS (MILLIONS)

1,250
1,200
1,100
1,000
900
800
700
600
500
400
300
200
100

'57 '58 '59 '60 '61 '62 '63 '64 '65 '66 '67 '68 '69 '70 '71

Fiscal Year

For what they are worth, Congress' cost estimates have always been at least twice as high as the Administration's. In an October 5, 1966, memo, Leon G. Billings, Senator Muskie's legislative assistant on the Public Works Committee, estimated that during 1967–72 about $20.25 billion worth of treatment facilities (based on 1966 construction costs) would have to be built to serve the population projected for 1980. He emphasized then that "this investment must be made now, water quality notwithstanding." In other words, even in places where the water was clean in 1966 it would have been cheaper, because of rising construction costs, to start building treatment plants immediately in anticipation of the pollution that would be caused by population growth.

The largest estimate of municipal treatment needs to date was prepared jointly by the National League of Cities and the U.S. Conference of Mayors at Senator Muskie's request. They collected data on the financial needs of 1008 cities, counties, and special districts responsible for sewage treatment and on July 6, 1970, reported that from *$33 to $37 billion* in public expenditures will be required between 1970 and 1976 to finance the sewers and treatment facilities that local authorities believe are necessary to meet *already identified* needs. The Mayors urged a Federal program of at least $2.5 billion yearly (which Muskie has proposed), but added that a "$3 to $4 billion a year Federal program can be easily justified in light of present needs." [5]

If Senator Muskie's scoop accomplished nothing else, it did prompt the Administration to take another survey. FWQA polled states and cities in July 1970 and came up with an overall cost of $12 billion—approximately $6 billion of it to be supplied by Federal funds—for fiscal years 1972–74 (compared to the $10 billion total cost estimate for the four-year period through 1974 which the Administration's 1971 budget request was based on).

The differences in the projections clearly illustrate the difficulty of making accurate economic predictions in the municipal construction area. The two biggest imponderables are the amount of inflation there will be in construction costs and the size of the rapidly growing industrial wasteload in municipal sewer systems.* Beyond that, it is important to recognize that the answer one gets depends on the question one asks. To come up with its $10 billion four-year projection in 1970, FWQA's economists asked the states what *facilities* their cities planned to build and computed the total by using statistical cost data and adding a fudge-factor for the heavy padding of contracts that typically takes place in municipal public works construction. The larger numbers—the Mayors' $37 billion and the Administration's $12 billion—were obtained by asking the cities and states how much *money* they thought they could use.

Given the current degree of state and local control over these Federal funds—FWQA typically makes no review of proposed projects to see whether they fulfill real pollution control needs †—there may be a perverse logic to this latter method of calculating costs. The construction industry exerts a powerful influence over local construction plans in many areas, particularly in the Northeast, where construction costs are more than twice the national average. These larger grants to the states should suffice to keep the construction industry busy and happy just about everywhere, with enough of a margin to insure that critical pollution control projects in places where construction interests are less powerful still get funded.

The Administration's $12 billion figure was so much lower than the Mayors' $37 billion cost estimate not only because the time spans were different (three years versus six) but also because the projected levels of treatment were different. The Administration is shooting only for secondary treatment—inadequate in many

---

* It should be noted that if the Federal government were to discontinue subsidizing that portion of municipal treatment capacity necessitated by industrial waste input, a hidden subsidy to industry, a greater measure of certainty would be restored to municipal grant needs. (See Chapter 17.)

† FWQA's construction grants section makes a detailed *technical* review of the plans to insure that the proposed project will be constructed properly. It does not, however, usually inquire as to whether the cost is excessive or whether the project will clean up pollution. In July 1970, FWQA published regulations (18 C.F.R. 601.32–33) requiring Federal evaluation of the pollution control value of any project. So far, however, FWQA does not have the manpower to perform this kind of review adequately, and old habits are hard to change.

locations—whereas the Mayors' survey for the Public Works Committee contemplated more advanced technology. The Mayors' estimate also included projects to which FWQA's money cannot legally be applied—the separation of combined sewer systems, for example. The Administration's calculation is limited to projects eligible for Federal aid under the present law. Finally, the Mayors asked the cities how much money would actually solve their pollution problem; the Nixon Administration asked them what they now "intend" to spend over the next three years. Whether or not the Administration's figures turn out to be accurate may thus depend on whether we make up our minds to demand of the cities more cleanup than they now "intend" to do.

On August 4, 1970—about a month after the Conference of Mayors had informed Senator Muskie that the cities could use $3 to $4 billion each year in Federal funding, and shortly after the Administration's survey showed that some $2 billion annually in Federal funds would be needed—Assistant Secretary of Interior Carl Klein appeared before Senator Muskie's Senate Air and Water Pollution Subcommittee. Klein told the Subcommittee that the most the Federal government could possibly use in fiscal 1971 was $985 million.[6] It would seem obvious to just about everyone in Congress and elsewhere that, when measured against all but the most patently fictional estimates of absolute need, Klein's maximum was absurd. How, then, have past and present Administrations been able to sow confusion in Congress and in the minds of the public with such remarkable success?

For one thing, the Administration's control over the rate of Federal spending can (intentionally or not) make its most ludicrous predictions come true. For example, by putting a "freeze" on pollution control construction funds until late in the fiscal year, as the Administration did in fiscal 1970 (funds were not released until February 1970),* the government can practically

---

* Pollution control funds were not invidiously singled out. They were part of a general freeze on government construction funds to combat inflation.

insure that there will not be enough time before the budget cycle closes (on June 30) to process all the outstanding grant applications. Administration spokesmen then can cite the fact that some of the appropriation was not "obligated" as evidence that future needs are less than imagined.* This was the approach Klein took in his August testimony before the Senate Air and Water Pollution Subcommittee. The fact that only $360 million out of the $800 million appropriation could be obligated during fiscal 1970 was the proof he offered that the Administration's new funding package for 1971 was adequate.

Apart from the fact that the $440 million "surplus" was created in part by the Administration's own funding freeze, such leftover funds are not a surplus at all. Under the distribution formula in the Pollution Control Act, it actually takes a "surplus" appropriation like this to get construction money to the states that really need it. Under this formula, most of the Federal pie is divided among the states roughly in the proportion that each state's population bears to the total U.S. population. Each state has until six months after the fiscal year ends to try to use up its allocation, at which time any "surplus" reverts to the Federal government for distribution to states that have used up their allocations and still need more money.

The fact that many legislators are not aware of the method for distributing leftover money has been used to good advantage by opponents of higher funding. During the October 1969 floor fight in the House, for example, Congressmen who supported the full $1 billion

---

* Another important factor that is easily forgotten when the Administration begins to play its "numbers game" is the effect of Administration budget requests on the number of grant applications outstanding. Both the number of grant applications the states *receive* from municipalities and the rate at which the states process them and send them on to the Federal government are roughly proportional to the size of the budget request (and the size of the previous year's appropriation). Applicants and state governments alike, in other words, accelerate their activities only when they are given reason to believe that accelerated activity may bear fruit. The best measure of the likelihood of their getting Federal grants in the immediate future is the Administration's budget request. Thus, if the Administration makes a low budget request, it can point to the fact that grant applications are being processed at a slow rate at the state level as "evidence" that no larger amount is needed.

authorization pointed to the $2.4 billion backlog in requests for construction grants as evidence of the need to boost the appropriation. Those who wanted to hold the level down explained that over half the backlog came from New York State, which under the population formula could get no more than $89 million even if the full $1 billion were to be appropriated. For this reason, they claimed, it was not possible to use any more than $600 million. Their argument carried the day, and they were right—up to a point. What they didn't mention— and what Assistant Secretary Klein never mentioned when he waved around the unspent $440 million from fiscal 1970—is that any "unused" money eventually goes to the states that *can* use it. It seems clear that, ultimately, the distribution scheme should be changed to reflect more closely the states' real funding needs.* But if New York and the other states that need the bulk of these grants † are ever going to catch up with their runaway pollution problems under the current distribution plan, it is *absolutely essential* that Congress vote an amount sufficient to create a large annual "surplus."

The large backlog of unapproved grant applications

---

* The Nixon Administration's 1971 legislative package includes a proposal that the formula be changed for this purpose (S. 1013). The proposal would set 10% of the grant money aside to be allocated by the Administrator of EPA on the basis of the most serious pollution control needs. Any grant money not obligated by the states at the end of the fiscal year would revert to the Administrator immediately (rather than six months later) and also would be available to go to the states with the greatest needs.

† The states with the largest backlogs of unfilled applications for construction grants as of June 30, 1970, were:

Applications pending June 30, 1970 at state agencies and FWQA regional offices	*Fiscal 1970 allocation*	*Estimate of backlog (applications pending minus fiscal 1970 allocation)*
(all figures in millions of dollars)		
New York	$592.3	$69.9	$522.4
Michigan	114.9	33.0	81.9
District of Columbia	49.9	3.8	46.1
Indiana	59.5	20.0	39.5
Maryland	52.1	13.6	38.5

It is important to remember that these figures do not measure the extent of need, since it has been found that many cities do not bother to apply for grants when funding levels are low.

at the state level has been only the most visible effect of the program's chronic underfunding. The Senate Subcommittee on Air and Water Pollution has heard witness after witness in recent years describe how municipalities that were going ahead with pollution control measures slowly but surely on their own before the Federal government began to make its shallow promises stopped construction entirely, waiting for the day when their grants may come through in the amount to which they are entitled.*

The situation was bad enough prior to 1967. Back then, there was a $1.2 million ceiling on the amount of any single grant; the government provided a maximum of 30% of the costs of any city's project, and annual authorizations were ludicrously low (averaging only $77 million annually for the entire U.S. for the 10 years prior to 1967). While the grants never got large enough to be of much assistance, at least few cities were tempted to wait for the peanuts FWQA might toss their way. But in fiscal 1968, when the annual authorization tripled at the same time that the individual grant ceiling was removed and the maximum Federal contribution jumped to 55%, there for the first time appeared to be enough money to make waiting worthwhile. The promised Federal support never materialized, however. As a result, municipal pollution control efforts ground to a virtual halt in many places.

When President Eisenhower vetoed a proposed boost in the Federal grants program back in 1960, he wrote, "By holding forth the promise of a large-scale program of long-term Federal support, it would tempt municipalities to delay essential water pollution abatement efforts while they waited for Federal funds." [7] Ike's prediction turned out to be accurate. But his hands-off approach to the sewage treatment crisis was no answer. If Wash-

---

* Some states try to spread their Federal money around as much as they can, giving less than the maximum authorized percentage of a treatment project's cost (from 30% to 55% depending on what conditions have been met) to each applicant. Polluters in these states thus have an incentive to wait until a better day to submit their applications and start construction. Other states give out only full-percentage grants, thereby limiting the number of recipients. Since the polluters who miss out have no incentive to build at all until there is money available for *them*, this slows matters down even more.

ington does not provide direct aid to the cities, they must rely to a greater extent on bond financing to cover their plant construction costs, driving bond interest rates higher and higher as they compete for funds. The only winner in the municipal bond market is the Wall Street investment banking community, which does the bulk of lending to cities. Because local money-raising mechanisms like water bills and property taxes take a bigger percentage bite out of the poor man's pocketbook than the rich man's (in contrast to the less regressive Federal income tax), the real financial losers in the local-bond method of funding pollution control are the cities' middle- and low-income citizens. Since the middle 1960's particularly, local property-tax bases have continued to decline, construction costs have skyrocketed, and treatment requirements have grown much stricter. For all these reasons, Federal participation in financing municipal pollution control is absolutely essential. But the experience with the grants program so far suggests that for the Federal government to have jumped in only ankle deep as it has up to now, promising to cover a higher percentage of treatment plant costs than it actually paid, may have caused damage to the nationwide pollution control effort outweighing whatever small benefits it has produced.

The Federal government's predilection for undernourished subsidy schemes has not only stalled local cleanup efforts in many places. It has also derailed the Federal enforcement program, although indications are that the Nixon Administration plans to keep the wreckage concealed, at least until cities start missing their deadlines under the Water Quality Act of 1965 in 1971 and 1972.

The long-standing, if informal, connection between Federal enforcement of water pollution laws and Federal sewage treatment grants to cities was acknowledged quite explicitly by Congress in relation to the Water Quality Act of 1965. In 1966 Congress passed four-year funding legislation authorizing $3.4 billion for fiscal 1968–71 for the purpose of helping the states begin implementing the water quality standard requirements in the 1965 Act. When the states began to set their deadlines for full compliance with the standards, they had to calculate how much of the $3.4 billion authorization would actually be available. The logical people to ask were the regional officials of FWQA. Just what their answers were has since been the subject of an on-going dispute between state and Federal officials.

FWQA's former Commissioner Joe Moore had an unusual opportunity to hear both sides of the story, since he came to the Federal government from a career in state government. He was chairman of the Texas Water Quality Board in 1966 when it was setting its construction timetables. "FWQA did not come out with any guidelines on how to make up an implementation schedule for a long time," Moore recalled to the Task Force. "But even though it's not in writing anywhere, I was personally told by FWQA officials that the 1972 compliance time we set was to be based on the assumption of full appropriations for the next five years. Of course, FWQA can always say 'we didn't *really* tell you that.'"

A former New England state official who is now part of the Administration team told the Task Force essentially the same thing, but reminded us that what he said had to be "off the record" because it conflicted with the "official" Federal position. That position has always been, basically, that no promises were ever made by anyone, but that if any had been, the states should not have believed them.

The states have reacted with varying degrees of enmity toward their sometime Federal "partners." Some have gone ahead and advanced their municipalities payment of the Federal share of the construction costs themselves (called "pre-financing" in bureaucratic jargon), hoping against hope that the money will someday arrive to reimburse them.* Some of these "pre-financ-

---

* The 1966 legislation included a provision that entitles state and local governments to refunds on projects constructed with partial or no Federal assistance. It also included a disclaimer, however, to the effect that approval of a project by the Secretary of Interior could not be construed as a "commitment or obligation of the United States to provide funds to make or pay any grant for such projects." All of the Federal money appropriated so far has been quickly used up on urgently needed new projects.

ing" states have threatened to sue the Federal government for the sums they believe are owed them. But most states have been neither generous, trusting, nor interested enough to pick up the Federal tab. The attitude toward the Federal government at the state level over most of the past four years has been uniformly one of contempt. As a Maryland official told the Task Force in 1969, "The Federal government is making a lot of demands, but it's not coming up with the money to meet them." Or as a New Jersey official commented, "FWQA should keep its mouth shut until they have the money for grants."

The Federal government has responded to this hostility by continuing to short-change the states while insisting that the original cleanup deadlines still hold. Typical of the Federal position were the remarks of Assistant Secretary of Interior Carl Klein before the 5000 members of the Water Pollution Control Federation (an association of water officials and engineers) at its October 1969 conference: "We can't have procrastination by anyone saying he's not getting enough money from the Federal government." [8] Much of the available evidence suggests, however, that the deadlines in a number of states are by now nothing more than a façade for the failure of the water quality standards program, propped up by the Federal government's wishful thinking and by little else.

But doesn't the government plan to enforce the Federal water quality standards when they are violated, as it continues to warn that it will? No one has yet been given any reason to believe that the Federal government keeps promises to enforce any better than it keeps promises to spend. In fact, even when the enforcement program has confronted a pollution situation head-on, it has always had to pay heavily for what little cleanup it got. As the Federal government scowls at a municipal polluter across the enforcement table, it typically slips a subsidy underneath the table at the same time. Because it is either unwilling or unable to *compel* polluters to clean up, the government has sometimes had difficulty even getting municipal polluters to accept their grants. Without the grants, municipal cleanup is hardly even a viable possibility.*

Former FWQA Commissioner Moore told the Task Force what *his* solution to this dilemma would have been had he not been ousted when the Republicans came in. He would have relaxed the deadlines. This alternative has been rejected by the Nixon Administration so far. If they continue to reject it, what will they do instead? Were there to be an improvement in Federal enforcement laws in 1971 coupled with a drastic change in the government's attitude toward enforcement, it is conceivable that the Administration's tough talk about enforcing the timetables might turn out to be more than bluster. True, such a policy would be unfair to some hard-pressed municipalities and states. But, for the most part, it would not be an unhappy possibility. The majority of the states have turned the Federal government's lack of commitment on the funding issue into a convenient smokescreen for their own inaction.†

But barring this unlikely turn of events, what can be expected? The Administration could simply announce that cleanup has been attained when the deadlines arrive. Or, better, the government could initiate a few token enforcement actions and make sure that it gives handsome awards to its few enforcement targets. But what most observers of the Federal scene told the Task Force they suspected would happen is that the dates would just slip silently by—with as little mention by anyone as possible.

In June 1971 the municipal grant authorization ap-

---

* It is possible that even were the government to try with all its might to enforce (something it has rarely, if ever, done), it would still be rebuffed. Since the Pollution Control Act stresses "economic feasibility" and the "equities" of pollution control, a municipality might be able to invoke successfully the government's failure to come through with the Federal share of the money as a defense in court. More important, since the real power of enforcement lies in public opinion, it's difficult for the Federal government to get any movement at all as long as everyone considers it (rather than the polluter) the villain of the piece.

† The states knew all along that authorizations were so low that, in any case, either state or local government would have to assume the bulk of the financial responsibility for seeing that municipalities met the water quality deadlines. Moreover, while the grant money legitimately expected from the Federal government has continued to go to most states slightly over a year late, many of the same states now are as much as two and three years behind in getting their municipalities to construct cleanup facilities.

proved in 1966 will end. When it does, the final tally for the four-year subsidy package will be: authorizations, $3.4 billion; appropriations, just over $2.2 billion. The Nixon Administration and Senator Muskie have both proposed a new funding package to take up where the current one will leave off. Senator Muskie wants $2.5 billion per year in Federal money for the next five years. President Nixon has come closer to Muskie than any President has ever come before; he now wants $2 billion annually for the next three years. One of these plans, or some middle ground between them, will no doubt be enacted. As the Congress contemplates adopting a new financing scheme and as the Administration begins to plan its budgets for the years ahead, they would do well to recall the lessons of the last four years. The cities need more Federal money than they have been given— even more than they have been promised—in the past. But promises broken are worse than promises never made.

17

Subsidies Abused, Sanctions Unused

More than 10 years after the Federal government first began dabbling in construction grants, Senator Muskie asked the General Accounting Office, Congress' official investigative watchdog, to find out what clean-water dividends had been realized from the program. The results of the GAO's study, which began in November 1968 and covered eight states in depth, were made available to the Senate's Air and Water Pollution Subcommittee in summer 1969 and later to the public.[1] Given the program's long history of underfunding and the anemic quality of Federal enforcement, the GAO's principal finding was a foregone conclusion: the water is no cleaner, but in a few places the construction grants program has kept it from getting much dirtier.

While the GAO mentioned in passing the obvious need for larger appropriations, its main task had been to find out exactly what the Federal government had bought with what it had spent (barely $2 billion since 1956). The findings were distressing. Much of the pitifully small sum appropriated has been wasted—either squandered on treatment plants that, for one reason or another, don't work well, or funneled off to pay the cleanup bills of private industry.

GAO's main whipping boy was FWQA, which has always been responsible for administering the grants program on a day-to-day basis. At first glance, the waste treatment section seems to be taking its job well in stride. Most of the work of processing applications is done in FWQA's regional offices, and though their numbers are few, the Federal personnel who administer grants in the regions are praised for their efficiency by the states. The program is a remarkable exception to the usual delays encountered by applicants for Federal grants. Under FWQA's procedures, the states are told well in advance of deadlines how much money they are entitled to. Applicants are told within a month after receipt of their applications whether they have been accepted or rejected.

While speed and efficiency are in themselves admirable, they unfortunately derive as much from the staffers' narrow conception of their role as from their competence. Grants personnel, according to several officials with whom the Task Force spoke, are considered "paper shufflers"; the program's efficiency has often been the efficiency of mindless, purely mechanical performance of a purely mechanical task—doling out a certain amount of money each year. After having been stung by several critical blasts from the GAO, the diligent staffers and their superiors in the water program are just beginning to comprehend, far too slowly, that this $1 billion subsidy program, more than 90% of the annual Federal water pollution budget, could be a powerful tool to promote water cleanup if it were used strategically. The heavy costs—in both financial losses and continuing pollution—of FWQA's mechanical approach to the distribution of waste treatment grants were spelled out in detail in the GAO's reports.

We Get What We Pay For: More Industrial Waste

Two of the largest municipal treatment plants in the United States built with substantial Federal assistance are in Nampa, Idaho (1960 population: 18,897) and Monsanto, Illinois, a railroad stop with a listed population of 324.[2] One plant was designed and built to serve the local food-processing industry, the other to treat the Monsanto Company's chemical wastes.

Wellston, Ohio (1970 population: 5410) also has a huge waste treatment plant. It was built in 1966 to attract a Ralston Purina Company turkey-processing plant to the area. Wellston Mayor James Reupert explained to the Task Force that the city had an "iron-clad contract" with Ralston Purina drawn up by "lawyers in Cleveland" so that the company would have had to reimburse Wellston for the full local cost of the waste treatment plant (over $1 million) had it ultimately decided to set up shop elsewhere after the treatment plant was built. But there would have been no one to reimburse the Federal government for the $805,836 it spent on this project.

In Senator Muskie's birthplace, Rumford, Maine (1960 population; 10,005), the Oxford Paper Company has found that the cheapest way to fund its waste treatment expenses is by tapping the Federal Treasury. Oxford Paper, Rumford's single industry, is the largest polluter on the Androscoggin River, one of New England's filthiest waterways. The Oxford mill employs, directly or indirectly, more than half of Rumford's work force and pays more than 54% of the town's taxes. (Oxford Paper also completely owns the Rumford Falls Power Company.) Local residents all agree that "the mill is Rumford, and Rumford is the mill." The mill's most pronounced influence on the town's policies has been through special study committees that recommend proposals for submission to Rumford's voters. Rumford's sewer study committee is typical. Back in 1967 and 1968, when the paper mill was first faced with a state requirement to give its wastes secondary treatment by 1976 or shut down, the sewer study committee was composed of engineers, 80% of whom were salaried employees of the mill. The members of the mill-dominated committee had little trouble persuading the town to solve Oxford Paper's problem by building a joint treatment facility, which will cost $6–8 million by the time it is completed. The mill agreed to pay 90% of the local costs of constructing the joint Rumford-Oxford Paper treatment plant (a fair share from Rumford's point of view, since the mill contributes about 90% of the town's effluent). But with state and Federal grant assistance, the local costs may be reduced by as much as 85%. In other words, Oxford Paper will be getting the benefit of a $6–8 million plant to comply with state pollution laws for the bargain price of $810,000 to $1.08 million.

The funding arrangements described above, whereby FWQA's Federal municipal grant money is being used to provide what are, in effect, disguised subsidies to industry, are neither unusual nor illegal. Although Congress has refused on several occasions to authorize subsidies explicitly designated for industry, the Federal

Water Pollution Control Act leaves FWQA a gaping loophole through which it can do indirectly what it may not do directly. The Pollution Control Act requires only that a treatment facility be publicly *owned* in order to qualify for a "municipal" grant. Thus if an industry and a city join together to treat their wastes, the government may pay up to 55% of the construction costs of both the city *and the industry,* provided only that the city retains nominal ownership.

Since the Act gives the EPA Administrator broad discretion in administering the subsidy program and commands him to insure that the grants serve the "public interest," FWQA presumably could, if it chose to, save grant money for the cities by means of an innovative grants policy. (FWQA might, for example, require cities that receive grants to promise to charge industries that use the treatment plants a fee which would cover their share of the construction costs; the grant amounts to cities might then be lowered, when appropriate, to reflect this additional source of revenue for some cities, leaving more Federal money to go where it is most needed.*) Federal officials have not been given to bold innovations, however, particularly ones that might displease corporate polluters. The result is that private industry, in effect, pockets a substantial portion of the money appropriated by Congress each year ostensibly intended to promote municipal cleanup.

Up through 1968, the GAO investigation showed, the Federal government spent or obligated over $80 million of the $1 billion total awarded during those early years to help build 400-plus "municipal" treatment plants designed to treat primarily—or, in some cases, exclusively—industrial wastes.[3] The GAO investigation only touched on the tip of the industrial subsidy iceberg, however. Since 1968 the municipal grants program has gone big-time (i.e., annual grant appropriations in the neighborhood of $1 billion) and all the available evidence indicates that industry is fast becoming the biggest beneficiary. FWQA's best estimates show that somewhere around half and possibly a great deal more of the BOD wasteload in the nation's municipal treatment

* It is clear from the law's legislative history that FWQA cannot refuse to fund the industrial portion of joint treatment plants altogether.

plants comes from industry.[4] This means it is likely that
somewhere around half and possibly more of the annual
$1 billion (fiscal 1971 appropriation) in municipal grant
subsidies—i.e., somewhere around $500 million or more
—is currently being funneled through municipal middle-
men to pay the cleanup costs of private industry.

These subsidies to industry not only rob the nation's
needy cities of scarce public funds to which they otherwise would be entitled, but, worse, actually encourage
the increased production of industrial waste. This is an
absolutely crucial point to understand. When corporations do not pay the full cost of cleansing their wastewater, they can afford to discharge more waste than they
otherwise could. Why change one's manufacturing process in order to keep toxic pollutants from going down
the drain—no matter how inexpensive the change might
be—when someone else will pay the treatment costs?
Why bother keeping uncontaminated cooling water separate from polluted wastewater when the whole mixture
will be handled at cut rates? The industry subsidy scheme
forces every taxpaying citizen to give his money to companies that pollute, rewarding their pollution with increased profits and decreasing the prices that consumers
who buy their products have to pay (thereby encouraging them to buy more and cause more pollution). Our
current subsidy policy is quite literally *subsidizing industrial waste production.* In the absence of financial deterrents, industrial wasteloads will go on increasing geo-
metrically, as they have in the past. Whole new rounds
of subsidized treatment plants will have to be built to
cope with the soaring overload. Unless this disastrous
giveaway can be ended, the pollution-subsidy spiral will
continue *ad infinitum.*

It should be emphasized that there is nothing objec-
tionable about joint municipal-industry waste treatment.
To the contrary, policing is easier when there is only
one system to police. And the fewer systems there are,
the more likely it is that each will have a competent
operator. Furthermore, joint treatment often makes for
both efficiencies of scale (i.e., bigger plants do the same
job more cheaply) and technological efficiencies (controlled amounts of certain types of industrial waste can

complement the biological process involved in secondary treatment).

More important than the question of who treats the wastes, however, is the question of who pays. It makes little sense to give subsidies to companies to encourage them to do what they would have good reason to do on their own without them. Where there are genuine economies in joint waste treatment, industries should not need a subsidy to take advantage of them—especially since municipal treatment of their wastes saves firms operation and maintenance headaches. If, on the other hand, a company can treat its own wastes at lower cost than the municipal treatment plant can, the Federal government should not be subsidizing the most inefficient solution to the industry's pollution problem.

It should be noted that public subsidies for industrial waste treatment rarely stop with FWQA's construction grants. Municipal sewer authorities typically finance both the local share of treatment plant construction (15–60% of the total construction cost, depending on whether or not the state supplements the Federal grant) plus all operation and maintenance costs through some combination of taxes and sewer service charges to their customers. For one reason or another—habit, ease of administration, or industrial pressure at the local level—most cities finance their municipal waste treatment expenses by methods that not only work out in industry's favor but lack any deterrent to industrial waste production. Neither property taxes, flat monthly sewer service charges, nor water bills based on the amount of water used bear any relationship to the burden of waste a user places on the treatment system.* There is only one pay-

---

* With enough citizen or Federal pressure, any of these methods of financing waste treatment could probably be manipulated so that industry as a whole would pay its fair share of treatment costs—but only temporarily. Industrial water rates could, for example, be raised so that the total amount of money collected in industrial water bills equaled the total cost of treating all the waste from industry at the time the new rate was set. Were this done, companies might be induced to cut down on the amount of water they used, but there would be nothing to prevent them from simply pouring more waste into less water. Industrial wasteloads would thus continue to rise even as industrial water use—and water bills—decreased, and soon industry as a whole would no longer be paying its fair share of waste treatment costs.

ment plan that not only divides the costs fairly between different kinds of users (residential, industrial, and commercial) but also offers the right deterrents to waste production by individual companies: a system of user charges based on the volume and strength of industrial waste, with periodic waste sampling and lower charges as a reward for limiting wasteloads. When Otsego, Michigan, started charging industrial users extra for above-average concentrations of BOD, for example, the total volume of BOD dropped almost 50%—from 28,000 pounds a day to 15,000—by the second billing period after the change was made. Cincinnati, Ohio, and several cities in Oregon have also reported reductions in suspended solids and BOD from industry after introducing similar surcharges.[5]

The reason so few cities have imposed industrial surcharges on sewer use is precisely the same reason that most cities and states have been unsuccessful in regulating direct industrial discharges into lakes and rivers. Industrial political and economic pressure on local government is too great. To prevent companies from playing one local jurisdiction off against another, Federal regulation is necessary in local charges to users just as it is with respect to direct industrial discharges.

FWQA needs no new authority to provide this regulation. It could insist on the imposition of local sewer-user charges as a condition to awarding Federal grants. But why, the reader may well ask, should a city not be permitted to inundate itself with industrial waste if it is willing to assume the ultimate moral and legal responsibility for preventing that waste from reaching the rivers? How does one justify the Federal government's dictating to local governments what method they should use to collect their revenues? The answer is simply that if the Federal government does not, one or both of two unacceptable consequences will ensue: (1) Federal tax money will have to be spent in increasing amounts to subsidize expensive advanced levels of municipal waste treatment that otherwise would not be necessary; * (2)

---

* This is an important point. If increased industrial wasteloads only created a need for additional municipal treatment *capacity,* any burden on the Federal Treasury could be avoided by simply refusing to subsidize the construction of industry's share of any capacity increase. In order to prevent the total pollution load from increasing as waste-

municipal plants will be unable to keep up with the unchecked growth of industrial wastes, and the contamination of our waterways will continue to accelerate. In the absence of local user charges, municipal treatment systems are, in effect, safe havens to which industries can flee to avoid state and Federal regulation. Shielded by their municipalities, corporations can either pollute with impunity or charge part of the cleanup bill to the Federal government. The Task Force believes that the Administrator's responsibility to administer the grant fund in the public interest leaves him no alternative but to attach the sewer surcharge string to his Federal subsidies. Such a policy is needed to save cities from their own industry-indentured status, which breeds increasing waste inundation.

FWQA made a feint in the direction of insisting on user charges in July 1970, but went no further. In response to GAO criticism of its acquiescence in local subsidies to industry, the water pollution agency published regulations that require grant applicants to "assure" FWQA that they will have "an equitable system of cost recovery" in effect once the plant starts operating.[6] Not only did the regulations fail to provide any sanction for non-compliance, however, but they carefully avoided telling the cities which "equitable" cost recovery system to choose. Any hope that FWQA might somehow plug that fatal loophole when it applied the new regulation

---

loads increase, however, an increase in treatment capacity is not enough. An increase in percentage removal rate—i.e., a more advanced treatment technology—is also necessary. The government could conceivably treat the funding of additional percentage removal ability in one of two ways. It could subsidize that proportion of the cost of the improvement which overall volume and strength of domestic wastes in the plant would bear to total wastes, on the theory that *all* wastes in the plant—both domestic and industrial—would in fact receive the advanced treatment. This, however, would amount to a Federal expenditure which would not have been necessary if a local user charge system had been imposed from the beginning to prevent industrial waste increases. Alternatively, the Federal government could refuse to provide Federal assistance for any portion of the cost of upgrading plant percentage removal ability, on the theory that, having caused the entire need for the expenditure, industry should bear the entire cost. This would be a difficult position to take politically, since the rationale for it is not at all obvious. Assuming, however, that the government *could* force industry to pay for the entire improvement, the cost to industry would in all probability be a great deal more than industry would have had to pay for treatment if local user charges had been imposed at the lower percentage removal levels.

has been disappointed so far. Not only has FWQA not plugged the loophole; the water pollution agency has not yet even begun to apply the regulation. Press releases went out in July 1970 heralding the change, and then the grants program simply continued as usual, mechanically handing out money as though the regulation had never happened.

When the Task Force inquired in January 1971 about the fate of the regulation, we were assured that it had not been forgotten. It would be enforced, grants program staffers told the Task Force, just as soon as specific implementing guidelines could be drawn up, possibly as early as April 1971. If the time when the cost recovery regulation would finally be enforced was uncertain, however, one thing *was* certain, they told us. Grant applicants will indeed be permitted, under the forthcoming guidelines, to use any system of cost recovery they choose—property tax, water bill, flat monthly sewer charge, etc. FWQA grants officials went to great lengths to make sure that the Task Force clearly understood that they had no intention of "interfering in local politics" by insisting that cities adopt the only cost recovery system that can bring a halt to increasing industrial waste pollution—the user charge.

Were FWQA to change its mind about local user charges, the industrial waste incentive problem would, unfortunately, still be only half solved. The stillborn July 1970 cost recovery regulation specified that only *local* costs had to be recovered from industry, leaving the direct Federal subsidy undisturbed. To meet the GAO's criticism on this count, FWQA's July 1970 regulations offered another cosmetic solution.[7] The regulation said nothing about the question that really matters— who should *pay* for the treatment. It specified only that industrial wastes must be "included in a waste treatment system treating the wastes of the entire community." What this means is that industries can still count on receiving their Federal subsidies, just as before. The only practical difference the regulation makes is that the Federal government may no longer fund "municipal" plants that are *100% industrial*. Since July 1970 companies have had to include a modicum of residential sewage in their treatment plants to make the subsidies legal.

In the final analysis, there is only one defensible rationale for underwriting industrial cleanup. It is better to pay companies to have their wastes treated in a "municipal" plant, several Federal grants officials argued to the Task Force, than to leave their wastes completely untreated. In other words, if one takes as given the fact that neither local, state, nor Federal governments have ever proved capable of enforcing pollution control laws against industrial polluters, it is better to buy them off than to let them continue to pollute. This may be true. But the price we pay for our failure to set high standards and enforce them is a heavy one in both inflated Federal budgets and inflated industrial wasteloads.

Treatment Plants that Don't Treat

The Task Force first became curious about the operating efficiency of waste treatment plants during our review of the Federal enforcement program. How was it possible, we wanted to know, that more than 1000 municipalities have constructed waste treatment plants in response to Federal enforcement conference cleanup recommendations and yet the water in most of the rivers those treatment plants discharge to is no cleaner than it was before the plants were built? We found that there are many answers to that question. But one of the first persons we talked to in 1969, a former FWQA enforcement official, paused for less than a second before he advised, "Take a look at the waste treatment plants." From his experience, he said, the plants just "don't work. If anybody looks, they'll find a national scandal in the waste of money on treatment plants that don't treat."

The "national scandal" did not escape the attention of the GAO, which devoted an entire report later that year to the chronic operation and maintenance difficulties the investigators discovered at most of the Federally funded treatment plants they inspected. Over and over again in their inspection tours the GAO investigators found four factors at work: (1) unsuitable treatment plant design; (2) understaffing; (3) inadequate training of and outright negligence by treatment plant operators; and (4) systems fouled up by industrial wastes they had never been designed to treat. When the GAO investigators showed up in Crawfordsville, Indiana, for example, they found all of these problems. In their effort to find out how much pollution the Crawfordsville plant had been getting out of its wastewater, they were somewhat handicapped by the fact that the plant superintendent had stopped keeping laboratory records some time before. He had been too busy, he said, training a new operator. The plant had not been built to accommodate the heavy rains the city periodically experienced, so when the first rains came the plant quite naturally bypassed raw sewage. And its biological treatment system had never operated efficiently, partially because of debilitating onslaughts of concentrated industrial waste.*

In 1971, more than two years after the GAO visit to Crawfordsville, things are only slightly improved. The treatment plant keeps records now, but the industrial waste problem is as bad as ever. A hog-slaughtering company that goes by the name of Porkland has never bothered to screen out its blood—which consumes much of the oxygen the treatment plant needs to decompose domestic sewage properly—and is flooding the plant with hog manure. Were not the Crawfordsville plant operators now careful to keep cleaning out the treatment apparatus, it would rapidly become jammed with Porkland's hog hair, toenails, other assorted pig parts, and even unborn pigs. "We have to take those pigs out and bury them like they [Porkland] should be doing," Crawfordsville's new plant superintendent, Dan Stutzman, complained to the Task Force. The six-foot-deep stream of *treated* effluent *leaving* the Crawfordsville treatment plant contains a moving mat of hog hair at the bottom six inches thick and another two feet of heavy black sludge from the treatment plant which coagulates around the hog hair. Since local officials do not now contem-

---

* The GAO reported several treatment plants where toxic industrial wastes entirely killed off the bacteria essential to the secondary treatment process. One plant the GAO was told about was shut down for a nine-month stretch in 1968 because toxic industrial wastes had stopped the operation of its sludge digestor.

plate imposing any legal requirement for pre-treatment on Porkland (nor charging the company anything more than the flat sewer rate for the added expense it causes, including more than $1000 per year in higher chlorination costs), it looks as if the Crawfordsville plant is one recipient of a $230,702 Federal construction grant that may go on operating below peak effectiveness for some time to come.

Behind every Federally funded treatment plant that turns out to be a lemon lies a failure by FWQA to carry out its clear statutory responsibility in administering the grants program. The Pollution Control Act states emphatically that "no grant shall be made for any project . . . until the applicant has made provision satisfactory to the [Administrator] for assuring proper and efficient operation and maintenance of the treatment works after completion of the construction thereof." Congress' mistake was expecting that FWQA would be willing and able to carry out this statutory mandate. The minimal steps FWQA has taken to assure that Federally funded plants will be maintained and operated properly have been distinguished by their almost total impotence in dealing with the basic problem.

One safeguard consists of a blank on the agency's construction grant application asking the applicant to "itemize number and type of employees to be hired, as well as amount per year for labor, chemicals, utilities, supplies, including those associated with laboratory operations, etc." During their brief review of the construction grants program, the GAO investigators examined 67 *approved* applications and found that almost 20% of the grant recipients had not even itemized this "assuring" information, much less done what they had promised.

Not that FWQA is likely to learn whether the promises have been kept. Inspections of municipal treatment plants are characteristically few and far between. The agency's handbook of procedures calls for a single inspection of each Federally financed project after it has been in operation for one year. The inspection is to be made by either an FWQA official or a state official or both. Under this flexible rule, FWQA's regional offices the practice, the GAO learned in 1968, at FWQA's Ohio Basin Regional Office, which required that the states make all the inspections. Sufficient commentary on the effectiveness of this approach was the GAO's discovery may decline to make any inspections at all.* This was that one Ohio Basin state, West Virginia, "had not performed any of the required inspections since 1965." (Since then the Ohio Basin region has been conducting occasional inspections, but still has only one man assigned to keep track of all state inspection efforts and conduct inspections himself in the region's four states.)

The GAO's revelations prompted FWQA to extract a promise from the states that they would try to do better. FWQA's July 1970 regulation prohibited grants except where the state agency "assures" FWQA that state officials will inspect the new facility at least annually for the first three years and periodically thereafter.[8] A clause to that effect is now included in every grant contract. It remains to be seen whether the state pledges will be honored when the projects which have been started since that time are completed, but once-a-year state inspections will be no panacea in any event. A state official from Maryland, where each plant is already given an annual spot check by the state, admitted to the Task Force that his "state inspection program is horrible" due to lack of personnel. He said that Maryland's 300 treatment plants *should* be inspected at least four times a year. Merely multiplying the number of inspections will not solve the operating efficiency problem either. One New Jersey treatment plant operator explained to the Task Force that in his state the state inspector "comes around a maximum of four times a year, between the hours of nine and five, in good weather." If an operator wants to shut down the plant awhile, he waits "until just after the inspector has gone."

FWQA's measures for improving the caliber of treat-

---

* When FWQA officials are able to make the inspection, it is generally done in half a day even though, in some cases, an accurate determination of whether a plant is operating efficiently may require as much as three weeks of observation. The FWQA makes no independent laboratory tests, but relies upon the states' or the plants' test results and records. Where such data is not kept, the inspector has no alternative but simply to make visual observations of how the plant is running that particular day.

ment plant operators have been somewhat more practical, though they have been restricted to the problems easiest to solve. Low status and low pay scales have traditionally made it hard for city and county treatment plants to attract the best-qualified people for operator jobs. The chemical or sanitary engineer who can get the most efficient waste removal from a large and complex plant will earn below the median salary ($16,490 a year) for his profession if he works for a county or city sewer authority.[9] The best engineers tend to get jobs with the Federal government, industry, or consulting or construction firms, all of which pay more. Pay scales in big Northeastern cities are among the highest. Yet Boston's biggest treatment plant, a primary system with 248 employees, averaging 299 million gallons of wastewater a day and serving 1.4 million people, pays its chief operator $12,870 tops after six years of service. The sewage treatment operator who runs the one-man plant in a hamlet of 1500 can earn more installing air-conditioners or commuting to an assembly-line job.

A good waste treatment plant operator has an exacting job. He has to know chemistry, biology, electricity, hydraulics, and math. He has to be at home with valves, motors, pumps, mixers, and generators, be able to keep them in good repair, and know how to read gauges and interpret charts. In many plants he takes the samples of incoming and outgoing wastewater and does the laboratory analysis on them himself to see what kind of waste removal he is getting. If too many solids are getting out, a competent operator gives the sewage more time in the settling tanks. If the effluent is low on dissolved oxygen, he gives it more time in the activated sludge. A vigilant operator watches over the incoming wastes, so if he spots a thick sludge from a gas station or factory riding in on some harmless sewage he can quickly turn off his trickling filter before the gummy oil knocks it out of commission. Above all, the operator has to be thoroughly familiar with his own treatment plant so he knows exactly what to do to avert catastrophes and, like a jockey with a thoroughbred, can urge it to peak performance.

Just how great a handicap operators had been laboring under in getting the utmost from their treatment plants was underscored by the regulations FWQA belatedly adopted in March 1969. Before the Federal government will make its final subsidy payment, the consulting engineer on a project is now required—for the first time—to prepare a manual describing in detail how to operate the plant. In addition, he has to stay at the plant four to eight weeks after start-up in order to supervise operating procedures and familiarize employees with the equipment. The Federal government has also started voluntary operator training and upgrading programs of its own to supplement the haphazard state courses, which average 31 hours in duration and reach less than half the eligible operators.[10] These Federal measures make eminent good sense as a starting point. But the best-instructed operator can't run a plant right if, for example, he takes the afternoon off.

"In my experience, it's not that the operator doesn't know what he's doing," one state pollution control official told the Task Force. "I usually find they've just gone fishing." One of these elusive operators was responsible for running a 350,000-gallon-a-day secondary treatment plant in Earlington, Kentucky, built with $118,429 in Federal funds. The first time an FWQA inspection team visited the plant, in 1968, "the operator, who was aware that we were to be in town, made himself unavailable." Later that year an inspector from the state pollution control agency also searched in vain for the man in charge. He found a broken flow-meter, pieces of blackened equipment from which a suspicious odor was emanating, gas forming in the chlorine contact chamber, and an effluent remarkable for its cloudiness and pungency—but, once again, no operator. On its inspection tour in February 1969 the GAO observed that the raw sewage coming into the plant was going out into the receiving stream virtually untreated. As usual, the operator was nowhere to be found.[11]

As precautions against this kind of gross negligence and just plain incompetence as well, Federal officials have considered imposing various kinds of operator qualification schemes—mandatory state certification (in effect in only 23 states as of 1969[12]), for example, or mandatory Civil Service status (to circumvent political hiring). Measures like these could hardly help making an improvement, but so far they have not been carried beyond the talking stage. In the final analysis, however, operator qualification requirements are no more the final answer than are any of the other halting steps FWQA has taken. What the Federal government needs most is not more empty "assurances" from the states or grant recipients but a way to make sure that the pledges it obtains will be honored.

FWQA's reliance on unenforceable promises overlooks the basic fact that cities, like industries, usually have every incentive to avoid spending money to clean up the water when people who live downstream—and outside city limits—receive most of the benefit. In Dayton, Ohio, a $134,000 chlorination unit that had been built to comply with state requirements for routine chlorination was used for a grand total of 60 days between 1966 and 1969. The reason was simple—the city saved $275 a day by not operating it.[13] In the same way, cities have little incentive to pay high wages for operators or, in places like Earlington, Kentucky, even insist they stay on the job during working hours. Most cities, like private industries, will not conscientiously control their pollution unless they suffer penalties when they do not.

While the Federal government is short on penalties to use against municipalities in the enforcement area, FWQA's construction subsidy could be filling a great deal of the void. Unless the Federal Water Pollution Control Act does not mean what it says, the "provision . . . for assuring proper operation and maintenance" that the Administrator is bound to insist upon with every grant not only could but should include enforceable contractual promises by grant recipients to suffer a prescribed financial penalty whenever they breach their agreements to run the plants well. (Cities might be required, for example, to post performance bonds, which would be forfeited to the government if specified operating standards were not maintained.)

This kind of sanction would, if vigorously applied, make a bigger difference in the condition of our lakes and rivers than a hundredfold increase in the number of inspecting personnel. (It should be noted that if the Federal government ever got serious enough about controlling municipal pollution to search for workable solutions in the grants area, it would also require municipal polluters to install continuous monitoring devices and periodically report the results of monitoring to FWQA and to the states.) A city could hardly afford not to demand that local industries give their wastes pre-treatment before dumping them into the sewers if the size of the city treasury depended on it. (Under a regulation adopted in July 1970, cities must now "assure" FWQA that they will require pre-treatment by industries.[14] Again there is no assurance that the cities will live up to their assurances.)

Up to now FWQA has been oblivious to the possibility that municipal construction subsidies might be used to actively promote water cleanup. The agency's grants section continues to plod along in its perfunctory way, handing out grants with little real assurance that they will be properly used. By not insisting on sanctions to accompany "assurances," FWQA has been conducting an elaborate exercise in self-deception at the public's expense. The lesson would seem to be clear. As long as it remains cheaper for cities to pollute than to control their pollution, we will continue to throw good money after bad by buying treatment plants that don't treat.

Industrial Footprints on the Freshly Mopped Floor

Two FWQA construction grants totaling $640,000 were awarded to Bogalusa, Louisiana, and Picayune, Mississippi, in 1960 and 1964, respectively. They were supposed to help clean up the Pearl River, which flows along the Mississippi-Louisiana border. Bogalusa built a plant that removed the equivalent of the BOD from 4500 people. The local Crown Zellerbach paper company, however, continued to dump in the equivalent of the BOD from 400,000 people. Picayune did a little better. Its treatment plant removed the equivalent of the BOD from 8000 people; but the nearby Cosby Chemical Corporation was discharging the BOD equivalent of the wastes from a population of 125,000. All in all, the two municipal treatment plants were keeping the equivalent of the BOD from 12,500 people out of the river, and just two of the industries that line the Pearl were dumping in the equivalent of the BOD from 525,000 people.

At a cost of $640,000 the Federal government had helped clean up less than 2.4% of the pollution in one stretch of the Pearl River.[15]

The situation on the Pearl, documented by the GAO in its 1969 report on the effectiveness of the construction grants program, is not unusual. Seven other examples included in the study showed a similar pattern: pollution gains from FWQA-supported municipal plants were completely canceled out by neighboring (mostly industrial) discharges. The primary lesson FWQA drew from the GAO's shocking findings was that grant spending should proceed in accordance with orderly regional plans to achieve maximum total cleanup on each stretch of water. FWQA published regulations in July 1970 which provide, in effect, that in awarding grants FWQA should consider whether neighboring dischargers will be cleaning up their wastes too.[16] It is difficult to argue with this approach in the abstract. But the Task Force believes that the main moral of the GAO's horror story is a different one.

The construction grants program has suffered less from its own flaws than from a massive failure of state and Federal enforcement agencies to compel cleanup of those polluters—particularly industry—who are not proceeding voluntarily. The situation on the Pearl River is illustrative. Two Federal enforcement conference sessions had been held on the Pearl, the first in 1963 and a follow-up session in 1968. Had Federal enforcement succeeded in forcing Crown Zellerbach and Cosby Chemical to reduce their respective torrents of filth by as little as half, the Pearl would be much brighter today even if Bogalusa and Picayune had never made a move toward cleanup. The conclusion seems inescapable. The Federal municipal subsidy program can accomplish little on its own, no matter how effective it becomes. It needs help—a great deal of help—from enforcement. We can eliminate all the municipal pollution in the country and this still will not prevent our rivers and lakes from getting progressively worse as long as we hesitate to enforce the law against corporate polluters.

18

Janus' Other Face:
The Federal Government
as Polluter

In the face of our environmental crisis, an aroused public's natural inclination is to turn to the Federal government for help. Yet the alarming truth is that the Federal government is one of the worst polluters in the U.S. Like a huge, grimy hand, its environmental exploitation stretches from Alaska, where giant Air Force bases befoul their still relatively pristine surroundings, to West Point, New York, where the U.S. Military Academy dumps more than a million gallons a day of poorly treated effluent into the Hudson River. In 1968 the Department of Defense alone was responsible for more than 335 million gallons of human waste per day, of which 25–30% was given less than secondary treatment. Defense's largest polluter, the Navy, contaminates our harbors and estuaries with waste from more than 700 U.S.-based vessels; only *three* experimental ships have any sewage treatment facilities at all.

Apart from conventional sewage, the Federal government is directly or indirectly responsible for such foul effluents as "red water" (caused by an unusable isomer of TNT), photographic processing wastes, chemical and biological warfare pollutants, aircraft washing wastes (from Air Force wash racks), heavy silting from road construction, high concentrations of pesticides and herbicides, salinity from unwarranted irrigation projects, and deoxygenated water from Federal dams. The government dredges up and relocates 7 million cubic yards of polluted spoil from the Great Lakes each year, contaminating new sections of the water; and it produces radioactive fallout and radioactive wastes from nuclear reactors. In short, the Federal government has not applied to its own operations the standards it demands of private and municipal polluters. The result of this schizophrenic stance was spelled out in a confidential Budget Bureau memorandum to the President in September 1969:

The government's position of leadership is greatly compromised so long as it is itself among the nation's greatest polluters. It is hardly sustainable that the Federal government should be bringing private firms and local governments into court to force compliance with air and water quality standards which the government itself chooses to ignore. This is not simply an abstract issue: local articles on pollution often carry photographs of Federal installations belching black clouds of soot into the air, or polluting the rivers.

No examples of governmental callousness could be more disturbing than the Defense Department's annual attempts to dump nerve gas in the ocean. In a way the Army's most recent scheme to give its leftover gases a burial at sea in August 1970 was one of the least controversial, although it provoked an unprecedented storm of protest. What made the outcome in that case a foregone conclusion was the Army's "almost unbelievable negligence," [1] as the House Subcommittee on Oceanography put it, in encasing a shipment of deadly nervegas rockets in concrete vaults years earlier. Once that had been done, there was no safe way to remove the rockets to deactivate the gas, and "due to the hazard to human life through continued storage, the only alternative," as the Subcommittee reluctantly concluded, was to sink the containers in 16,000 feet of water off the Florida coast—regardless of the damage to aquatic life that would almost surely result.[2]

Far more reprehensible was the Army's attempt to deposit nerve gas and live explosives off the New Jersey coast in spring 1969, since in that case deactivation and land disposal were clearly the less dangerous alternative. On May 8, 1969, Congressional hearings uncovered the Army's plans to dump massive quantities of unused and unneeded poison gas into the Atlantic Ocean 200 miles off Atlantic City. Also included in this 27,000-ton surprise package for unwary seagoers were numerous other chemical warfare agents and live bombs. The chemical agents included 5000 *tons* of mustard gas, the virulent agent that causes skin and eye blisters, shock, bone-marrow damage, ulceration of the gastro-intestinal tract, and possibly cancer or genetic damage.

The most lethal of the chemicals to be dumped, however, was 2000 tons of the nerve gas GB, an acetylcholinesterase inhibitor that destroys communication between nerve cells. Since GB kills in amounts as small as 3/1000 of a gram, the total to be dumped was enough to do in every human being on earth 200 times over.* A public outcry led by Congressman Richard D. McCarthy (D.-N.Y.) spurred hastily convened hearings by the House Subcommittee on International Organizations and Movements.

The Army claimed everything would be under strict control. The gas was to be loaded onto ships, hauled out to sea, and sunk in 7200 feet of water. At that depth, said military spokesmen, the poison would rest below the range of marine life and would decompose long before reaching the surface. The Subcommittee was also assured that at a submerged temperature of 4°C., mustard gas would freeze solid (and thus presumably be slow to diffuse). But then Dr. Stephen L. Teitlebaum of the Committee on Environmental Information, Chemical and Biological Warfare Subcommittee, gave testimony that proved somewhat disconcerting. By the Army's own figures given in testimony, he said, the gas containers would, if airtight, rupture from the pressure at 1000 feet, and if not airtight, would leak. Either way, such a release of gas would cause potentially massive genetic damage to all marine life in the vicinity.[3] Under close questioning, the Army admitted it had no idea *where* the nerve gas containers might burst:

CONGRESSMAN H. R. GROSS [R.-Iowa]: Am I to sit here and believe that you have adopted a method of burial, but you have not sent down a container, a metal container of the

* These figures are theoretical and assume perfect distribution. According to Dr. Stephen L. Teitlebaum of the Committee on Environmental Information, Chemical and Biological Warfare Subcommittee, efficient military delivery of the GB could kill half a billion people. Presumably, accidental distribution of the gas would result in a killing power below this one.

average strength of the containers that carry this material? Am I to believe that you haven't experimented to find out at what depth those containers will rupture?

DR. FROSCHI [Assistant Secretary for Research and Development, Department of the Navy]: The containers then full, there would probably be a great variability in the rupture depth. I think this is fair to say that this is one remaining piece of information that I would like to see developed. (Laughter) [4]

Other pieces of information should also have been "developed." Dr. Teitlebaum noted the need for documentation of just how long it would take GB to decompose in sea water. The Army had made no tests to determine the effects of three earlier ocean dumpings, but it claimed no harm was foreseeable so long as the tanks burst below a depth of 1000 feet. No evidence was presented to support this view. It is difficult for laymen to believe that a huge killer cloud of gas at 1000 feet would have no adverse effect on the myriad forms of ocean life—including shark, barracuda, lantern fish, and many species of herring and salmon—found at that depth, especially when the testimony has not been supported by a single specialist in this field.

It is clear now that the Army did not possess the scientific evidence it should have had before making a decision to dump poison into the ocean. What information the Army did have—supposedly provided by experts on ecology—proved to be grossly insufficient. Among the authorities it claimed to have consulted was the Department of Interior. But in the Subcommittee hearings the performances of Interior and its FWQA could only be described as catatonic. Interior, which was to be the *only* independent source of scientific support for the Army's dumping plan, was represented by Dr. Leslie I. Glasgow, Assistant Secretary for Fish and Wildlife, Parks and Marine Resources. He began by delivering an impassioned speech against ocean dumping, denouncing it as a shortsighted measure that causes great long-term problems:

We are looking to the ocean as a very important source of food in the future. *We are opposed to the use of the ocean as a dumping ground.* [Emphasis added.]

In response to subsequent questioning, however, he *agreed* with the Army's assertion that dumping was the best solution at hand:

CONGRESSMAN CORNELIUS E. GALLAGHER [D.-N.J.]: Sir, do you concur in the Army's evaluation of the extent of contamination which would result in carrying out the current proposal?

MR. GLASGOW: I have been advised by very knowledgeable people that the method of disposal that they propose is probably the safest.

CONGRESSMAN GALLAGHER: I want to ask you later whether or not that is the safest. I am talking now about the ecology of the ocean to which your statement is directed. Do you concur in their evaluation of the extent of contamination in the ocean?

MR. GLASGOW: I think that the contamination under the conditions they propose would be minimal.

CONGRESSMAN GALLAGHER: Therefore, you concur with them?

MR. GLASGOW: *We are opposed to pollution anywhere. This is the posture of the Department of the Interior.* [Emphasis added.]

Still trying to determine the gentleman's position, the puzzled Congressional committee chased up and down:

CONGRESSMAN GALLAGHER: Doctor, you concluded your statement by saying that the Interior Department is opposed to the use of the ocean as a dumping ground. Are you opposed to the Army's proposal on the basis of present evidence submitted?

MR. GLASGOW: *This is a general position we take. We are opposed to pollution whether it is land or ocean. I repeat that this is the best thing that is known for disposal at the present time.*

CONGRESSMAN GALLAGHER: Then this does not violate your general principle of using the ocean as a dumping ground?

MR. GLASGOW: *Again, we are opposed to any form of pollution wherever it might be.* [Emphasis added.] [5]

If that opposition were so categorical, why did Interior condone the dumping of 7000 tons of poisonous gas into the ocean? And who were the "very knowledgeable people" acting as advisors? Roland Smith, of Interior's Bureau of Commerical Fisheries, responded with shocking clarity.

CONGRESSMAN DONALD M. FRASER [D.-Minn.]: What you were given was the conclusions that the Army had reached on these questions, not the basic technical data from which you would make your own judgments. Is that true?

MR. SMITH: That is true.

CONGRESSMAN FRASER: So you were confronted essentially with the same kind of statements that we were given in the testimony yesterday?

MR. SMITH: That is true.

CONGRESSMAN FRASER: You are not then in a position to go behind that and make an independent evaluation; is that true?

MR. SMITH: That is true.

CONGRESSMAN FRASER: Thank you.[6]

FWQA provided the most dramatic revelation. FWQA's Kenneth Biglane, Director of the Technical Support Division, testified that the Department had never even *heard* of the dumpings before May 9, 1969, one day *after* the start of the hearings and after the story appeared in *The New York Times*.

The first day of the hearings had been adjourned "with some disappointment" on the part of the Chairman, Congressman Gallagher, because the Army had requested a recess until May 13 "to get some additional technical support." How literally that phrase can be taken! FWQA obligingly showed up to help, but never bothered to check the Army's assertions:

CONGRESSMAN FRASER: Is it true then up until Friday [May 9] since you had no knowledge of the proposed dumping, that you have not been able to make a technical analysis? Since Friday have you been able to make a technical review of the proposed dump?

MR. BIGLANE: Not of this particular exercise, no sir.[7]

Both Defense and Interior had plainly flouted the general standards of an executive order signed by President Johnson on July 2, 1966, which stated that

(d) No waste shall be discharged into waters if it contains any substances in concentrations which are hazardous to health.

(e) No waste shall be discharged into waters if it contains any substances in concentrations which will result in substantial harm to domestic animals, fish, shellfish, or wildlife, if methods of treatment or disposal are available that will remove or render harmless such pollutants. . . . *A determination that such methods are not available or cannot reasonably be developed will not be made without the concurrence of the Secretary of the Interior.* [Emphasis added.]

Not only had the Secretary not *concurred* in the determination of the safety of the dumping, but his department had not even been asked to participate in the decision. And when Interior and FWQA finally found out what was going on, they did not conduct an independent investigation but parroted the Army's position in an equivocal rescue attempt, and thus prostituted both themselves and their mission.

The nerve gas incident spotlights the inadequacy of the government's approach to its own pollution problems. While the Army eventually decided to deactivate the gas and dispose of it on land,* this reversal came only after news of the original dumping plan sparked a public furor. The government's failure to police its activities in this instance was no accident. It is symptomatic of a general lack of respect for pollution control throughout the government which extends even to the most mundane day-to-day activities—the treatment of ordinary sewage, for instance.

The cleanup program at the U.S. Military Academy at West Point is illustrative. It was revealed at a Federal enforcement conference in 1965 that the latrines of the Long Gray Line were seriously polluting the Hudson River system, contributing about 1.2 million gallons of partially treated effluent every day from the Academy's two major treatment plants. In 1968 the Army received authorization from Congress for a $2,137,000 program to construct new treatment facilities. FWQA certified the project and gave it high priority, and presumably expected construction to begin by late 1967. One can understand the chagrin, then, that two FWQA inspectors expressed when they got back from a routine check of West Point in April 1969: "We expected to find a secondary treatment plant under construction but in-

---

* The Army, incidentally, originally dismissed the idea of "demilitarizing" the gas, claiming it would take nearly six years of dangerous work. At the hearings, however, the chief engineer at Edgewood Arsenal testified one year would be sufficient.

stead found that they had not even put the job out for bid, reports an internal FWQA memo. It turned out that the Army's construction program for 1968 at the Academy had run into severe "cost overruns." Rather than postpone the construction of a new academic building or new dormitories, the Army did the logical thing— given its priorities—and sacrificed the sewage treatment facilities.

The Army's attitude is typical of government agencies. Like private corporations, agencies do not see pollution abatement as part of their job. Each one has its own special interests, its own narrowly defined mission. Just as the search for profit tends to blind industrialists to the environmental "side-effects" of their activities, so the bureaucratic mission tends to blind agency personnel to the environmental consequences of *their* activities. As Chairman Murray Stein observed at the December 1968 Calumet River enforcement conference:

If you have a man in charge of an Air Force base or one of these operations, the first thing he will tell you, he is not in the waste treatment business; he is in the business of the defense of our country. And you wouldn't want another kind of man there.* [8]

Perhaps we wouldn't, but someone has to clean up the mess. The need for waste treatment is a part of life at any human installation, no matter what its primary purpose is. Yet, out of an operating budget that approaches $80 billion, the Defense Department asked for only $19.5 million, and received only $12.3 million (less than one-fourth the cost of a single C-5A aircraft), for pollution control in 1969.†

In the face of a widespread "pollution is not our job" attitude, it has long been apparent that the anti-pollution cause needs a vigorous advocate of its own within the government. The logical choice has always been the

* FWQA officials have found that officials at the Defense Department sometimes refuse to provide information on pollution problems at military installations on the grounds that "national security" would be violated.
† Perhaps the DOD was out of breath in 1969 from its $40.4 million effort the year before (FY 1968). In FY 1970, DOD's pollution control request was $77.9 million, an improvement over 1969 but still a paltry one-tenth of one percent of its total outlays.

Federal government's water pollution agency, and that responsibility was bestowed upon FWQA formally in 1966. Under the terms of Presidential executive orders that have been in effect since then, the agency or department housing the water unit is charged with providing not only "technical advice . . . to the heads of agencies" but "leadership" in combating pollution from Federal government activities. Unfortunately, FWQA has proved unequal to its task.

FWQA has taken its responsibilities so lightly that it has relegated the office with the job of policing Federal installations to its most inferior organizational status. A branch is the lowest generic entity on the bureaucratic totem pole, and FWQA's Federal Activities Branch is appropriately regarded as a second-class unit.* FWQA staffers told the Task Force that the Branch was a "dumping ground for those out of favor."

The Federal Activities Branch has not only been out of favor; it has also been out of touch with both FWQA and Interior, not to mention other agencies. When FWQA received a letter of complaint concerning the Army's dumping of nerve gas, the Branch, in drafting a reply, wrote that "Mr. Hickel [was] very concerned about the problem." The staff member who wrote the letter, when asked his source of information, replied, "I know Mr. Hickel was concerned, because I read it in a press release." Questioned by the Task Force on many occasions, Federal Activities Branch officials repeatedly cited the mass media as their prime reference for the information they provided.

Unable to keep tabs even on its own bureaucratic superiors, this small branch has a larger assignment that is crushing in its futility. There are some 37,000 Federal installations in existence, about 7000 of which are important enough from a pollution standpoint to warrant regular inspection. FWQA field personnel in the Federal Activities Branch make about 550 inspections per year (and submit only about 150 reports to the central office). Theoretically, then, each major facility can be

---

* Prior to their 1968 downgrading, the employees of this office made up the Federal Activities *Division*. As of this writing it had not yet been decided what the Federal Activities unit would be called in the new EPA, but the Task Force was informed that its lowly status was not likely to change.

inspected only once every 13 years. What happens in practice is that some of the more visible installations are inspected regularly, while most are never visited at all. There is no guidance from Branch headquarters on whether or when or how to investigate a particular installation. "That is the result of an engineering decision out in the field," says one headquarters staffer. But the Task Force was assured that "if our people see something in the newspaper—that there is a complaint—then they jump right in." Is the Branch a surveillance unit or a clipping service?

Even when it finally gets the news, FWQA has shown it lacks the backbone a small agency would need to call more powerful Federal agencies to task. In the West Point case, for instance, staffers in FWQA's Federal Activities Branch dashed off an indignant internal memo when they first discovered the treatment project's cancellation. After noting "it is now questionable whether or not the facilities will be in operation by 1972 as required by the second session of the Conference on the Hudson River," the memo concluded:

There is no way that [FWQA] can or should justify this sort of action. The report for the Third Session of the Conference should state the facts and indicate that the switch [of West Point's pollution control funds to other programs] was made without consulting [FWQA] and that [FWQA] would not have concurred with the USMA in this action.

But when the third conference session was called to order, it was obvious that FWQA had capitulated. The Academy presented four pages of testimony justifying its decision. Except for a one-sentence reference to the postponement buried deep in its formal submission for the conference record, FWQA said not a word.

Rather than making the inevitable embarrassing backdown, FWQA's senior officials would evidently have preferred that the Federal Activities Branch shut its eyes to Federal pollution problems altogether, if we are to believe these comments in a 1969 internal report from FWQA's Office of Operations (the Branch's administrative overseer within FWQA):

The [Branch's] major resource-consuming effort appears to be field surveys and follow-up visits to identify pollution problems and induce their correction. *Why should* [FWQA] *be so deeply involved in the affairs of other agencies?* [Emphasis added.]

The scanty information that FWQA has available on Federal polluters reflects this ostrich-like approach to its assignment. The Federal government is capable of generating some rather bizarre pollutants; yet when FWQA does look for pollution, it looks almost exclusively for sewage problems. Accordingly, FWQA has no information available on the extent of pesticide use by Federal agencies, for example. When the Task Force examined FWQA's file on "Hazardous Materials" in late 1969, all it contained was some scattered references to oil pipeline problems. A Task Force query on this subject drew the reply: "Some things would be nice to keep track of, but . . ." An inventory form prepared by FWQA for soliciting information from Federal facilities about their wastes has 90 questions, but only one of them deals with industrial-type wastes. There are no questions relating to hazardous substances, nor any on whether an installation uses advanced treatment methods that might remove them. Because of its lack of information, FWQA is often unable to pinpoint even the most frequently recurring pollution problems. Nearly every time the Task Force inquired into a specific case of pollution by a Federal installation, FWQA was unable to provide adequate information from its headquarters files and had to call a regional office or another agency.* What little data FWQA did have could not be accepted with full confidence, especially after the Task Force was told by a Federal Activities Branch staffer, "I assure you, I wouldn't trust those files."

His warning was soon borne out. In gathering pollution data on Redstone Arsenal in Huntsville, Alabama, for instance, the Task Force got a different estimate of daily discharges from the arsenal every time we checked a different FWQA source. Three different inquiries

* We noticed one exception to the generally incomplete condition of FWQA's information. The files in FWQA's Federal Activities Branch contain a dossier, evidently well filled, on Ralph Nader. Curious about Mr. Nader's contribution to water pollution from Federal facilities, the Task Force requested access to this file. We were informed that it contains mostly internal working memoranda and is, therefore, not for public inspection.

yielded three different quantities, ranging from 500,000 to 5 million gallons per day, not an insignificant margin of error. None of FWQA's figures came anywhere near the Defense Department's own figures: a whopping 1,822,500 gallons per day of sewage and 7 million gallons per day of industrial waste.

At one point a skeptical Task Force member was assured by an FWQA official that he was getting just as much information from the agency as a member of Congress would. This apparently was no exaggeration. The House Subcommittee on Power and Natural Resources, for example, said it had

been much disappointed by the [Federal Activities Branch's] apparent inability and repeated failures to furnish adequate information concerning the status and progress of water pollution control efforts at Federal installations.* [9]

FWQA has been listening to Congressional criticism like this for a long time. It was apparent to the House Subcommittee on Power [now Conservation] and Natural Resources as early as 1964 that Federal pollution control authorities had "not adequately exercised [their] authority to maintain surveillance over the waste disposal practices of Federal agencies" and had "failed to provide leadership." [10] In 1966 FWQA was urged by the House Conservation and Natural Resources Subcommittee to "exercise its powers vigorously, rather than as a passive consultant who responds only when asked to do so." [11] The tone of the Reuss Subcommittee's most recent report, in 1969, was one of despair.[12]

Were the Subcommittee to look again today, it would find no cause for rejoicing. The February 1971 reply of one regional official in the Federal Activities Branch to a Task Force question about the amount of improvement that had taken place in the two years of the Nixon Administration sums up what our observations indicate is still the general situation: "The truthful answer would

* A long-awaited computer inventory system is now in the throes of birth—six years after it was first suggested by the Power and Natural Resources Subcommittee. Undoubtedly the system will speed the flow of data, but without a concurrent improvement in FWQA's ability to gather information, this will be but a means to discover nothing very quickly. As one FWQA staffer noted enthusiastically, "You can feed something you've just read in the newspaper right into the computer."

be that we have not changed our operating procedures at all."

FWQA's perennial listlessness has disappointed just about everybody but the Federal agencies whose polluting activities it has ignored. While the Reuss Subcommittee was castigating FWQA in 1969 for "not performing its mission anywhere near the level that it should" and for providing "inadequate aid and coordination," [13] the Defense Department was telling the Task Force that its relations with FWQA were cordial and satisfactory. "I don't think [FWQA] has ever pressed us on anything," reflected John Heard, the director of construction programs in the office of the Deputy Secretary of Defense for Installations and Logistics. "We have provided more leadership than they have."

Criticism of FWQA's chronic passivity falls short of the mark if it does not take into account the inadequacies of the agency's basic mandate. FWQA's powers are purely advisory, and success depends upon the willing cooperation of the polluting agencies. The notions that lie behind FWQA's meager grant of authority are familiar ones—that conflict should be minimized, that solutions should be coaxed and "worked out in conference" rather than sought out with determination, that problems should be solved through "partnership" and "consensus" rather than through aggressive advocacy and even enforcement when necessary. Accordingly, FWQA is expected to do its job through education and friendly persuasion of its sister agencies. Its failure should not be surprising. For several reasons, education and persuasion alone are no more likely to alter government behavior than they are to produce industrial and municipal reform.

In the first place, agency officials will always be faced with choices to spend money either on pollution control or on something they consider more in keeping with their primary mission. What of the unforeseen pollution situations—like the need to get rid of some nerve gas or some other highly dangerous substance, for instance —that seem to crop up in the Federal government with disturbing regularity? Expenditures for pollution control in such unexpected situations must come out of an agency's regular operating funds.

There is really no guarantee that routine pollution problems will be handled any better. A military base may have a treatment plant; but how do you make the base commander assign his best personnel to operate or maintain it, or spend time inspecting it, or even bother to operate it when all these annoyances benefit only the civilians downstream and at the same time take men away from jobs his superiors consider more important? At the 1968 Calumet River enforcement conference, the Coast Guard station at Indiana Harbor was found to have been operating its sewage treatment plant only part of the time. The rest of the time it kept the plant shut down—saving the operating costs—and dumped its raw sewage into the harbor.[16] Government employees have weathered platitudes and paper policies long enough to know that if exhortation is not accompanied by sanction, it is intended mostly for public and not for internal consumption.

And finally, there are several agencies whose pollution is not merely incidental to their activities, but virtually inevitable. The U.S. Army Corps of Engineers, for example, specializes in huge construction projects intended to alter the environment—building dams, digging canals, straightening out rivers, diking and deepening lakes and harbors, filling marshland for real estate development, and so on—which destroy aquatic life and leave pollution in their wake. The Engineers are not about to be "persuaded" or "educated" out of any part of this billion-dollar-per-year business. Neither are the many Senators and Congressmen sitting on key Public Works and Appropriations committees who count on these "pork barrel" projects to bring new industry, money, and jobs into their districts.

At the base of FWQA's powerlessness are the weak and ambiguous executive orders from the two most recent administrations. President Johnson's 1966 order not only made following FWQA's advice on pollution matters entirely optional, but did not specify who was to review the various agencies' compliance with water pollution control standards. The result was that each

agency made its own determination of what abatement
steps were necessary in its activities.

The Nixon executive order, issued in 1970, goes the
Johnson order one better. Whereas the old order was
ambiguous as to what surveillance and enforcement
powers the FWQA had, the new order clearly denies
FWQA those powers, leaving that organization a bureaucratic eunuch. Nixon's order stipulates that "heads
of agencies" are to "maintain review and surveillance to
ensure that the standards set forth in . . . this order
are met on a continuing basis." Federal agencies have,
in effect, been placed on their honor to do their best
not to pollute. Given the ignorance and hostility of
many agencies toward pollution control, an abatement
program that has no provision for independent surveillance is unrealistic, to say the least.

FWQA can inspect only by requesting an appointment
with the agency whose facilities are to be inspected.
There is typically a lag, up to a month, between the
filing of the request and the tour itself. (According to
FWQA enforcement officials, even private industrial
polluters rarely keep FWQA cooling its heels that long.
Polluting agencies obviously can conduct crash cleanups
in the meantime and temporarily curtail their polluting
activities. Furthermore, the inspecting itself may be
so limited as to yield no useful information. The General Services Administration in Denver took an FWQA
inspector on a grand tour over all the grounds *except*
to the one ditch that was polluting the water with grease
and oil from its motor pool.

Even had FWQA's inspectors somehow stumbled into
that ditch, Presidents Johnson and Nixon have insured
that FWQA could do nothing about it but complain.
For, in the end, to abate water pollution costs money,
and under the terms of the executive orders, FWQA
plays only the most trivial part in determining where
money for pollution abatement goes within the government. The critical voice at every step along the tortuous
path from conception to completion of a waste treatment facility has been given to persons who neither
know very much nor care very much about water quality and how to enhance it. The net result is a budgetary
process that stacks the deck against pollution abatement.

The long journey can begin only if a polluting agency becomes aware that it is polluting and decides to correct its abuses. That agency seeks whatever technical assistance it thinks it needs to come up with an abatement plan and then submits a request for funding (together with a cost estimate) to the Office of Management and Budget (OMB). FWQA is nothing but a passive technical consultant at this stage, and the other agencies' lack of expertise in this area may be reflected in the fact that in the past some 60% of these cost estimates have turned out to be too low.

At this point FWQA enters the picture for what may be the first—and is usually the last—time. OMB sends all the requests it has received over to FWQA, where Federal Activities Branch staffers spend hour upon painstaking hour meticulously assigning rankings to the projects. FWQA then returns the list to OMB, which proceeds to attack it with all the sensitivity of a jackhammer operator putting the finishing touches on a china figurine. OMB makes the decision on both how much to spend and what to spend it on. (In preparing the President's budget, OMB negotiates directly with the various individual agencies concerned and can swap one proposed abatement project for another, regardless of FWQA's suggested priorities.)

Thoroughly mutilated by this point, the budget requests then must run a gantlet of four Congressional committees—one for authorization and then one for appropriation in both the Senate and the House. As a confidential Budget Bureau (now OMB) communiqué to the President noted in September 1969, "It . . . [is] difficult enough to secure necessary appropriations from Congress, especially from the military appropriations committees"; for a number of years, an OMB official told the Task Force, Congress has been appropriating only about two-thirds of the funds the President has requested for abating pollution in Federal facilities. FWQA officials have calculated that only some 75% of the requested funds for fiscal 1971 were granted by Congress during 1970. They told the Task Force they were hopeful the figure might struggle its way up to 85% for fiscal 1972. The random assortment of pollution abatement projects that survive this budget process are a far cry from the orderly list of priority projects so sedulously compiled by Federal Activities Branch staffers many months before.

The public pressure that made pollution a national issue in winter and spring 1970 has also made an instant environmentalist of almost every politician. Accordingly, there has been the predictable flurry of legislative and Executive activity ostensibly aimed at revitalizing the Federal government's effort to control its own pollution. The amount of hustle and bustle is a hopeful sign that something substantive may follow, and there has even been, on occasion,'some real reform. But a number of regressive measures have also slipped in under the banner of progress. And a disappointingly large proportion of the "reforms" are simply reformulations of old policies that are already proven failures.

On January 1, 1970, Congress passed the National Environmental Policy Act establishing the Council on Environmental Quality. Congress failed, however, to give the Council any authority that existing agencies didn't already have. The Council is merely another advisory body, an information center, policy advisor, and technical researcher. The Act's most significant provision was a stipulation that Federal agencies must include "in every recommendation or report on proposals for legislation and other major Federal actions significantly affecting the quality of the human environment" a detailed statement on the possible and expected environmental impact of the proposed action, plus a list of alternatives to that action. In preparing these statements, agencies are to consult those agencies with special expertise in the field. Presumably, this means EPA's Water Quality Office must now be consulted whenever any Federal agency contemplates taking any major action affecting water quality. However, EPA's role remains simply that of technical advisor.

The Environmental Policy Act was followed by President Nixon's executive order. The official hoopla accompanying its announcement gave the impression that bold new measures were being taken to combat pollution caused by Federal agencies. In reality, the order was a cautious blend of forward and backward shuffles; it not only left the fundamental defects in the Federal cleanup plan—lack of sanction and independent surveillance authority and misplaced control over abatement funds —unaltered, but was weaker in certain respects than President Johnson's order, which it replaced.

On the plus side, Nixon's order stipulated that Federally owned vessels, as well as fixed installations, be covered by minimum standards; that water pollution control facilities at Federal installations be operated by trained personnel; that major water resources projects be reviewed by the Administrator of EPA for possible impact on water quality; and that heads of individual agencies no longer be empowered to reallocate funds initially budgeted for pollution control. In addition, the Nixon order has improved on the system for setting waste treatment requirements at Federal facilities. The old order had required simply that Federal installations give their waste "secondary treatment." The new order, on the other hand, mentions no specific standard, but instead requires Federal installations to conform to the Federal Water Pollution Control Act, under which each state sets its own standards subject to the approval of the Administrator of EPA. Although the states have no way to enforce their standards at Federal facilities (because of Constitutional limitations on state power) nor even the right to check and see if they are being met, standards for Federal installations have at least gained a flexibility they did not have under the old order. Over the long run, they will automatically rise with state standards.

The most significant pledge in President Nixon's February 1970 announcement of his new executive order was his promise to set aside $359 million ($110 million for air pollution and $249 million for water) to construct waste treatment facilities at Federal installations. True, a one-shot $249 million expenditure is small potatoes compared to, say, the $1 billion the Federal government now spends each year on sewage treatment grants to municipalities, or to the total Federal budget of $200 billion. And it is clear that unless the kitty grows, it will quickly disappear. The $249 million represents only a "cautious estimate" (since agencies lacking expertise in pollution control typically underestimate the cost of facilities they need) of "already identified" needs (the price tag for projects already proposed to EPA's Water Quality Office). It does not include such new proposals as installing treatment devices on U.S. Navy vessels, for example. Officials from both OMB and the Water Quality Office warned the Task Force that the President will have to "escalate" this amount considerably if all Federal installations are to meet the construction schedule he promulgated. Still, it's a quantum jump from the spending levels of the past (averaging about $43 million annually for air and water together in fiscal 1968–70). If the President spends the money he has pledged (and all indications are that he will), a realistic start will for the first time have been made toward meeting some of the goals in a Presidential order on Federal pollution.

So much for bold new initiatives. The remaining departures President Nixon made from the Johnson policies were designed to bridge the traditional gap between promise and performance in an unusual but admittedly practical way—by substantially cutting back on the promises. In the first place, the Nixon order eliminated a provision that the old executive order had contained directing all Federal agencies to use their "market influence" to reform those private polluters who do the Federal government's polluting on a contractual basis.

Under Section 7 of President Johnson's executive order, agency heads were to scour the ranks of private firms with whom they did business for polluters of various shades. When "significant potential for reduction of water pollution" appeared, they were to develop "appropriate recommendations for accomplishing such reduction," with FWQA lending the usual helping hand on technical matters. Federal agencies were encouraged to forbear from dealing with serious polluters entirely.

Section 7 had problems (e.g., how to determine whether a firm's pollution was serious enough for the order to apply), but they were hardly insurmountable. And the potential of this directive for bringing about nationwide industrial reform was staggering. The Federal government will buy an estimated $55 billion worth of goods and services in fiscal 1971. There is not a single major industrial polluter in the country who does not peddle his wares directly or indirectly to the Federal government. Were Federal agencies to implement fully a don't-buy-from-polluters policy, polluting corporations would not only have to worry about losing their lucrative government contracts. Once the public learned who pollutes more than whom, and who doesn't pollute at all (information that up to now the government has either studiously neglected to gather or kept secret), many private consumers would undoubtedly make the government's preferences their own. For the first time there would be a real incentive, a profit incentive, for industrial polluters to be the first—instead of the last— to clean up.

But even in its lifetime Section 7 was taken no more seriously than most of the Johnson order's other provisions. It was ignored entirely for about a year. Then in May 1967 the Department of Interior seems to have remembered—but only briefly—that it had been assigned the "leadership" role in carrying it out. In that month Assistant Secretary Frank DiLuzio appointed a departmental study team to review the effectiveness and potential of the "market influence" provision. The study group compiled a comprehensive report, and its conclusions were encouraging. The report compared Section 7 to analogous provisions now in government contracts to promote worker safety and equal-opportunity employment. It identified as the major obstacle to Section 7's successful implementation the natural tendency of government agencies to pursue their primary program activities without much attention to pollution provisions in their contracts with private firms. But this problem could be overcome, the report noted, by a strong and effective FWQA review procedure.

Once the DiLuzio report had been completed, the whole idea went back into hibernation. The study sat untended on various desks during the remaining two years of the Johnson Administration and on into the Nixon term. And not a single Federal agency, not even FWQA or the Department of Interior, ever used the authority it had under Section 7. Why not? Mainly because it looked like too much trouble. One FWQA official who had participated in discussions on Section 7 told the Task

Force he was sure the reason the Johnson Administration never pushed it was that they "were afraid of another quagmire like the civil rights issue [the equal-opportunity provisions in government contracts]." The Budget Bureau offered another explanation in a memorandum sent to President Nixon in September 1969 recommending that Section 7 be revived. "It is generally agreed," the memorandum noted, "that the Federal government could use its contract and purchasing power to make substantial contributions to reducing a wide variety of environmental problems." But "there would be strong resistance in the departments to any major change in purchasing or operations procedures, not so much because of extra costs as the inconvenience of changing traditional practices."

The various departments have of course never been reluctant to violate Presidential executive orders on pollution when to follow them would be an "inconvenience." But Federal agencies were free to ignore Section 7 without even having to violate the order. While they had been required in that provision to examine their practices, they were merely *encouraged* to do something about them. In putting together his new order, President Nixon could have given life to Section 7 by making it mandatory. But he was apparently less anxious to confront established bureaucratic purchasing patterns than the Budget Bureau had hoped he would be. The President's new order insured that Section 7 would no longer be ignored by eliminating it entirely.

President Nixon has also scrapped his predecessor's June 1972 deadline for compliance with treatment requirements, moving the target date back six months. By itself, this extension would be hard to quibble with, because neither President Nixon during his first year in office nor President Johnson before him had allocated the funds that would have been necessary to keep to the earlier timetable. The situation had deteriorated so badly by the time President Nixon decided to turn a sympathetic ear to the environment that, even with a crash construction program, it probably would have been physically impossible to have fully operating treatment plants at every Federal installation by the original date. Under the circumstances, it was probably necessary to slide back the deadline in order to inject some realism into a program that Federal officials had long since stopped taking seriously. Unfortunately, however, President Nixon tampered with more than the target date. The new order specifies only that projects be *under way* when zero hour arrives. Given the Budget Bureau's five-year allowance for completion of projects, this means that Federal installations may not be cleaned up until 1977.* Thus it is entirely possible that the Federal government will not have completed construction of the facilities necessary to treat its routine wastes until 11 years after President Johnson first pledged the government to that goal—more time than it took the nation to put a man on the moon.

Were the possibility of further delay in treatment plant construction the most serious defect in the government's cleanup scheme, there would still be ample grounds for optimism. But the absence of surveillance and sanctions in the Nixon order guarantees that pollution will survive the day when the last outstanding Federal installation celebrates the arrival of its waste treatment facility. Moreover, no performance specifications are being set for these new treatment plants at Federal facilities. In this case, the oversight was not entirely the President's. The Nixon executive order directs agency heads to submit "performance specifications" for proposed abatement facilities to the Administrator, and to adjust those specifications, if necessary, to meet EPA's requirements. Water Quality Office officials in the field were at first encouraged by this provision, reading it as a virtual mandate to set strict effluent standards for each facility. Enthusiasm has since worn off somewhat, and for good reason. By March 1971, more than seven months after the August 5, 1970, deadline for all the "performance specifications" to have been submitted, FWQA has only seen a negligible number of them. Some of the Water Quality Office regions—the Boston-based Northeast Region, for example—have not yet received a single specification.

---

* Even the 1977 date is not firm. The Budget Bureau was given unlimited power in the order to extend the compliance date even further if the Bureau's director deems a project "not technically feasible or immediately necessary."

When the Task Force asked the Boston region's Wallace Stickney about this miserable record, he was quick to concede that the Water Quality Office was not blameless. "I guess it's our fault in part because they [the specifications] have not been proposed," Stickney said, "and we haven't gone out and hit them [i.e., the other agencies] on the head about it." But, as usual, the root of the problem lies as much in the executive order as it does in the Water Quality Office's characteristic nonfeasance. Like all the other empty imperatives in the Nixon order, the requirement to submit "performance specifications" for treatment plants contains no sanctions for non-compliance. It looks like Federal facilities are building a bevy of treatment plants that won't treat.

Even should this not prove to be the case, the most basic problem will still not have been met. Human sewage and other routine pollutants are only the most benign of the Federal government's wastes. For those exotic wastes that shrug off regular treatment, shortcut disposal methods will continue to be just as tempting as they have always been to Federal agencies as long as there continues to be little risk that any outsider will be the wiser. No Chief Executive will be able to solve this problem without stepping on bureaucratic toes, something that all executive orders have carefully avoided doing. Each agency still has the exclusive right to scrutinize its own polluting activities and deal with them as it sees fit.

Thus, the most serious flaw in President Nixon's executive orders remains their failure to confront the issue of bureaucratic authority. The President may issue all the directives he wants; he may use the toughest language he can summon; but so long as he refuses to grant EPA, or some other independent body with expertise and interest in pollution control, the authority it needs to see that treatment goals are not only *set* but also *met,* then the directives and the procedures they outline will be essentially irrelevant. The necessary changes are hardly radical. The House Conservation and Natural Resources Subcommittee has been pleading with the Executive branch since 1965 to empower FWQA to maintain active surveillance over Federal installations and to assume the responsibility for coordinating water pollution control plans that OMB now has. But, more important, if either the President or Congress took their pronouncements seriously enough to see to it that anyone would follow them, they could make sure that Federal employees know not only what they are supposed to do but what will happen to them if they don't do it.* Congress could, for example, make Federal employees who knowingly fail to report pollution problems to EPA subject to criminal sanctions—fines and/or jail sentences. The President could forbid the promotion for a specified period of time—one year, for instance—of any individual who violated key provisions of the executive order. Fear that he might receive a black mark on his service record would make any man bucking for General a militant environmentalist.

The glaring inadequacies of existing legislation and executive orders dealing with Federal installations have not only left those installations free to pollute. They have also rendered EPA's Water Quality Office impotent to use even the limited authority it has been given to influence the decisions other agencies make on pollution control. The executive orders sketch out for the Water Quality Office a role as advocate, or ombudsman—calling problems to the attention of the public and the decision-makers, persuading, criticizing, yapping at the heels of more powerful Federal agencies whenever a pollution problem at one of their installations goes unattended. If an ombudsman is to perform his job, however, he needs something the Water Quality Office has never been able to get—information. No legal requirement telling Federal employees to report suspected pollution problems to EPA subjects them to a specified

* The problem with abating pollution in any large governmental unit, particularly the Federal government, is that those who know about a violation often have no control over the funds that can or should be used to take care of it. A mechanism needs to be set up which gets information on a violation quickly to a decision-maker who *does* have control over funds that can be used for pollution control. Thus, one way to create a workable deterrent to pollution by Federal installations would be to set penalties for persons who knowingly fail to report pollution problems to the proper authorities. The Task Force does not wish to imply that it would be impossible for an agency to set up an internal pollution enforcement system. But we do believe that an essential component of such a system would be a requirement that information about pollution control violations be transmitted immediately to pollution control officials *outside* the regular agency command channels.

penalty if they don't. Since the Water Quality Office also lacks the authority to inspect Federal installations without notice and at will, it cannot even get the basic data it needs to track down a violation unless it has the polluter's full cooperation. And because the water pollution agency has no real control in the disbursement of pollution abatement money, other Federal agencies tend to bypass the Water Quality Office and head straight for OMB when they think they have a pollution problem. For many of these agencies, exchanges with the Water Quality Office are no more than courtesy calls. The Water Quality Office is in much the same position with other Federal agencies as it has been with private industrial polluters, and for the same reasons. It has been forced to cultivate the good will of the polluters it hopes to reform in order to preserve what little access and influence it has.

The frustrations of trying to cultivate this good will and at the same time work for pollution abatement became clear to the Task Force when we sought access to files in FWQA's Federal Activities Branch in 1969. We were not allowed to see the files, and after some discussion the reason emerged. A Defense Department official had called the Federal Activities Branch and warned that future cooperation would be "seriously impaired" if the Task Force were allowed to see the files. The Federal Activities Branch staffer told the Task Force he had spent years building a close relationship with the Defense Department so that it trusted him and kept him informed. Anything that jeopardized that relationship, that impaired FWQA's ability "to consult with them and to get our advice bought . . . I will fight against that." And they told you allowing us access to the files would erode that confidence? "That's what they said, yes."

Within the very narrow confines of his particular job, it is difficult to find fault with this staffer's rationale. To cooperate too closely with the Task Force would have impaired his communication with the Department of Defense, a disaster in an organization with as little authority and political support as FWQA. What if a staffer picked up a newspaper and read that the Army was planning to dump tons of surplus defoliants into

Chesapeake Bay? The lines to the Pentagon would be closed. And for the same reasons that FWQA felt constrained to conceal Federal pollution problems from the Task Force, it has shown itself ready to conceal or play down the potential effects of Federal pollution to the public at large, whether from nerve gas or ordinary sewage—a curious role for an ombudsman.

FWQA is placed in the same dilemma every time it discovers Federal pollution. Should it sound the alarm to the public? Complain vigorously to the culprit agency? Take forceful action? If it does, FWQA fears it may never learn about the *next* violation. It is easy enough to quarrel with the way staffers resolve each particular predicament. But until the agency no longer has to rely on amicable relations with other agencies to get its information, FWQA's lack of knowledge, lack of power, and lack of aggressiveness will feed on one another in a trilogy of mutual causality. And whenever Federal agencies are discovered in violation of the standards the Federal government asks other polluters to follow, the government's hypocrisy will be fair game for private and municipal polluters casting about for an excuse to stall.

Innovation and Obfuscation: The Uses and Abuses of Pollution Control Research

What progress has been made toward the eradication of water pollution as a result of FWQA's research effort? Non-scientists often hesitate to ask a question this simple about so complex a subject. Yet it is the most important one that could be asked about FWQA's Research and Development (R&D) program. The Task Force confronted this issue and arrived at the following conclusion: after 10 years, and the expenditure of over $240 million, the research effort at FWQA has failed to provide a single significant technological innovation that has been widely implemented and is now being used to control water pollution.

Congress has little political incentive for scrutinizing the agency's research budget. Large research budgets are at worst relatively painless and at best a political asset. They give government officials a program to point to as proof that they are doing something about a problem like water pollution, without the nasty head-on confrontations that regulatory activities produce. If it produces no concrete results, it eases the pressure on industry and local government to control water pollution, since they can argue that suitable technology doesn't exist. If the research does result in workable technology, then they have gotten the benefits of free research. And as long as research continues, they can argue that it makes better sense to postpone controlling pollution until the "breakthroughs" that are sure to come make abatement less expensive and more effective.

Accordingly, FWQA's research program is left to

spend its funds without meaningful inquiries from Congress or the public. Typically, testimony by research administrators is scientific pablum, specially prepared to satisfy the non-scientists' taste for impressive terminology. The appearance of Dr. Leon Weinberger, then FWQA's Assistant Commissioner for R&D, before the Senate Air and Water Pollution Subcommittee in 1965 exemplifies the technique and its success.[1] Dr. Weinberger dwelt at great length on the problems of water pollution; finally, he provided Senator Muskie, who had received his vague prologue appreciatively, with a description of the agency's research to meet these problems. Eutectic freezing, peroxide oxidation, chemical oxidation, reverse osmosis, electrodialysis, carbon adsorption were flung untranslated at the committee, accompanied by Dr. Weinberger's misleading description of FWQA's role in developing these wondrous processes. To conclude his testimony, Dr. Weinberger produced two sample bottles, one containing tap water, and one of water from sewage treated with one of the magic processes. The bottle containing the tap water was darker; by the logic of television commercials, then, it was dirtier. The ability to produce one small bottle of water clearer than tap water does not justify a research program, since anyone can make distilled water on a small scale. But Senator Muskie could only comment politely, "It is a very interesting before and after picture, isn't it?"[2] The result of this casual surveillance is a self-perpetuating program without adequate controls to insure real results. Control has been exercised by those who fail to understand why research is needed, and funds have been granted by those who fail to comprehend to what end the funds are being used.

FWQA's research budget in recent years has been some $40 million annually, about 50% of the agency's total budget if construction grant appropriations are excluded. The R&D program consists of a large in-house research effort at eight laboratories around the country (six more are planned), many pilot plants (small-scale model facilities where a new technology is tried out), and fewer demonstration plants (full-scale demonstrations that technology really works). In addition, numerous contracts and grants are awarded to industry, municipalities, universities, and private research organizations to study specified subjects and also to run pilot and demonstration projects.

Primary responsibility for directing FWQA's research program lies with Dr. David Stephan, Assistant Commissioner for Research and Development. Though largely unknown to the general public, his decisions have more impact on water pollution control than those of many better-known government officials and members of Congress. A chemical engineer, Stephan has been with the research program since the early 1960's and has headed the program since 1968. He is an excellent scientist. This contention was not disputed by anyone the Task Force talked to during our investigation. Dr. Stephan's technical competence is widely respected by individuals in FWQA and by others associated with water pollution control research.

Stephan's demeanor is one of a capable scientist-administrator, a hard-headed pragmatist. He describes his program as oriented toward solving practical problems of practical significance rather than solving theoretical problems of interest only to scientists. As Dr. Stephan put it, "The FWQA is trying to conduct a payoff program in R&D. A problem-oriented effort is needed rather than research for the sake of research." So the Task Force took Stephan at his word and looked for a "payoff." We didn't find one.

The baseline from which to measure any pollution control payoff is the knowledge that was available when FWQA's research began. The sanitary engineering profession's technological base has historically lagged far behind the discoveries of the natural sciences or the technology of other engineering fields. The profession's psychology has been conservative, and its technology is a grab bag of stopgap measures used to control pollution in the early 20th century.

Traditional municipal sewage treatment systems are known as either primary or secondary treatment. As indicated earlier, primary treatment screens sewage, thereby removing large objects, and then holds the screened sewage in tanks or ponds to allow other gross insoluble pollutants to settle out. Such a system is inefficient, since the materials to be removed, called settleable solids, settle very slowly. Long holding times are thus required, huge holding tanks have to be built, and heavy capital expenditures and large amounts of land are used in the process.

Secondary treatment employs biological processes like trickling filters * or activated sludge.† Both depend on small organisms (bacteria, insect larvae, sludge worms, etc.) to convert organic pollution into energy, carbon dioxide, water, and cell material. The cell material—the digesting animals themselves and their excretions—must then be removed by settling.

Once settled, this leftover "sludge" (which is actually a highly concentrated form of the original wastewater's organic pollutants) must itself be disposed of. It is sometimes dumped at sea, where it kills bottom life and in some cases has eventually commenced landing operations on ocean beaches. Some cities burn it, causing air pollution, and a few have used it for landfill. Milwaukee sells it as a low-grade fertilizer, but it has already become a glut on the market.

Primary treatment removes 30–40% of organic pollution, measured as biochemical oxygen demand (BOD); secondary removes 50–90%. It has been estimated that most secondary treatment plants, which serve about 67% of the people who have sewage treatment,‡ remove about 65% of BOD. The secondary plant at Washington, D.C., averages 50–70% removal. Assuming an average of 60% removal for 67% of the nation's treated sewage, and 35% removal for the 33% given only primary or "intermediate" treatment (between primary and secondary),[3] the overall average is 48–50% removal of organic pollution. This means that every other time you flush

---

* Trickling filters are large beds of some aggregate material such as gravel which can provide a site for biological organisms to grow on. The sewage is sprayed across the top of the filter and trickles through it. The filter is porous so oxygen is also available. The organisms in the filter then use the organic pollutants in the water and the oxygen as a food source for their growth.

† Activated sludge plants, first introduced in 1913, use a different approach to effect the same ends. The "sludge" in such a system is a mass of biological organisms. Air is pumped through this mixture, and the organisms consume both the oxygen and the pollutants. The sludge is then settled out of the water in a settling tank beyond the activated sludge tank.

‡ A third of the population has no sewage treatment at all.

an average toilet in an average city with average sewage treatment, the waste goes straight to the river.

Clearly, there is need for improvement. When the Federal R&D program began in 1960, these primitive techniques—primary and secondary treatment—were the only methods being applied in water pollution control. The problem was not that better treatment methods had not yet been discovered. There was a large backlog of unused sophisticated technology theoretically available for water pollution control. These chemical processes had already been widely used in the solution of other water quality problems—notably, in municipal drinking water purification. In effect, the necessary research and much of the development on these techniques had already been completed. All that was necessary before widespread application to pollution control could take place was a minimum of development—perhaps a few years' work at the outside.

Beyond this, however, new ideas were needed. The old chemical engineering technology promised to be extremely expensive when applied to pollution control. Drastic cost reductions in any of these chemical processes were unlikely, since the technologies were "mature." They had been used for so long (outside the pollution control field) that they were well understood and probably not susceptible to much further change. And there were other problems that the chemical techniques would not solve—where and how to better dispose of the waste sludge left after treatment, for example. As a first step, of course, all the available unused knowledge had to be put to work. But what was really needed was *innovation*.

FWQA's research program was created to fulfill this need. In order to determine whether, and to what extent, the R&D effort has contributed to water cleanup, the Task Force began by asking two questions: (1) what *new* ideas has FWQA brought to practical pollution control? and (2) which, if any, of the control techniques that the research program has been working on have come into *widespread application* toward the solution of the nation's pollution problems? The Task Force expected some positive answers to these questions when we spoke with Dr. Stephan, because four years earlier, in the September 1965 issue of *Civil Engineering,* he had given the following description of the agency's research work:

The water renovation field is still in its infancy, but already a variety of *significant findings* have been made. In fact, *full-scale application* of several *"advanced"* waste treatment processes has now been achieved, full-scale application of other "advanced" processes is planned, and a variety of pilot-plant studies are underway. [Emphasis added.] [4]

But in 1969 all Stephan could point to when the Task Force asked him to identify those "significant findings" were some relatively obvious adaptations of the technology that existed before the research program got under way. And *none* of the treatment processes, old or new, that FWQA has done research on has yet been applied to any significant degree in practical pollution control situations. One of the advances Stephan claimed, for example, was phosphate removal technology. He admitted that its application has hardly been widespread. (In 1969 there was only one operating full-scale "demonstration" plant; * by March 1971 that number had grown to three operating plants using phosphate removal, two of them FWQA "demonstrations." †) The question is, is this technology really "advanced"?

Dr. Stephan's phosphate removal technology is chemical precipitation, a process that converts dissolved materials into insoluble solids which are then removable from water by settling. Precipitation is one of a family of similar techniques, including also coagulation, sedimentation, and flocculation. A book on industrial waste treatment published in 1952 ‡ mentions that a prominent organization specializing in the development of practical and economical methods of waste treatment had a list of precipitants, coagulants, and flocculating agents that they would ordinarily consider as a solution to every waste problem presented to them. Iron chloride, one of the three precipitants typically used in FWQA's phos-

* At Lake Tahoe, California, treating 7.5 million gal./day.
† Another 15 plants are currently in the design or construction stage of installing phosphate removal technology. This number is a significant leap from one or three, but it assumes its proper perspective in comparison with the 12,565 treatment plants in the U.S. as of 1968.
‡ Edmund B. Besselievre, *Industrial Waste Treatment* (New York: McGraw-Hill, 1952).

phate removal, was on that list in 1952. So was sodium aluminate, another of FWQA's precipitating agents. Out of 59 research projects dealing with coagulation and precipitation underwritten by the FWQA as of 1969, six demonstrate the efficacy of sodium aluminate. You cannot call a research program successful because it duplicates the discoveries of 20 years ago. The same can be said for Dr. Stephan's third innovative coagulation process, involving lime. A reading of a book copyrighted in 1954 * would have provided the basic answers to that adventure into advanced waste treatment research.

Dr. Stephan also pointed with pride to FWQA's much-heralded "development" of "tertiary" treatment. Tertiary treatment is the collective name for various processes that may be used to treat wastewater after it has already received primary and secondary treatment. The treatment techniques used in FWQA's tertiary treatment system are lime precipitation (already discussed as a phosphate removal process), rapid sand filtration, ammonia stripping, and carbon adsorption. Has any of these processes come into substantial practical use in pollution control? Does any of them represent real innovation or —as Dr. Stephan put it in his 1965 article—"significant findings"?

Rapid sand filtration is the passing of water through a bed of sand. The technique was patented by John W. Hyatt of Newark, New Jersey. The patent causes no problem, since it was granted in the early 1880's and the rights have expired. This venerable process has been used extensively in municipal drinking water treatment and industrial water treatment since the 1890's. It was being applied in municipal sewage treatment, on a limited basis, before FWQA's research program started.

Ammonia stripping (currently being used full-scale in two FWQA "demonstration" plants) is the special application of a common engineering technique—the use of air to strip a volatile component from a liquid. Any engineering textbook on transport phenomena will

---

* Shepard T. Powell, *Water Conditioning for Industry* (New York: McGraw-Hill, 1954), particularly p. 263. This source discusses the removal of hardness in water by using phosphates to precipitate calcium out of solution. The precipitate is calcium phosphate. The lime process developed by FWQA for phosphate removal does just the reverse.

discuss this technique and provide the theoretical basis for designing a large-scale system.

Carbon adsorption is used for "polishing," or removing residual organic material from treated water. It has been used in laboratories for purifying water since the late 19th century. As early as 1961 this process was hailed in the research program's Summary Report as "probably the most likely contender for a position in the technology of advanced waste treatment." The Summary Report went on to note that carbon adsorption was "a *proven technique* for removing [from] water the same soluble organic contaminants that are refractory toward conventional water and waste treatment. *Right now,* tastes and odors *are being removed* from many [drinking] water supplies by activated carbon" (emphasis added). But after a decade of FWQA research on this already "proven technique," carbon adsorption is currently being employed in only one full-scale treatment system (at Lake Tahoe) and is slated for eventual use in only four others, some of which are as far along as the construction stage.

A number of waste treatment processes that FWQA has been researching are listed in a 1969 report to FWQA Commissioner Dominick concerning the agency's Municipal Treatment Research Program. In addition to the "advanced" techniques already discussed, this report lists three others: electrodialysis, ion exchange, and reverse osmosis.* Nothing new—these

---

* *Electrodialysis* is a separation technique utilizing the properties of semi-permeable membranes and electrolysis. A semi-permeable membrane is a membrane that will allow some but not all substances to pass through it. Electrodialysis is used to separate salts from water. Salts exist in water in the form of positive and negative ions. Membranes that are permeable to only the positive or negative ions are arranged in such a pattern as to effect a separation of the salts and water under the influence of an electric field.

Ion exchange makes use of large organic molecules known as ion exchange resins. These molecules have ionic parts; that is, there is a positive or negative site on the molecule that has an associated ion of opposite charge. The resins are chosen so that the ions in water that are pollutants will have a greater affinity for the charged sites on the ion exchange resin than the ion originally associated with the resin. When water flows past this resin, the polluting ions in the water are exchanged for the ions on the resin.

Reverse osmosis takes advantage of a common phenomenon in reverse. This process again makes use of a semi-permeable membrane. Normal osmosis is the transfer of water through such a membrane from an area that has a high concentration of water to an area that has a low concentration of water. This concentration gradient pro-

processes have appeared in the Advanced Waste Treatment Research Summary Reports since 1961. Ion exchange was applied commercially to water quality improvement problems in 1905. Yet after all this time none of these three well-studied techniques has been used in a full-scale operating municipal waste treatment plant anywhere.

One technique that FWQA has worked on deserves mention as a genuine innovation: biological denitrification. In biological denitrification, a special microorganism in the activated sludge converts the nitrogen (in the form of ammonia) in fecal waste into nitrates. This change reduces pollution because ammonia adds to the BOD load of sewage. Nitrates, however, can stimulate algal growth. In FWQA's denitrification process the nitrates are then separated from these bugs and put in with a second kind of bug with different tastes. In this stage the bugs transform the nitrates into nitrogen gas, which escapes to become part of the atmosphere. What FWQA has added to the water treatment technology that was already available is the idea of *segregating* the two kinds of bugs and letting them do their work in consecutive stages under optimal conditions (temperature, etc.), rather than leaving them jammed together in a single batch of sludge where neither kind could do the job to the best of its ability.

The idea is a good one. However, its applicability may be limited. Phosphorus, not nitrogen, is the key nutrient for algal growth in lakes and streams and the one that needs the greatest immediate reduction in most places. And eliminating ammonia makes only minor inroads on the BOD in municipal sewage. As of March 1971, FWQA's biological denitrification process has not been used in a single full-scale operating waste treatment plant. Only two communities (Tampa, Florida, and Chicago) are making active plans to adopt it in the near future.

In 10 years of FWQA municipal waste treatment re-

---

duces a pressure known as osmotic pressure. In reverse osmosis a physical pressure in excess of osmotic pressure is applied to a solution with a relatively low concentration of water—that is, a solution of polluted water. Water is forced through a semi-permeable membrane by this pressure while the pollutant, usually a salt in water treatment applications, is left behind because it cannot penetrate the membrane.

search, no real progress toward the scientific solution of water pollution has occurred. Very little new information, no breakthroughs, no application of knowledge. When pressed by the Task Force, Dr. Stephan could only reply, "I guess that there have been no widespread applications of technology we have developed. I will have to check with Middleton [Director of FWQA's Taft Research Center in Cincinnati] to confirm this, however."

How could FWQA have devoted 10 years and more than $240 million to water pollution research and added so little to either the knowledge available or the technology in use before the program began? The most obvious answer is simply that no one—in Congress or at the agency itself—has ever seriously demanded anything more of the research effort. The Task Force believes, however, that both the lack of scrutiny which FWQA's researchers have enjoyed and the failure of the program itself derive from a more basic weakness in our nation's approach to pollution: *there is no commitment to apply existing technology* to pollution control. There is, similarly, no impetus to apply the results of research. Under these conditions, even the most *productive* research would be, from a practical standpoint, just what Dr. Stephan told the Task Force he most abhors—research for the sake of research.

It should be emphasized in this regard that FWQA's under-achieving researchers are not incompetent. Virtually all of them are skilled in their specialty, many are both dedicated and genuinely impressive. But research cannot operate in a vacuum. New ideas are harder to come by when there is no strong demand to use them. As long as no one on the outside is in the market for a developed technology, the government gets no automatic signal that it is time to stop developing and move on to another idea. In the absence of this feedback, the temptation is great in a research program to believe that the reason technology is not being used is that it is not sufficiently refined. And if there is a large lag time between development of a technology and enforcement pressures to put it to use, the technology may never be used at all. In short, when the results of research don't really matter very much, chances are good that there *may not be* many results. That is precisely what has happened at FWQA.

The R&D effort started out well enough. FWQA's researchers realized that the inefficient primary and secondary technology in use had to be *completely replaced*. Most of the highly innovative waste removal processes they investigated in those early years worked just fine— in the chemistry laboratory. But it is an elementary tenet in the world of R&D that if the output of a research effort is to be usable, it must remain within, or at least somewhere near, the limits of engineering capability (not to be confused with pure science capability). No one would have seriously suggested that society possessed either the engineering wherewithal to build or the money to buy these products of unconstrained scientific inspiration on a large-scale basis.

Hydration, freezing, electrochemical degradation, solvent extraction, emulsion separation, electrodialysis, reverse osmosis, eutectic freezing, ozone oxidation, corona discharge—all technically elegant but physically impractical processes—were among FWQA's early pursuits. Ten years later only three of them—ozone oxidation, electrodialysis, and reverse osmosis—are still believed to have any potential for future use, and even these processes are nowhere near implementation. As an unusually ludicrous example of FWQA's fascination for the exotic during the R&D program's infancy, consider the 1965 research report, *Waste Disposal on Spacecraft and Its Bearing on Terrestrial Problems*. After discussing two processes, wet combustion and a closed-cycle digestive process like activated sludge, the report suggests "terrestrial" applications, oblivious to the fact that their utility in space adds no new dimensions to their application on earth.*

The Water Quality Act of 1965 dramatically changed the emphasis of the program by providing for greatly increased funds for demonstration. But it was becoming

* Despite a turn in the practical direction in the overall program since that time, FWQA's interest in outer space persisted. In 1969 a company called Arde Incorporated held a $70,000 contract on the "Applicability of ARDOX catalysts to the oxidation of Municipal Sewage Effluents and of Wastes Produced During Manned Space Flight." The catalysts proved to be unsuccessful.

increasingly obvious that FWQA had nothing which it could point to as usable research progress. Faced with the need to demonstrate *something,* FWQA turned to the most expedient alternative. Rather than continue to do the research necessary to *replace* traditional control technology, the R&D program would simply employ the chemical engineering technology of drinking water treatment as a final "tertiary" stage to be *added on* after primary and secondary. Applying sophisticated tertiary treatment to the half-treated effluents from primitive biological plants makes about as much sense as putting a jet afterburner on a Model T Ford. But because the tertiary treatment idea represented an easy route to superficial progress, it was useful in buoying a sinking program.

Since that time FWQA has broadened its research effort to include promising studies on the long- and short-term effects of various pollutants on streams and on aquatic life. FWQA has also begun research on industrial pollution control. (The Task Force believes this policy is largely mistaken, no matter how competently carried out [see below].) But FWQA's long-standing effort to develop municipal control technology continues to consume time and talent while adding very little to the nation's cleanup potential. Since the passage of the 1965 Act, the orientation of the program has shifted away from *research* to *demonstration.* The purported rationale of the new focus is that by building and operating visible hardware which employs tried-and-true pollution control techniques, the R&D program can prove to the most cynical polluter that such techniques exist. Once these techniques have been demonstrated over and over again, and refined to the point of absolute optimal performance, then polluters may begin to use them.

The research program's budget reflects this massive shift in emphasis toward demonstration after 1966.

While FWQA has been demonstrating, however, the polluters have been looking the other way. And neither the Federal government nor local control authorities have tried to make polluters take notice. The Federal enforcement program has only recently begun to recommend high levels of treatment in places that should have had them 10 years earlier. Similarly, FWQA's construction grants office has never made a special effort to promote the use of advanced treatment technology.

FEDERAL WATER POLLUTION CONTROL RESEARCH EXPENDITURES
(in millions of dollars)

	In-house operations	Research grants and contracts	Demonstration projects	Total
Before–1966	16.7[a]	18.3	2.7	37.7
1966–1970	46.6[b]	50.5	111.9	209.0
Total	63.3	68.8	114.6	246.8

[a] This figure includes administration and overhead costs and wages of in-house personnel who both perform research and supervise demonstration projects.
[b] In-house personnel spent a much greater proportion of their time during this period supervising demonstration projects.

But hope springs eternal at FWQA's R&D office, and R&D continues to demonstrate. To the extent that the current focus on demonstration can be justified at all, it is premised on a fundamental misconception of the reason that polluters do not use existing pollution control technology. It is not that polluters—or pollution control authorities—do not know how to control pollution. Nor is it that polluters are waiting only until the available techniques become more refined. It is rather that control officials have not been willing to set high water quality goals and demand that polluters take whatever measures are necessary to meet them.

Given FWQA's current preoccupation with demonstrating old inadequate technology, what chance is there of real pollution control progress in the foreseeable future? The Task Force believes that unless someone supplies the main missing ingredients—high standards for pure water and a willingness to enforce them—there is virtually no chance at all. Should we ever make up our minds to demand clean water, however, a number of good ideas are available that could take us beyond our present primitive state.

What follows should by no means be considered a definitive list of the directions waste treatment practice might profitably take, but rather a sampling of some of the original thinking going on in pollution control. Many of the ideas described below, though not all, have received limited testing in FWQA's research program, buried deep down on the priority list below "tertiary treatment." Additional research emphasis is needed to put some of these ideas to work. Others need only to be picked up and used.

One scheme with high potential is using FWQA's "tertiary" treatment system to treat raw sewage rather than effluent that has already undergone conventional secondary treatment. The city of Rocky River, Ohio, was the first to opt for this plan. Rocky River, an affluent area where real estate prices are high, wanted to save money by building its new treatment plant on the smallest possible site. So it sought FWQA's help in designing a 10-million-gallon-per-day plant that would eliminate the secondary (biological) stage. The Rocky River plant, soon to be constructed, will rely instead on a chemical process (lime precipitation) followed by carbon adsorption (a physical process that takes the place of biological secondary treatment). This process, dubbed "physical-chemical" treatment, not only saves on real estate costs but is also more reliable than secondary treatment because it has no living organisms that can be knocked out of commission by industrial wastes. And whereas it takes around a week to grow a new culture in a secondary plant after shutting the plant down, "physical-chemical" can be shut down and started up again with no lag time.

There is a moral to the Rocky River story. "Physical-chemical" has never been "demonstrated" by FWQA. (Rocky River will get an FWQA research grant and be the first full-scale "demonstration.") Yet "physical-chemical," unlike the much-demonstrated "tertiary" treatment, looks as if it will catch on in several places; FWQA officials told the Task Force that, as of March 1971, some 10 cities have begun planning to adopt it without the lure of an FWQA demonstration grant. Good ideas apparently do not need much "demonstration."

Another idea with obvious potential is improving on the poor performance of existing treatment systems. One reason municipal plant performance is so erratic is that flow rates of city sewage fluctuate from hour to hour. A common suggestion for eliminating this uncertainty and increasing treatment efficiency is to install holding tanks to retain sewage during high-flow periods and release it for treatment during low-flow periods. One high-ranking FWQA scientist explained it to the Task Force this way:

You should get rid of primary treatment, of settling tanks, and convert them into holding tanks. That is what is done in any other process where flow rates are variable. This is not a new concept. If holding tanks were used, a steady flow of water through the treatment plant could be maintained. This would markedly improve performance.

Bob Smith, a researcher at FWQA's Cincinnati lab, told the Task Force that making this modification would improve the average removal efficiency of a typical plant by 20%. Yet despite the potential of this seemingly simple change, FWQA has proceeded no further with it than a single field evaluation study of a recently completed 100,000-gallon-per-day holding tank in Newark, New York.

FWQA has also given short shrift to the likelihood that the secondary treatment process itself could be radically improved by making some fundamental changes. It is widely acknowledged that the only reason biological treatment is not more effective is that oxygen is not transferred in enough quantity to the little creatures that do the work. But out of a total of approximately 260 FWQA research projects on improved municipal waste treatment, only 20 are designed to help the bugs breathe better. Among the 20, however, are some promising prospects. One is a demonstration plant in Batavia, New York, that will substitute liquid oxygen for air in the activated sludge process (suggested by the Linde Division of Union Carbide Company). Three of the 20 projects are investigating the use of a device that the Allis-Chalmers Manufacturing Company suggested called a "rotating biological contactor." The bugs get to ride on the contactor, a drum that revolves like a ferris wheel half submerged in the sewage. Their sagging appetites are stimulated each time they come up for air.

There are also a number of ideas for improving the

"trickling filter" method of biological treatment, some of them quite old. As early as 1948 a sanitary engineering consultant named August Vorndran tried to interest his colleagues in meeting the problem of limited oxygen transfer. In the old-style "trickling filter," the bugs cling to rocks or gravel packed in a large box (perhaps eight feet or so high) and feast on sewage that has been poured over the top and is seeping downward. Vorndran suggested that contact of the sewage with the air would be increased if the gravel were replaced with vertical flat-surfaced planks placed close together and one above another. His ideas were praised in 1956 by M.I.T.'s prominent sanitary engineering professor Ross E. Mc-Kinney.

Dow Chemical has come up with an idea similar to Vorndran's (though independently conceived) that looks even more promising. Instead of making the filter of gravel or thin panels, Dow now manufactures corrugated vinyl "cores," little plastic honeycombs that can be stacked up like building blocks. (B. F. Goodrich and Ethyl Corporation are also now marketing similar plastic filters.) The sewage flow down through an old-style gravel filter typically seeks out certain narrow pathways after a while, giving some bugs more waste than they can handle while those in other parts of the filter go hungry. The plastic blocks are more efficient because the water seeps down through them evenly. And because they are lighter than gravel and don't jam together at the bottom of the box where the pressure is greatest, they can be piled much higher—to 15 or 20 feet. The October 1967 issue of *American City* magazine reported that a plant in Cedar Rapids, Iowa, had converted to the plastic filters and discovered that two plastic filters could do the same work as 14 gravel ones. They were cheaper to operate and cost less to buy initially—$1,846,000 for 14 gravel filters to $1,577,000 for two plastic ones. And that doesn't count the tremendous saving in real estate.[5] Using the new filters is like replacing a community of one-story buildings with a smaller number of two- and three-story buildings. Although Federal research personnel gave Dow a great deal of encouragement in the early development stages (back in the 1950's), including, for several months, the full-time assistance of two government researchers, neither Dow, Ethyl, nor B. F. Goodrich needed an FWQA research grant to come up with these practical improvements in pollution control technology.

The most neglected ideas in pollution control come from ecologists, who view attempts to control pollution that do not conform to natural life cycles as ultimately futile. They have aptly characterized modern sewage treatment, even at its best, as ecologically shortsighted: it merely transforms wastes into other dangerous compounds—the sludge left over after waste removal—and relocates them, causing pollution in a more distant location. Furthermore, taking water from underground supplies, using it, and then dumping it into the streams continually depletes groundwater. Groundwater loss can be even more serious when wastewater is not treated and dumped locally but piped to large distant treatment plants for waste removal. Ecologists have suggested a return to natural treatment cycles, using wastewater to reclaim and fertilize land and using land as a filter to purify the water. In the process, the water would return to the underground stores from which it came.

A good example is the work of Louis Kardos, environmental scientist at Pennsylvania State University's Institute for Research on Land and Water Resources. In the March 1970 issue of *Environment* magazine, Kardos describes experiments that his research group conducted during 1963–68 (with the help of FWQA demonstration grants totaling $168,091 in 1965–67), using the earth as a "living filter." The Penn State team piped the city's treated effluent out to experimental plots nearby and sprayed it over the ground. They found out that the wastewater more than doubled the crop yield, replenished groundwater levels, and became drinkable in the process (by Public Health Service standards). The experiments showed that to dispose of the wastes from a city of 100,000 in this fashion would require only about two square miles of land.

Despite the demonstrated success of the Penn State experiment, an exciting 1968 proposal for a far more sophisticated version in Muskegon County, Michigan, was greeted with a conspicuous lack of interest by pollution control officials. The same narrowness that has held back water pollution control nationwide came out in the Muskegon experience—myopic assessments of treatment costs and limited aspirations for water quality. The revolutionary Muskegon County plan could conceivably spell the beginning of the end for many stopgap treatment methods and add-on devices that are the current stock-in-trade of the sanitary engineering profession. Had it not been energetically pushed by people outside the water pollution fraternity, the Muskegon proposal would just be a good idea that never made it.

The basic elements in the Muskegon County system—lagoon-like treatment "cells" and spray irrigation—are not revolutionary. But they have never before been combined in this country into a system to treat and recycle the waste of an entire county—some 13 communities with a combined population of 170,000, and five industries (paper, chemical, engine manufacturing, and metal casting and plating). If everything goes according to schedule, by 1972 all these wastes will flow through two pipelines to two isolated tracts in the barren eastern part of Muskegon County. On the larger 10,000-acre tract three "treatment cells"—huge earthen embankments covering eight acres each—will receive the wastes piped from communities and industries as far as 15 miles to the west, on the shore of Lake Michigan. Ten big propellers floating on the surface of each enormous vat of wastewater will continually beat oxygen into it, helping colonies of bacteria to decompose organic matter and, if the theories work out, keeping odors to a minimum. After up to three days in the treatment cell, where settleable solids will also settle out, the liquid effluent will be piped to two huge storage basins, each some 18 feet high and covering 850 acres. The effluent will be held in these artificial sewage lakes whenever the ground is frozen. On warm days the effluent will be piped to rotary spray rigs and used to irrigate the soil.

According to Dr. John Sheaffer, a 39-year-old geographer with a light in his eye and a background in natural resources management who is the moving spirit behind the Muskegon project, the effluent from even this intermediate stage in the Muskegon County process will be superior to that of a conventional activated sludge (secondary treatment) plant. And the huge treatment cells and storage basins improve on the normal performance of secondary treatment plants in other ways as well. Their capacity is so immense that they need never bypass raw sewage—as many conventional plants do—when the waste flow is swollen by a heavy storm. And there are so many colonies of helpful bacteria growing in each treatment lagoon that no lethal slug of toxic industrial waste is sufficient to kill them off.

The final stage of the Muskegon County treatment process, the "living filter" of the earth, will outperform any technology now in existence. Experiments have shown that virtually all the phosphates in domestic sewage are removed by the time the water has moved only 12 inches downward through the soil.[6] The "living filter" also intercepts what are probably the most persistent and dangerous pollutants in domestic wastes: viruses. The numerous viruses in raw sewage are the suspected cause of many of the country's most serious diseases. Conventional secondary treatment does nothing to remove them; neither does chlorination or municipal drinking water treatment. But early indications are that the "living filter" may provide 100% removal of viruses. Research has shown that many viruses are transformed into innocuous proteins as they pass through the soil.[7]

The biggest imponderable in the plan is how much nitrate will show up in the final "treated" product. The earlier Kardos experiments at Penn State indicate that the soil-filtered water will be low enough in nitrate to fall well below Public Health Service levels for drinking water. (Nitrate concentrations in "renovated" wastewater in Penn State samples taken six inches below the ground ranged from 0.7 to 3.1 parts per million [ppm], depending on the type of tree grown on the soil.[8] The PHS limit, considered high by many authorities, is 45 ppm.) And there is a strong possibility that water disposal researchers will, in any event, discover workable ways to limit the amount of nitrate in the water before it is sprayed on the land or encourage the growth of "denitrifying" bacteria in the soil. But the Muskegon planners have taken no chances that the groundwater will be contaminated by nitrate or any other harmful chemical. A series of wells and drainage ditches will be dug to lower the natural groundwater level and make it possible for water that has passed through the "living filter" to be diverted back for further treatment if the planned monitoring program turns up any impurities. It will also be possible to route the treated water into depleted surface streams.

If a closed waste treatment system of this sort, using the soil as a natural filter, can achieve such a high level of treatment, why had no one tried to put one together on a large scale before? That was one of the questions Sheaffer puzzled over in 1968 when he was asked by Roderick Dittmer, director of planning for Muskegon County, to suggest a policy for meeting a 1972 state and Federal deadline of an 80% reduction in phosphate discharges into Lake Michigan. Sheaffer came to the conclusion that a system like this had never been set up "because no set of elected officials had ever laid down a sufficiently demanding policy to suggest the notion to public appointees responsible for municipal water and sewage works," as he later wrote in the November 7, 1970, *Saturday Review*. Sheaffer then very logically put the horse before the cart and wrote out a demanding policy. His statement, far more ambitious than any handed down by state or Federal pollution control officials, effectively banned the dumping of most pollutants into lakes and rivers.

County planning director Dittmer endorsed the stringent policy; the Muskegon County Planning Commission followed suit. The first serious opposition came from the state water pollution agency. The trouble with the policy statement, the Michigan State Water Resources Commission noted disapprovingly, was that it was designed to solve pollution problems for the year 2000. It would be expensive enough just to tackle the problems of the next decade.

Whatever the Michigan Commission's real objections to the plan, it seems clear that if the system works as expected, costs are high only from the most narrow short-term perspective. The projected capital costs of the Muskegon County scheme are unquestionably enormous—somewhere around $30 million. Operation and maintenance costs are low, however. If the Muskegon County planners are correct, the average annual cost of the Muskegon plan spread over its 20-year financing period is actually $3 *less* per capita than the cost of meeting the inadequate requirements the Federal government is now giving most polluters in the Great Lakes area—secondary treatment plus 80% phosphate removal.

Apart from long-run cost saving, the Muskegon planning team expects natural recycling to produce economic fringe benefits. After irrigation, it is anticipated, the now barren acreage will support a corn crop that will put the Penn State harvest to shame; the corn is likely to bring in as much as $740,000 in annual profits. As the corn springs up in the east of Muskegon County, the county's three now-putrefied lakes—Lake Muskegon, Mona Lake, and White Lake—will be spared the stinking primary and secondary effluents they now absorb; recreational and boating interests are expected to flock to Muskegon County to ply their trade. And the bumper corn crop could give birth to a variety of local by-product processing industries (e.g., cattle feed, corn oil, etc.). It has been estimated that up to 1200 new jobs may be triggered by the switch to a natural disposal system in Muskegon County.[9]

Likely advantages notwithstanding, however, opposition to the Muskegon project within the state's pollution control establishment—not only the Michigan Water Resources Commission but other state and local bodies as well—persisted. FWQA remained skeptical until Michigan's Republican Congressman Guy Vander Jagt took on the Muskegon proposal as his pet project. After meeting with the Muskegon County planners in late 1969, the Congressman began lobbying in Washington to get Federal funding for the proposal. In January 1970, FWQA approved a $42,250 grant for an engineering feasibility study; it reportedly came as a surprise to some FWQA officials when the report [10] concluded that lagoon treatment and spray irrigation on the scale envisioned in Muskegon County were definitely feasible. But Vander Jagt had still more prodding to do before

FWQA would agree to help fund the actual construction of this "feasible" project.

In the spring of 1970 Congressman Vander Jagt met with Russell Train, Chairman of President Nixon's Council on Environmental Quality, and John Ehrlichman, White House advisor on domestic affairs. He asked both of them to speak to FWQA Commissioner David Dominick about the Muskegon system. They did. At a meeting between Dominick, Vander Jagt, Sheaffer, and Muskegon County representatives in May 1970, the FWQA Commissioner is said to have concluded with some excitement, "My R&D people had told me there were no new breakthroughs, but there are." In September 1970 Secretary of Interior Walter Hickel announced that FWQA had awarded the Muskegon project a $1,083,750 research and development grant and a $981,650 construction grant for the initial year of the seven-year, $30 million project.

Whatever reluctance FWQA may have had about the Muskegon project before Congressman Vander Jagt interceded, state and local opposition proved a tougher nut to crack. The city of Muskegon clung steadfastly to its original plan to upgrade its own conventional treatment plant, until a powerful industrial resident in the county, the S. D. Warren Paper Company (a division of Scott Paper), pressed the city to change its mind. The prospect of an economical once-and-for-all solution to their pollution control needs (not to mention the usual hidden Federal subsidy in joint treatment, which may in Warren Paper's case amount to as much as $8 million) was too good for the paper-makers to pass up. The state agencies, however, remained unimpressed.[11]

Their conservatism in this regard was, if nothing else, predictable. Both the Michigan State Water Resources Commission (which allocates Federal construction grants among competing state projects) and the Michigan Public Health Department (which issues construction permits for sewage treatment facilities) are, like all state agencies, staffed mainly by old-line sanitary engineers; these are men whose natural bias in favor of the old technology is reinforced by their close ties to the engineering consultants who normally supervise sewage treatment construction. Local consultants have a powerful vested interest in perpetuating the only kind of treatment technology they know how to build. The Muskegon project was an affront to the locals on several counts. It was designed by an out-of-state engineer with special qualifications for this kind of work.* But the project's worst sin is that it represents a permanent solution. If it lives up to expectations, projections of an ever expanding market for the old-liners' services will go up in smoke.

In any event, it took Congressman Vander Jagt's intercession and pressure from Michigan Governor William Milliken before the state agencies finally succumbed. The State Water Resources Commission gave its official go-ahead in June 1970. The Public Health Department granted its construction permit in the spring of 1971. If all goes as scheduled, the nation's first large-scale natural waste disposal system will be in action by 1972.

Recent developments now make it more likely that a number of places in the country will be faced with the same decisions Muskegon County faced in its fight for clean water. The Army Corps of Engineers, a long-standing villain to environmentalists, has a bold strategy for salvaging its environmental reputation. The Corps is the government's recognized expert in large-scale regional planning. Having enticed Dr. Sheaffer into taking a high-level Corps position in late 1970, the Engineers have proposed that their planning ability be put to use designing regional land disposal treatment systems like Muskegon's. The Corps scheme is simply to do the planning in various locales around the country, calculating costs and benefits of alternative strategies for waste disposal (both land and water), and then let the cities take it from there, picking the plan they like, if any, and seeing it through under their own arrangements to final completion. An agency of no mean aspirations, the Corps plans to christen its entree into the waste treatment planning business by taking on, for starters, Chicago, Cleveland, Detroit, San Francisco, and Boston.[12] At this date, the only remaining question is

* Bauer Engineering Company of Chicago.

whether Congress will go along by providing the requested funds, some $4 million in new appropriations for fiscal 1972.*

It remains to be seen, of course, whether a natural waste recycling system will, when the detailed figures are in, prove as attractive an alternative for all areas of the country, for all types of waste, and for all population sizes (e.g., big cities), as it now appears to be for Muskegon County's problems. Answers to the unanswered questions will be coming in in the months and years just ahead. One element of the Muskegon County experience, however, unquestionably holds true for the country at large. We will only begin to outgrow our fixation on the halfway treatment technology handed down to us by the pollution control profession when we stop being satisfied with the moderately polluted rivers and lakes which that technology offers and begin to demand truly clean water.

The main political advantage of research is that it excuses inaction. These problems are complicated and need much study, the concerned citizen is told. No solutions have been found for acid mine drainage pollution, although the problem is under study; for agricultural runoff pollution, although the problem is under study; and so on. "Under study" is a finely developed response at FWQA. When confronted with a pollution situation, the standard reflex action is to give out grants and contracts for development and demonstration and then to explain to whoever asks that no more can be done until the studies are completed. FWQA applies the "under study" technique to virtually every pollution problem. But it is most dangerous from a pollution control standpoint when the pollution "under study" is industrial.

FWQA currently gives over half of its demonstration grants for advanced waste treatment technology directly to industry ($10.1 out of $19.1 million in fiscal 1969, compared to a total Federal enforcement budget of $3.8 million). Additional millions are given out in contracts for research on industrial pollution problems. The public gets nothing for these expenditures but continued

---

* The $4 million will be added to $2.7 million in funds that the Corps plans to switch from another project.

industrial inaction and continued industrial pollution. Unlike the cities, industries can do their own research when they have to. According to Dr. Leon Weinberger, head of the research program when it was first organized, the Federal government began subsidizing industrial research precisely because "it was surmised that industry had a greater capability in the research area than [FWQA]." So why not simply enforce the water quality laws and let *industry* pay for whatever innovation it takes for them to comply? On those occasions when the Federal government has overcome its instinctive tendency to hand over a development grant, the results have been encouraging. At the second session of the Animas River enforcement conference in 1959, for example, the conferees recommended that the Vanadium Corporation of America, a uranium refinery in Durango, Colorado, provide adequate treatment for the radioactive tailings it had been dumping into the Animas River, a tributary of the Colorado. Vanadium claimed that there was no technology available for them to meet the recommendations and that if the government persisted in enforcing them, the company would have to shut down. When threatened with the Hearing Board stage of the conference procedure, however, Vanadium developed at a cost of only $50,000 a process change that not only met the requirements but provided another marketable product. The new process refined radium in addition to the uranium that previously had been the only extract from their ore.

Federal and local enforcement requirements have resulted in similar advances in the pulp and paper, sugar beet, and oil industries, to mention just a few. P. Nick Gammelgard, spokesman for the American Petroleum Institute, explained to the Task Force how his industry developed the technique it currently uses for disposing of the salt brine that is pumped from underground with the oil—high-pressured injection back into empty underground wells.

About twelve or so years ago our oil people used to let their brine run right out into the streams. As you know, when they do this, the water isn't much good for anything. The background for the change to underground injection came from state water control districts in Texas where we had most of our big fields. They simply told us, "After a certain date, you're just not going to be able to do it anymore." Once they had imposed some enforceable deadlines, we got busy and did the research necessary to come up with this new technique in time to meet them.*

Pollution control officials sometimes find (and often suspect), upon enforcing against industry, that the research has already been done and the technology is simply not being used until it becomes necessary. If there is no enforcement, one company can hold the answer to a pollution problem in its hip pocket while other companies—and the Federal research program— duplicate the first company's research and development efforts. Only vigorous law enforcement brings the fruits of industrial research out of the pocket and into the plant.

Contrast the enforcement approach with the philosophy of the Federal research effort. When an industry claims there is no technology available for cleanup, it is awarded a development grant. The rationale, which is embodied in the Federal law, is that the government is entitled to demand cleanup only if it can first demonstrate "technological feasibility." Government researchers rarely object to conducting industrial research programs. Because industrial research requires virtually endless funds, it guarantees research officials an ever expanding influence over the government's control effort, whether they seek it or not. As Dr. Stephan pointed out to the Task Force, doing research for industry is more difficult than researching municipal pollution. Municipal pollution is very much the same from one location to another, and research applicable to one city is applicable to the others. But a single demonstration project in a given industry, Stephan explained, might not prove the feasibility of pollution control. Plants in different parts of the country, may be so different that a grant for each plant would be "necessary."

The Federal industrial research program raises the specter of eternal prevarication by every industrial establishment in the country, each arguing that the fine

* The brine injection technique is an improvement on former practices. But it is now recognized as a major cause of groundwater pollution. The oil industry will have to be pushed again.

distinctions and unique aspects of its operation justify yet another demonstration project. It makes government researchers the bureaucratic enemy of government enforcement efforts—because enforcement could destroy the entire *raison d'être* of the lucrative industrial grants scheme. It creates inequities in the private sector—between firms that can make profitable process changes at government expense and those that must develop control techniques without Federal assistance. And it perpetuates the conventional wisdom that abatement is going slowly because not enough public funds are being spent to develop technology. Congress could do pollution control an immense favor by eliminating Section 6(b), which permits research subsidies to industry, from the Pollution Control Act.

If Federal research simply excuses inaction, why conduct it at all—even on municipal control problems? Is there not a danger that municipal pollution, like industrial pollution, will be "under study" in perpetuity? There is. But the Federal government cannot so easily turn its back on the cities as it could on industry if it chose to. Unlike industries, municipal polluters can neither shut down nor move. Nor do their citizens respond to an increase in waste treatment costs the way industries do—by reducing their waste output. The local costs of municipal waste treatment fall heavily on the poor. The rest falls on the Federal government. So there is every reason for Washington to want to ease the burden on the nation's cities and on itself by making pollution control less expensive.

But if the past 10 years of Federal research have taught us anything, it is that action must not be deferred in the hope that lower-cost solutions will be found. Another decade of R&D like the last one will bring us no closer to that mythical "breakthrough" than we are today. Not until there is pressure to *use* pollution control technology and not just *study* it will any technological improvement take place. There is no technological solution for a society that does not use existing technology.

Part Six

REDRESS

20

Conclusions and Recommendations

The major problem in pollution control is the vast economic and political power of large polluters. Water pollution exists, in large part, because polluters have more influence over government than do those they "pollute." As long as this disproportionate influence persists, so will the pollution. It is a mistake to suppose that new laws with higher cleanup requirements and tougher penalties will ultimately succeed in eliminating environmental contamination; unless new laws also tip the scales of influence over government in favor of the public, the requirements they set will be consistently violated and the penalties rarely used.

Just as elementary physics tells us there are three ways to bring a simple lever into balance, there are essentially three ways to alter the existing political imbalance. One can move the fulcrum (i.e., government) or change the weights on either end (i.e., the polluters' or the people's). The Task Force believes that in order to restore the environment to equilibrium, it will be necessary to do all three: (1) structure the laws to make government less susceptible to special interests; (2) place more power in the hands of the people; and (3) strike at the very sources of illegitimate private influence.

As an essential first step toward making the nation's pollution control effort less vulnerable to political sabotage, pollution control officials must be deprived of the discretion to enforce or not, as they choose. Discretion invites pressure from polluters to see that it is exercised in their favor. Federal officials in the field must be charged with a mandatory legal duty to investigate and issue abatement orders immediately upon receiving notice of a violation, to impose civil sanctions, and to seek criminal sanctions from the courts.

Removing governmental discretion would itself provide people with a right they do not now have—the right to go to court to compel pollution control officials to carry out their assigned duties. To make the removal of discretion fully effective, the people should have additional power at their disposal. Public authorities who knowingly acquiesce in pollution should be subject to more than court command to do their job. By violating their public trust they become co-conspirators in environmental crime. Officials should be subject to the same penalties for conscious nonfeasance as the polluters are for pollution.

Aggrieved citizens need more than legal authority to force the government to move against polluters. They need authority to move against polluters themselves. Government's resources are limited. As long as political power gravitates toward private wealth, large polluters will always find ways to make government's resources for pollution control even more limited. The environment is much too precious for its protection to be left in the hands of just a few. The right of citizens to sue polluters directly for compliance with Federal and state requirements is an indispensable element in any realistic environmental reform package.

To help citizens in their individual and collective efforts to secure pollution abatement, the Environmental Protection Agency should be required by law to take an activist public information stance—sending out regular status reports on the compliance of individual polluters with state and Federal cleanup requirements, collecting and publishing a broad range of facts on pollution (e.g., which companies in each industry pollute more and which ones pollute less, where the water is worst, which cities subsidize industrial waste treatment and by how much, etc.),* preparing evidence for

* The Federal water program's Public Information section has for several years published a weekly compendium of press clippings from around the country on water pollution called *Water Pollution News*—for internal agency distribution only. *Water Pollution News* is an invaluable source of up-to-date information on the state of the nation's water and of new ideas in pollution control. In addition to systemat-

use in court and training citizen groups to prepare their own, supplying professional guidance and technical expertise, providing expert witnesses for state and local hearings on pollution as well as for lawsuits against polluters. Congress should consider establishing government-funded environmental assistance offices similar to the community legal aid under the Office of Economic Opportunity. Unless the citizens' right to an environment free of deadly contamination is fully protected, their other legal rights will, in the end, mean very little.

But just as the people should not have to depend on government to take action against polluters, they should not have to rely on the government for technical expertise. If the citizen effort is to be a viable countervailing force against polluters, it will ultimately have to be self-sustaining. The people's cause must have money of its own, not only to support lawsuits and purchase technical assistance comparable to that which corporations can buy, but, more important, to wage the long-run political battle for clean water against the well-financed industrial opponent.

The income that corporations generate while they pollute has, up to now, served only to perpetuate pollution. It has paid for obfuscating propaganda in the media, backed political candidates who represent narrow business interests, and flooded legislatures across the land with lobbyists for environmental destruction. Pollution must henceforth be made to pay for its own defeat. Citizens who report pollution violations to the government or citizens who sue polluters on their own should be compensated for their public service by the same polluters whose illegal discharges make the service necessary. The compensation scheme should provide for automatic fine, in the case of either citizen suit or government abatement order, at least as great as the amount it would have cost a polluter to avoid his violation in the first place and, in any event, a specified minimum. (A requirement like this would place polluters in a dilemma unlike any they have faced before: by complaining about excessive cleanup costs, they

---

tically collecting and disseminating technical and factual information on pollution, EPA's Water Quality Office should make subscriptions to *Water Pollution News* available to the public.

would be naming their own penalty.) A specified proportion of the fine should go to the citizens whose efforts brought it about. What is left over should be placed in a trust fund to assist citizen environmental efforts all over the country. In other words, polluters should be made to finance the citizen fight against pollution. Polluters' dollars should pay not only for the expense to citizens of individual legal actions against polluters but also for the full cost to the people of matching industry's force on every front with equivalent force of their own—combating corporate propaganda, backing environmental candidates, and sending professional people's lobbyists to state legislatures and to Washington.

Finally, we must face up to the fact that we will not be able to restore our environment unless we also alter the economic arrangements that underlie private industry's political power. Just as some have sought to justify war, hot or cold, as a permanent national way of life by reasoning that the country could not afford the economic disruption that would inevitably accompany the transition to a peacetime economy, industry has always justified its war on downstream water users on similar economic grounds. Pollution, the business ethic tells us, is a necessary evil, the price we must pay for our television sets, our automobiles, and all the other material blessings that spell progress. Those who advocate making the transition to a pollution-free economy have been cast by industrial spokesmen as irresponsible "ecomaniacs," wild men who would sharply curtail production and plunge the country into widespread depression just to save a few fish. The nation can have either employment or its environment, we are told, but it cannot have both. This is an unacceptable choice. Both material security and a healthy environment are essential, and they are ours to claim provided we recognize what has kept them from us up to now.

Industry's control over the economic well-being of its employees is its ultimate weapon against the people and their environment. Both Federal and state control officials, in speaking to the Task Force, excused their failure to press polluters for more rapid cleanup on the grounds that people might conceivably be thrown out of work if they did. The courts have been equally mesmerized by the smell of the payroll; under the common law, the employment that industries provide has traditionally been regarded as a favor not quickly to be forgotten when cases are brought against polluters. Most modern-day pollution control statutes share the common law's bias against the environment, typically instructing the court to consider the "economic feasibility" of abatement. Congressmen rationalize their financial dependence on corporate constituents as being a legitimate part of the close relationships they form while seeking jobs for the home district; the legislatures are, predictably, permeated with the business perspective. And in industrial centers and factory towns across the country, fear for their livelihood turns workers into industry's allies against the environment. The power of the payroll has so weakened the nation's will to control pollution that companies have not, until very recently, had to use it overtly in confrontations with control authorities. It has been sufficient to hold it in reserve.

That no longer is the case. As public pressure for a decent environment mounts, undisguised corporate threats to the nation's economic security promise to become a regular occurrence. Since mid-1970 a number of corporations, most notably Union Carbide and U.S. Steel, have conjured up the ugly specter of massive unemployment in a last-ditch effort to stave off environmental regulation. Environmentalists and pollution control officials across the country have begun to discern what is shaping up as a nationwide country-club conspiracy to frighten blue-collar workers into open hostility toward the cause of clean air and clean water.

The lesson is clear. We cannot clean up the water by simply tinkering with one part of our complex economic and political system—the part that now allows companies to dump their wastes in the rivers. We cannot have business as usual, only stop polluting. If we are going to protect our environment from abuse by uncaring corporations, then we must also protect our working population from abuse by an uncaring corporate system. Industry must no longer be permitted to starve the public into silent submission.

In this regard, three problems require immediate

solution. First, stiffened control requirements will almost certainly be the straw that breaks the camel's back in that small percentage of plants so marginal that they are now profiting only from their pollution. Second, the environment promises to become, if it has not already, a convenient scapegoat for many of the some 10,000 businesses that go under each year at the hands of the marketplace, pollution regulation or no. Taking a cheap shot at the conservationists in passing is a good way for a firm to conceal the real reason for shutdown and at the same time feed a potential environmental backlash. A third and even more disturbing possibility as abatement requirements are toughened is that large corporations waging psychological warfare with cleanup authorities may find it expedient to make their employees hostages for environmental blackmail, falsely claiming that layoffs will be "necessary" if the authorities do not back down.

It should be noted that economic upheaval, real or imaginary, due to pollution control will cease to be a problem once stringent discharge requirements are in force everywhere in the country. New industrial plants will presumably come equipped with whatever it takes to comply with the law. Likewise, it is important to recognize that warnings of a national economic downturn as a result of stricter controls on discharge obfuscate the real issue at hand. The costs of effective pollution control in most industries would barely show up on the budget sheet, and even if they did, no company could claim a competitive advantage by avoiding them. Cleaning up pollution should, overall, create *more* jobs because there is more work to be done. The real problem for the immediate future is not depression or recession but rather *dislocation*. Some places will get more than their fair share of layoffs; other places will get more than their fair share of new jobs. And for each layoff that pollution control requirements actually make "necessary," the environmental movement can expect to be falsely accused for many times that number.

To nip the environmental backlash in the bud, Congress should first of all make it possible to expose publicly the real reasons for job reductions by establishing broad disclosure requirements concerning companies' wage and hiring practices. Power to enforce the requirements should be given to an environmental labor board modeled after the National Labor Relations Board and perhaps located within EPA or, alternatively, within the Department of Labor. Just as the National Labor Relations Act (otherwise known as the Wagner Act) makes it an "unfair labor practice" to use dismissal or the threat of dismissal to discourage union activity, so should it be an "unfair environmental practice" to dismiss or threaten to dismiss workers as a result of pollution control requirements. To implement the scheme, the environmental board should be given authority to subpoena company financial records, analyze them, and publicize both the data and its conclusions whenever a company suggests by word or deed to its employees, its stockholders, or to any member of the public that layoffs have been necessitated or will be necessitated by cleanup orders. The threat of exposure alone should help to discourage firms from making fraudulent claims about the importance of abatement costs in their employment decisions.

Public review of corporate financial data will not, however, buy bread or pay rent for workers who have lost their jobs. Those workers should not be made to suffer for their employers' shortsightedness while the corporate owners and executives responsible pull up stakes without ever having spent a cent on pollution control and proceed to the next profitable venture. Industries that have been permitted to pollute have been receiving a public subsidy. They should now be made to pay it back. Congress should require companies that reduce their work forces as a result of environmental cost pressures to continue to pay the wages of those employees who lose their jobs for some specified period of time—say, six months—after their dismissal. If a firm folds up entirely, payment for the workers should come out of the company's assets before they are distributed. Where insufficient assets remain to give the workers their due, the Federal government should make up the rest, perhaps out of a special corporate tax surcharge to be assessed nationwide. (A tiny fraction of 1% of corporate sales revenues would be more than enough.) Should a company's workers find other employment during the compensation period, the amount earned could be set against the company's obligation. This would induce corporations to become, in effect, employment brokers for their former employees. A plan like this would have two constructive effects beyond easing the burden on the environmentally unemployed. It would make companies think much harder before dismissing workers in response to pollution regulation. And it would free the nation's working population to come out of industry's corner and into the environment's.

The government should also have the alternative, when the public interest requires it, of taking over and operating a closing plant for the benefit of its workers and the public. Having run up a huge debt on the public with its pollution, the firm should be declared in a state of environmental bankruptcy. A public receiver should be appointed and the firm's profits used to pay off its debt—i.e., to install pollution control equipment and operate it for some specified period of time. Another alternative that should be available is providing long-term government loans (perhaps through the Small Business Administration) to companies willing to keep operating but unable to raise the capital necessary to finance pollution control equipment.

Finally, it should be an "unfair environmental practice" for a company to take economic reprisal against an employee because he reports his employer's pollution violation to the public authorities. Congress should insure that workers who "blow the whistle" on their employers' environmental crime will be fully protected.*

Apart from having these remedies available to use on a case-by-case basis, the Federal government should also be actively planning on a broad scale for the clean economy of the future. We could, if we tried, predict well in advance what industries will have to cut down their operations, where, approximately when, and what the effect will be on the various locales around the country. A good indication of how serious we have been up to now about restoring a healthy environment is the fact that no one has even made a serious attempt to dis-

---

* A bill (S. 523), introduced in February 1971 by Senator Muskie, would provide this kind of protection.

cover—let alone to deal with—the economic consequences everyone professes to be so worried about. A
high-level commission should be established by the Administration to begin planning for whatever change will
accompany environmental improvement. (Senator Muskie's Air and Water Pollution Subcommittee might well
commence the dialogue by holding hearings on the
topic.) The study should proceed industry by industry,
location by location, to plot the way to a pollution-free
future. Will the people have to move to find non-polluting industries or should non-polluting industries be
going to the places where the old polluting industries
are on their way out? Should the government be retrain-
ing workers for jobs in clean industries? Can the government predict and publish in advance where and when
labor pools will be available? It is difficult to overstate
the importance of reliable economic data. If we are to
avoid problems, we must first understand them. We
can no longer afford to play blind man's bluff with the
fate of our workforce and our natural resources.

Summarized below are the Task Force's conclusions
and policy recommendations in the key areas we investigated. (Most of our more specific recommendations
are made—or are implied in the shortcomings we documented—in the chapters dealing with each topic.)

Standard-setting and Enforcement

What little there has been of the Federal enforcement
effort has produced more tangible motion in the right
direction than any other Federal program dealing with
water pollution. (See Chapter 10.) It has not, however,
cleaned up the country's water. The irony of the government's position is to be found in the following fact:
Federal enforcement is needed because the states are
often too weak to stand up to their large polluters; yet
Washington's pollution control laws, particularly the
ones enacted in the last 15 years, are far weaker than
the laws that many of the states have to work with.
Metaphysical jurisdictional restrictions (Chapters 11
and 14), lack of information-gathering authority (Chapter 12), mandatory waiting periods (Chapter 13), and
debilitating instructions to the court (e.g., consider the

"economic feasibility" of abatement) have so hamstrung abatement efforts under the Federal Water Pollution Control Act that they have merely slowed the pace of continuing deterioration of our lakes and rivers. The Refuse Act of 1899 (Chapter 15) is the best law the government has to work with against industrial polluters. (It is not applicable to municipal sewage.) It has been given only minimal use so far, however, because, like the later laws, it gives Federal officials excessive discretion not to enforce it. (Any discretion at all has proved to be excessive.)

Some 70 years late, pollution control should now be brought into the 20th century. As a starter, Congress should not only eliminate the discretion in the Federal Water Pollution Control Act but eliminate its other obvious defects as well by incorporating into the newer regulatory scheme all the powers the Federal government now has with respect to industry under the Refuse Act. Until this is done, President Nixon should transfer, by executive order, the Corps of Engineers' permit-granting authority under the Refuse Act to the Environmental Protection Agency to eliminate the confusion and possible serious legal snarls inherent in the scheme the Administration has announced it plans to follow. (As it now stands, EPA will approve the polluter's discharge limits and the Corps will perform the mechanical task of handing out the licenses.*) The President should also reconsider the announced Administration policy, under its forthcoming permit system, of acquiescing to moribund state pollution control standards on intrastate waters. Unless the Administration changes its mind on this crucial point, the Refuse Act permit system will provide industries on the majority of the nation's waterways with Federal "licenses to pollute." (See Chapter 15.)

Beyond that, the law should no longer require the government to go to court to seek a civil fine against polluters. The Administrator should be given the authority—indeed, the duty—to assess fines himself. The

---

* For a discussion of the legal problems involved, see the testimony of William H. Rodgers, Jr., associate professor of law at the University of Washington, at the hearings of the Senate Commerce Committee's Subcommittee on Energy, Natural Resources and the Environment on February 19, 1971.

burden of proving pollution should be shifted so that the government does not have to prove each separate day of a continuing violation in order to award a penalty for each day. Once an illegal discharge has been shown, the polluter should have the burden of demonstrating to the Administrator's satisfaction that he has come back into compliance. Until he does, the penalty fines would continue to be added on automatically each day.

In addition to penalty fines against polluting companies and municipalities, the government should be required to bring sanctions to bear against individuals knowingly responsible for violations. Sanctions which attach to the individual are especially needed because of the oligopolistic (i.e., dominated by a few large noncompetitive firms) structure of most heavily polluting industries (e.g., steel, oil, etc.). When a company is not in active price competition with other companies in the industry, it can much more easily pass on a fine to the consumer in the form of higher prices rather than absorbing it as a loss out of profits. Under such circumstances, companies may even use pollution penalties as an excuse to raise prices by more than the added cost, thereby actually increasing profits. This is because noncompetitive firms know that they can often count on other firms in the industry to match a price increase by one firm with price increases of their own. The best way to guard against this phenomenon is to impose penalties that cannot be passed on. Company executives who are guilty of repeated violations of pollution laws should be barred from working in the same capacity for any company and in the same industry for, say, three years. (There is precedent for such a penalty in several laws, including Landrum-Griffin, which mandates removal from union office of labor officials found guilty of certain corrupt practices.) We should no more allow pollution-prone corporate officials to remain in office and endanger the public than we would permit a repeated and reckless motor vehicle law violator to retain his driver's license or a union official to continue to damage his union members.

As for the intricate water quality standards approach to setting effluent requirements (Chapter 14), it should

ultimately be made irrelevant (except as a check on our
ability to prevent secret dumping and control runoff
pollution) by instituting a "no dumping" policy everywhere. Land disposal of wastes and natural recycling
based on ecological principles is already a feasible
alternative for domestic wastes in many, if not all,
locations, and for many, if not all, industries. (See
Chapter 19.) As a short-term interim measure, the Task
Force believes that the water quality standards approach
to regulation can be made workable provided that unanswered scientific questions are resolved in favor of
the public rather than in favor of the polluters. The
Pollution Control Act should state clearly that polluters
wishing to deposit a given material in the water have
the burden of convincing the Administrator of EPA
that their discharge will not damage present or possible
future desired water uses or degrade the quality of the
receiving waterbody. The Administrator of EPA should
be empowered to designate certain waterbodies as de-
serving special protection—water of exceptionally high
quality, for example, like Lake Superior or Lake Tahoe,
and water whose degradation may become irreversible
if it proceeds any further, like Lake Michigan. No
dumping whatsover should be permitted in these waters
as an immediate policy, and the government should
move toward implementing the zero discharge requirements on as rapid a timetable as is reasonably possible.

Whether or not Congress grants the Administrator
this authority under a strengthened Federal Water
Pollution Control Act, he has a positive obligation to
go as far as he can go toward protecting the nation's
most sensitive and critical waterbodies on his own. He
should recommend to the Corps of Engineers that no
discharge permits under the Refuse Act be granted to
industries located on these waters. Administrator Ruckelshaus should then take the unprecedented step of asking
the Justice Department that the Refuse Act be regularly enforced.

As a supplement to, but not a substitute for, the water
quality standards regulatory scheme (i.e., setting discharge limits or prohibitions and enforcing them), the
government's abatement plan should also include a regu-

lar monetary fee, which polluters would have to pay. The fee for each polluter should be keyed directly to the amount of waste he discharges. Such a charge scheme would bring about a more efficient allocation of pollution control burdens, thus making the same amount of cleanup cost a great deal less. And it would reduce polluters' demands to be permitted to discharge large quantities of waste, and thus reduce the pressure on public agencies to permit more pollution and to extend cleanup deadlines.

The best charge mechanism, the Task Force believes, would not be the oft-proposed flat national effluent tax, because it would (1) not take account of widely differing local water conditions and needs, and/or (2) involve heavy administrative expense, since public agencies would have to continue to compute and adjust the charge, using complex and unreliable economic data, as new or different industries came into each area.[1] A better plan would be to settle on a maximum total pollution load—a tiny one—which could be permitted on each portion of each stream consistent with preserving water quality and then to place that load, in effect, in a regulated market, renting it out in short-term permits to the highest-bidding polluters. (Special provisions could insure against monopolization of the entire permitted amount as a predatory business practice, and other provisions could be made to reduce the costs for municipal polluters.) Polluters would be allowed to sublet their permits to others, under carefully specified conditions. Such a scheme would be a powerful stimulus to technological progress. As industry expanded in a given area, the demand for the available small supply of permits would increase and the price would rise. It would be to each firm's advantage to find ways to treat its wastes more cheaply to avoid the increasing cost of polluting.[2]

Charge schemes, like those discussed above, have often been opposed by environmentalists on the grounds that it is wrong to allow polluters to discharge their wastes into the water if only they pay for it. Those criticisms miss the point. We should proceed as rapidly as we reasonably can to the day when there will be no dumping—when all wastes will be recycled and reused. But as long as we *do* permit polluters to dump anything at all into our public waters, as we do now under the water quality standards approach, we should make them pay for it.

It is important to recognize that, under the water quality standards scheme, the key to improving waters which are now polluted lies in upgrading assigned water uses and setting high criteria for water quality. As it now stands, citizen involvement in the process of setting standards is confined to making presentations at public hearings, on the basis of which a state or Federal administrative body assigns water uses and sets criteria. The water belongs to the people, and the people should have the final say. Citizens should have a right to intervene in the standard-setting process on the side of the environment. Standards should not only be subject to periodic mandatory review for possible upgrading. In addition, people living within specified distances of given stretches of water should have a statutory right to petition and vote for cleaner water when they feel the administrative process has sold them down the river.

As a final note on Federal water quality standards, the Task Force would like to remind Mr. Ruckelshaus and Commissioner Dominick that 22 states still do not have their standards fully approved more than 3½ years after the June 1967 deadline for their final submission, and point out that the Pollution Control Act contains the authority needed to get these disputes resolved: the power to hold Federal standard-setting conferences and promulgate Federal standards. Administrator Ruckelshaus should call standard-setting conferences for all the outstanding states without further delay. If standard-setting proceeds at its present rate after the standards jurisdiction is expanded to cover all navigable waters, we may find the Federal government still arguing with the states over their unapproved standards far into the 1980's or beyond. Congress should save EPA's Water Quality Office from itself by making it mandatory that Federal standard-setting conferences be called in all states that have not submitted approvable standards by six months after the final submission date.

Construction Grants

The Federal government is guilty of colossal incompetence and dereliction of duty to the American taxpayer in the management of billions of dollars of subsidies to municipalities for waste treatment facility construction. EPA's Water Quality Office could be using its grants as a lever to obtain proper operation and maintenance of treatment plants and to insure that the cities don't subsidize industrial waste treatment costs and thereby encourage increased industrial waste production. But it does not. Instead of using its subsidies to reduce pollution, the Federal government's policy has been to *promote increased pollution*. Not content merely to acquiesce in local subsidies to industry, EPA's Water Quality Office kicks in roughly half a billion dollars each year to subsidize industrial waste treatment costs—and industrial waste production—directly out of the Federal Treasury. (See Chapter 17.)

EPA Administrator William Ruckelshaus should:

1. Publish a regulation immediately requiring industries discharging into municipal plants built with Federal funds to pay user charges that reflect the *full* cost of treating their wastes, including that part of the construction costs furnished by the Federal government.*

2. Insure proper operation and maintenance at the local level by requiring grant recipients to post performance bonds or make similar enforceable agreements prior to their receiving Federal subsidies. Under these agreements, penalty fees would be forfeited to the Federal government whenever the city did not meet performance requirements. A variation which should be explored is requiring recipients to agree that if their plants are not properly operated and maintained, state or Federal government may supply a trained operator for the plant and bill the city for his salary.

Congress, for its part, should expressly forbid the

---

* It should be noted that the water pollution control legislation which the Nixon Administration proposed in 1971 would eliminate the Federal subsidy to industry in essentially this way. It would not, however, require user charges to cover local operation expenses. We believe that, whether or not user charges become explicitly mandatory under the statute, the Administrator of EPA has a positive obligation to administer the grant fund in the public interest by taking this step on his own.

Water Quality Office's continued violation of its public
trust by *requiring* that EPA take the foregoing measures. Unless Congress or EPA can muster up the courage to cut industry off, the industrial subsidies will
continue to promote inflated Federal budgets and inflated industrial pollution at the same time. Congress
could also profitably boost the Federal share of construction costs to municipalities from its present maximum of 55% to a much higher figure, perhaps 80–85%.
This change would make pollution control expenses fall
less heavily on the poor (because the Federal income
tax is less regressive than most local funding schemes),
who now bear a disproportionate share of the financial
burden for water cleanup. It would also reduce pressures on the municipal bond market and give greater
legitimacy to stiff Federal requirements for treatment
plant operation. (Leaving a small percentage to be
funded locally would maintain some incentive for local
authorities to procure the best buy for their money—
and the Federal government's money.) Finally, Congress should make subsidies available for some of the
costs of *operation* as well as the cost of construction,
thereby encouraging municipalities not to avoid needed
treatment (phosphate removal, for example) because
its operating cost is high (in the case of phosphate removal, the cost of purchasing chemicals) and its capital
cost (which the government now subsidizes) low.

Pollution from Federal Facilities

Pollution is good for business in Federal agencies just
as it is in private industry. (See Chapter 18.) Pollution from Federal facilities will not be controlled until
Federal employees are subject to laws with teeth and
not merely the eloquent exhortations they have been
subjected to thus far. Both Congress and the Administration could supply effective sanctions for non-compliance; they both should. In addition, citizens should be
granted a right to sue the Federal government for pollution violations and obtain a specified fine. This arrangement would bring all the power of the Office of
Management and Budget down on the offending agency
for having caused an unnecessary Federal outlay. The
Administration should bring EPA's Water Quality Of-

fice out of isolation by granting the pollution control unit the broad information-gathering power it needs to carry out its assignment and a central role in allocating pollution control funds to Federal agencies. To short-circuit the cumbersome budgetary process, EPA should be charged with administering a central omnibus fund for handling unforeseen pollution problems at Federal installations. Finally, the Federal government should stop paying for pollution in its role as a consumer. A Federal agency is not really buying from the lowest bidder when it buys from a firm that pollutes. The cost to the nation of each private company's pollution should be added onto the asking price whenever the government shops around for the best bargain. President Nixon should dust off the directive to that effect which he excised from the old Johnson Administration executive order and begin implementing it immediately.

Research

The best government research program is a strong enforcement program to generate a demand for practical new pollution control ideas. (See Chapter 19.) We cannot afford to stand still waiting for the ultimate solutions to descend upon us—or for the Water Quality Office's R&D program to discover them. The commitment to control pollution must come first; only then will the solutions follow.

Federal spending on *demonstration* projects (as opposed to research and development) should be sharply curtailed; control techniques should be *applied* to practical pollution control once they have been developed, not simply demonstrated *ad infinitum*. It makes even less sense for the government to lavish large sums of money to make a select few private and municipal polluters (chosen on what basis?) models for the latest in control technology when so many members of the Federal family are still sporting pollution control hand-me-downs (or, dare we say it, running in the raw). The Water Quality Office's demonstration effort should be focused on Federal facilities that pollute, a powerful way to shift Washington's traditional "do as I say, not as I do" stance to one of innovative leadership.

The government's research on industrial pollution

control techniques should henceforth be confined by Congress to studying methods for treating municipal and industrial waste jointly in municipal plants. The Task Force would recommend that there be only one very limited exception to this general rule: when the Administration makes an affirmative public finding that to enforce discharge limits for a given industry without regard to a lack of availability of effective control technology would result in nationwide (as opposed to local) curtailment of production so serious that national security or the country's overall economic well-being would be damaged. In that case, the Federal government might give a massive assist in research to that particular industry to find new control techniques at the same time it searches for non-polluting alternative ways to meet the national economic or security threat. It should be mandatory that such a finding be periodically reviewed for its continued applicability. With that single exception, the Federal government should let industry do its own research.

It is important to note, however, that it is appropriate to curtail research on industrial pollution control only in conjunction with strong laws which do not require the government to prove that abatement is "practicable" or "technologically feasible." Otherwise, a Federal pullback in the research area would simply award industry a monopoly position from which it could claim, "We can't do it," and thereby hold up progress. It is time to rid ourselves of the notion that the public must show industrial polluters how to clean up before making them do it. The price of making a private profit must include the full cost of controlling—and the cost of figuring out how to control—pollution. Private companies come and go, but we have only one environment to lose.

Non-point Source Pollution

We may eventually succeed in limiting municipal and industrial pollution and still lose our lakes and rivers to the polluters no one remembered to control—farmers, ranchers, the mining and real estate industries, and other non-point sources of pollution. (See Chapter V.) The Water Quality Office should make a start by setting water quality standards implementation schedules for non-point polluters wherever possible. Congress should require that minimum soil erosion standards be met on all cleared farm and forest land subject to Federal jurisdiction and on construction sites subject to Federal control (those on which Federally financed building is taking place, for example). Because pesticides and fertilizers often make their way into our waterways, a sane policy of screening these chemicals before they reach the market and wide use is mandatory. The myopic federal reclamation program should be revised so that ecological considerations receive prominent consideration. The nation's overall need for irrigated land should be re-evaluated in light of extensive land retirement under the Soil Bank Program.

None of these measures are politically expedient. They would limit powerful and heretofore sacrosanct private interests. But political expedients notwithstanding, we must recognize that our nation's environmental destruction will not be halted until these interests are held publicly accountable.

Phosphate Detergents

We cannot wait for the detergent industry to decide when it will stop destroying our rivers and lakes with its phosphate cleaning products. Several steps should be taken without further delay. (See Chapter 4.) Federal legislation should be passed requiring, at the very least, an immediate substantial reduction in the phosphate content of all detergents. Congress should explicitly authorize states and localities to enact local bans on phosphates.

It is important to recognize that the question of whether or not a *national* ban on phosphates in detergents *everywhere* should be imposed turns on the meaning of a single two-word phrase that crops up in virtually every industry statement on the problem: "adequate substitute." Washing clothes with non-phosphate products, we are told by the detergent-makers, would leave them stiff (and uncomfortable) or dirty (and unhealthy), and it might wear them out faster as well. To guard against these dire possibilities, imaginary or not, most of the national ban legislation proposed so far has provided that if the industry can make a showing of "good faith effort," failure to come up with an "adequate substitute" within the allotted time period would justify an extension of the deadline. What constitutes an "adequate substitute"? The legislation does not say. In the absence of clear objective standards for determining what is an adequate substitute for phosphates, the definition could only be supplied more or less arbitrarily by a government official under tremendous industry pressure, or else by the industry itself. What is an "adequate substitute" is the key question in the phosphate pollution problem, and up to now the detergent industry has maintained a monopoly on the right to answer it. We must find a workable way to make the *public* and not the industry the judge.

Congress should require that the Administrator of EPA publish clear criteria and minimum performance standards for evaluating any phosphate substitute, based on tests of products on the market. The minimum standards set for any criterion—cleaning power, for instance—should in no case be higher than the performance of detergents on the market after a substantial phosphate reduction has already been imposed. The law should then specify that whenever a non-phosphate product is produced that meets the minimum standards, a national ban on phosphates would go into effect within a certain period of time—say, two years or less.

As an initial approach, a ban tied to clear-cut criteria would be better than a ban after a given time period, for several reasons. It would eliminate any discretion to extend the deadline. It would guarantee that there *was* an adequate substitute before the ban took effect. It would provide industrial researchers with a clear-cut standard to aim for and a financial incentive to be the first to reach it. And it would eliminate phosphates at the earliest possible moment—possibly much sooner than they would be eliminated if a mandatory period of, say, two years were set. By piercing the industry's inflated claims about the wondrous properties of phosphates, we might very well discover that the ban could go into effect immediately without ruining our wash. Should an "adequate substitute" for phosphates not quickly appear, however, and the phosphate detergent pollution problem still remain, there should be no doubt about our course of action. We must be prepared to lower our definition of what constitutes an "adequate substitute." Stained clothing can be discarded and replaced. There is no way to replace a stained environment.

Eliminating phosphates from detergents will not solve the problem, however. The phosphate in human excrement would remain to take an increasing toll. We also need a national commitment to provide phosphate removal in municipal sewage treatment. But since the phosphates in detergents do, in any event, add to the cost of effective sewage treatment, they should be made to pay their way. Phosphate content in washing products should be taxed and the proceeds used to help finance phosphate removal at sewage treatment plants. A higher charge on higher-phosphate detergents would speed the industry's search for a non-polluting alternative and encourage consumers to purchase the lower-phosphate products.

If the Federal Trade Commission does not impose labeling requirements on the detergent industry, Congress should. The label must include not only all ingredients but also a clear warning that phosphates damage our environment. The FTC or Congress should go on from the narrow problem of phosphate pollution to demand that any advertisement of a product found to contain environmentally hazardous ingredients include a prominent message warning the consumer of the environmental harm he may cause by using it.

The most important lesson our experience with the soap and detergent industry teaches us is that it will not suffice to act *after* environmental harm has already been caused. Restrictions on the phosphate content in detergents would do more harm than good if they were not accompanied by legislation forbidding the introduction of any new substances into commercial products without adequate pre-testing for their environmental dangers.

Money and Manpower

The water pollution program has never had the financial or personnel resources it needs to do its job. As an explanation for the program's failure, however, the Task

Force believes that money and manpower shortages are secondary in importance to—and in many cases simply manifest—the political and policy constraints on the Federal effort. Two examples should suffice to make the point:

1. *Research, Development, and Demonstration*

The program hardest hit by personnel shortages is the Water Quality Office's research effort. Federal agency manpower ceilings imposed in the mid-1960's to counteract the inflationary impact of the Vietnam war caught the R&D program in the middle of a period of rapid expansion to meet new statutory responsibilities. Senior scientists had already been recruited and the water unit was just beginning to look for junior researchers, non-professional technicians, and administrative personnel when the hiring crackdown came. The R&D program was originally supposed to have been 1000 strong by 1970 (professional and clerical combined); it has never had more than a top-heavy 500. Interdisciplinary research teams—comprised of chemists, engineers, mathematicians, and biologists, for example—are only as strong as their weakest links. Many of the links were never inserted or have long since departed the Federal program in frustration. The Water Quality Office pays expensive overhead on eight R&D labs well stocked with research equipment still waiting to be used by scientists who never arrived, increasing the cost of an unproductive program. Short on junior personnel, the research section's overworked senior scientists do a great deal of their own routine lab preparation and paperwork—and consequently less research.

The heaviest drain on R&D's time and talent is not paperwork, however, but the Water Quality Office's misguided demonstration program. (See Chapter 19.) Supervising demonstration projects and outside research contracts is an "extra-curricular" activity that consumes anywhere from 25% to 75% of the average Federal research scientist's work time, leaving little left over for their own research assignments. The heavy stress on demonstration of available control techniques also requires that personnel slots which otherwise could be filled with people working on new solutions must instead be given over to the traditional engineers who make supervising demonstration work their primary assignment. The Task Force would thus not describe the R&D program's *main* problem as a *shortage* of manpower, serious though that shortage is, but rather as a *misallocation* of available men and money into an overextended demonstration effort.

2. *Enforcement*

Federal enforcement has been, if nothing else, cheap. From 1966 through 1970 the nation's annual enforcement budget averaged only $3.6 million (compared to an average $298 million per year during the same period for treatment plant subsidies, $41.8 million annually for research, and $4.5 million each year just for administration and management in the water pollution unit). Up until late 1970 the Water Quality Office's enforcement section did not have enough people in some of the regional offices even to check the states' files and records on pollution, let alone get out and look at real-life polluters very often.

The Water Quality Office's Middle Atlantic Region,* for example, would need an estimated 56 full-time enforcement personnel (counting both laboratory and field people) to begin enforcing the "shellfish clause" (see Chapter 6) in its coastal areas, overseeing water quality standards compliance, and documenting what regional officials described to the Task Force as "six or seven clear-cut cases of interstate pollution" on the region's numerous rivers. The region would need eight more people to handle its oil pollution responsibilities and 38 on top of that to maintain adequate surveillance of industrial polluters subject to the Refuse Act of 1899. The Middle Atlantic Region now has only 23 persons total working in enforcement. (Three more positions are authorized but currently unfilled.) Before the move to EPA in December 1970, the Water Quality Office had

* The Middle Atlantic Region includes the states of Pennsylvania, Maryland, Virginia, North and South Carolina, plus the District of Columbia. When the EPA regions become effective in June 1971, the Middle Atlantic will drop North and South Carolina and pick up West Virginia and Delaware to become EPA's Region 3.

only one enforcement officer to police all the pollution in the region's five states. Prior to 1969 the region had no one working in enforcement.

To take another example, it would require an estimated 40–50 persons to prepare the potential enforcement conferences and oversee standards compliance in the Missouri River Basin states of Iowa, Nebraska, Kansas, and Missouri. Water Quality Office regional officials could not estimate exactly how many extra enforcement employees would be needed to enforce the Refuse Act against Missouri Basin industrial polluters except to say that the number would be "significant." Were EPA to make the 66,000 animal feedlots in just those four states comply with the Refuse Act of 1899, some 200 or more additional enforcers would be needed. There are now only 12 persons assigned to Federal enforcement in those four states,* an increase of two over the previous year's total.

The enforcement program's flimsy financial underpinning simply reflects the lack of emphasis on enforcement in high-level policy. Given the heavy political constraints on Federal abatement activity up to now, additional field personnel could only have been a source of additional embarrassment to Administration officials trying their best not to discover where the pollution is. It should suffice to say that if the Federal government ever intends to begin enforcing even the laws that now exist, let alone take on any new authority, it will need a tremendous increase in surveillance and enforcement manpower. Laws with no one to enforce them are no better, indeed are usually worse, than no laws at all. As long as citizens are not under any illusion that they are being protected, they can at least take action in self-defense.

By the time the environment reached its apogee as a political issue in spring 1970, Congressional hoppers had long since been filled to the overflowing with new pollution control offerings. When 1970 drew to a close along with the 91st Congress, all these proposed bills

* Regional officials told the Task Force that as many as four or five persons formally funded under other programs actually worked on enforcement whenever the need for them arose.

remained approximately where they had been on the day they were introduced—no runs, no hits, no errors, none left on.* (The Senate Public Works Committee at least held hearings on new water pollution legislation in 1970; its counterpart in the House did not.) The 1971 legislative season opened more or less officially on February 8 with the President's Message to Congress on the Environment. Many of the same proposals are back again, warmed over for another try, along with several new ones. More than ever before, however, the field is dominated by the environmental offerings of two well-known public figures—President Nixon, on the one hand, and the Democratic Party's current front-running prospect to oppose him for the Presidency in 1972, Maine Senator Edmund Muskie, on the other.

The Administration comes into this early confrontation with Muskie as the visiting team in the Senator's home arena, the Senate Public Works Committee's Subcommittee on Air and Water Pollution. Subcommittee Chairman Muskie has had a corner on the upper chamber's water pollution business ever since he first carved out his "Mr. Pollution Control" niche in relative obscurity back in the mid-1960's. Since that time both the Senator and the environment have taken their separate paths into the political big leagues, and Muskie now finds himself matched against the President in a test of leadership on a key national issue. Given this highly charged setting, one might expect that if good water pollution legislation is ever going to be proposed by either Edmund Muskie or Richard Nixon, now would be the time. Unfortunately, the public can only hope that one or both will later move individually or be pushed to a stronger position in the water area than either has taken in his 1971 draft legislation. For if the contest between them were to be decided on the basis of their weak initial entries, the environment would probably turn out to be the real loser.

Take the Nixon proposal for example. There is no

* The Water Quality Improvement Act of 1970 (P.L. 91–224), which deals primarily with pollution from offshore oil spills, was no new proposal. It was the culmination of attempts ranging over several years to give the Federal government an enforceable oil pollution law. House and Senate versions of the bill had been passed in 1969. The two chambers finally got together on a common version in April 1970.

doubt that it eliminates the most prehistoric limitations in the old Federal Water Pollution Control Act. It would expand the Federal water quality standards jurisdiction to navigable waters and intrastate pollution; it would telescope the old law's strung-out abatement process; it would explicitly authorize the setting of precise effluent limits for each polluter; it would give the government most of the information-gathering authority it needs (including the right to require polluters to install and maintain their own effluent-monitoring devices); and it would provide civil penalties against pollution law violators ($25,000 per day for first offenders, $50,000 per day for repeaters). So far so good. From that point on, however, the bill is a disappointment, especially in view of the unprecedented public clamor for clean water. At a time when fundamental reform is needed, the Administration has proposed making what are essentially only rudimentary first moves in the direction of creating a truly effective water pollution law. Many of the old legislation's most basic defects have been transplanted intact, or modified only slightly, into the new Nixon plan.

Polluters can, for example, take solace—though just how much is not clear at this point—in a single seven-word phrase that is sprinkled strategically throughout the Nixon bill: "taking into account the practicability of compliance." Neither the Administrator nor any court may order a polluter to comply with the law without doing that. The phrase "taking into account the practicability of compliance" retains just enough of the flavor of a similar phrase in the current version of the Water Pollution Control Act—"giving due consideration to the practicability and to the technological (or physical) and economic feasibility of complying"—to suggest that its intended effect may be very much the same: to give judges the authority to second-guess cleanup requirements on the grounds that they may have unacceptable economic consequences for the firms involved.

This phrase cannot be dismissed as the uninspired product of incompetent lower-level staffers. As it was originally drawn up at EPA, the bill did not contain these words. They were inserted later during interagency meetings on the new legislation, reportedly at the insistence of the Department of Commerce, home of President Nixon's government-sponsored organization of top industrial leaders, the National Industrial Pollution Control Council.[3] (See Chapter 12.) Keeping the Council contented has become high-priority business for the Nixon Administration during this period when its new environmental program is being unfolded. On February 10, 1971, the day the Administration's water bill went to Congress, President Nixon was busy entertaining the Council's industrialists and their business affiliates, some 200 strong, at a White House reception. He took the opportunity to promise his guests that they would not be made the "scapegoats" of the Administration's effort to clean up the environment. "The Government—this Administration, I can assure you—is not here to beat industry over the head," the President told the executives in his warmly applauded remarks.[4] It is difficult to avoid concluding that the "practicability" phrase which was added to the Nixon bill was intended to provide the industrialists with added assurance.

Polluters can find additional consolation in the fact that the Administration's draft legislation would, like the old Act, permit the Administrator of EPA to enforce or not, as he pleases. There is only one mandatory "shall" in the bill's enforcement scheme and, like the "shall" in the present law (Chapter 6), it may not really be mandatory, after all. Whenever he "determines that any person is in violation of water quality standards," the Administrator "shall" notify the violator and the state agencies involved of the offense and of the remedial action required to achieve compliance. The Administrator is not, however, under any obligation to conduct the investigation that almost certainly would have to precede such a determination. After the Administrator delivers notice of violation, if he does, from that point on he faces nothing but discretionary "mays." He "may" order compliance or he may not, and so on. Consistent with the stale policy of deference to state action embodied in the old Pollution Control Act, the Administration bill actually *forbids* any penalties being given any polluter for his violations, no matter how flagrant or serious they may be, if the states involved "are taking appropriate and sufficient action . . . to secure compliance sometime in the future."

When the Administrator of EPA is not prohibited by this clause from taking Federal action at all, the Nixon enforcement scheme gives him his choice between two routes to cleanup, both of which contain unnecessary roadblocks. The Administrator cannot, in any case, immediately order a polluter to obey the law, nor can he assess penalties for past violations on his own. The Administrator can, however, ask the Attorney General to seek abatement orders or fines in a court trial against the polluter, with all its attendant hazards and delays. (Imagine a law that required the Police Department to try the bank robber in court before it could stop a bank robbery in process.) If the Administrator wants to sidestep the court route, the Nixon bill provides an alternative administrative path that could only have been designed by someone with a lingering affection for the present Water Pollution Control Act's tortuous scheme. Having delivered notice of violation, the Administrator must wait 30 days before he can order the polluter to obey the law. During this time the polluter gets to pollute for free while he makes up his mind whether or not to stop breaking the law "voluntarily" (or while the state decides to act). Under the Nixon plan, pollution control is still no profession for the impatient.

The Administration bill gives citizens the right to sue polluters for compliance with standards, but then nearly takes it all back again in the fine print. Citizens are barred from suing if either a state or the Federal government is suing the polluter. Thus if a state or the Federal government is willing to settle for a weak judgment (one with an unreasonably lengthy compliance time, for example), the citizen is left with no recourse. The Nixon bill also shields polluters from irate citizen litigants with 60 days of mandatory waiting after the offending discharger receives "notice" of his violation (as though he didn't know of it already in most cases) from the prospective plaintiffs. (The fact that the polluter may already have been given "notice" several times by state and Federal officials makes no difference; the citizen must give "notice" again and then wait.) The most worthless part of the citizen suit provision is one that gives any person a right to sue the Administrator to force him to perform any "duty . . . which is not discretionary." The obvious problem, of course, is finding any meaningful authority that is not wholly discretionary under this discretion-ridden bill.

Finally, the President's proposal is distinguished by its conspicuous failure to set any final deadline for all polluters to be in compliance with the new water quality standards that would be set in the areas of expanded Federal jurisdiction.

The Nixon bill's saving grace is its section on municipal treatment plant subsidies, which would eliminate the present pollution-producing Federal subsidy to industry. (See Chapter 17.) Each industry discharging into municipal treatment plants built with Federal funds would be required to pay a user charge that covers the share of construction costs attributable to its wastes. The Federal portion of the fee would revert to the Federal Treasury.* Unfortunately, however, the Nixon subsidy provision stops short of requiring that grant recipients assess a user charge to cover local operation and maintenance costs. The Administration would retain discretion to leave local treatment cost subsidies— and the inflated industrial wasteloads they encourage— undisturbed.

One final word to be said in favor of the Nixon water pollution proposal is that it is much improved over the

---

* There is one problem with this provision in the Nixon bill. It would permit the Administrator to choose to allow the municipality which collected the industrial user fee to keep the Federal portion and apply it to present and future treatment needs. This provision would, in effect, give a tremendous bonus to cities which could attract a great deal of industry into joining into the municipal system. Aside from the temptation such a setup would provide for cities to find ways to give industries hidden subsidies, it would be an extremely inequitable way to distribute billions each year in Federal subsidy money. Citizens in cities with a great deal of industry would receive a very large Federal subsidy and have to pay minimal, if any, waste treatment costs on their own. Persons in cities without much industry would not get this "extra" Federal subsidy and would have to pay much more for their waste treatment. We believe that the user charges collected from industry should all revert to the Federal Treasury and be redistributed to the cities on a more equitable basis as an additional subsidy.

almost unbelievably inadequate measure the President submitted as part of his environmental package the year before. Senator Muskie, on the other hand, would appear to believe the pollution crisis is not as bad this year as it was last, if one can judge from his February 1971 draft legislation. It is actually weaker in many significant respects than the bill the Senator submitted in 1970. The 1970 Muskie draft, for example, made Federal enforcement orders mandatory, no longer to be subjected to political scrutiny. Federal representatives in the field, it provided, "shall" issue cleanup orders immediately upon learning of a violation of Federal standards. The latest Muskie proposal abandons the mandatory "shall" at the field level for the familiar format of high-level discretion: whenever the Administrator of EPA finds a violation, he "may" issue an abatement order. Or then again, he may not. Nixon's bill at least reminds the Administrator to notify the state agencies about the violation. Muskie's does not even go that far.

The 1970 Muskie draft specified that when effluent requirements were being violated, the polluter was to be given a *maximum of 72 hours* to come back into compliance. Now it's a time "which the Administrator determines is reasonable."

The 1970 Muskie draft stated explicitly that cleanup orders were not to be suspended while the polluter appealed, but were to remain in force from the beginning. In 1971 Senator Muskie reverses himself completely. An order "shall not take effect until the person to whom it is issued has had an opportunity to confer with the Administrator concerning the alleged violation." * The effect of this provision would be approximately the same as that of a statute giving every lawbreaker in Los Angeles a right to have a personal chat with the Chief of Police before he obeyed any policeman's order to

---

* It is doubtful that the Administrator could legally delegate to subordinates the power to "confer" with polluters, since it is clearly intended as a protection to polluters. This interpretation is reinforced by the fact that the bill spells out clearly "the Administrator or his representative" in other places when it is intended that the power may be delegated. It should be noted that orders to abate "hazardous substances" violations were exempted from this conferral provision in the 1971 Muskie bill.

stop committing his crime. The only way the Administrator could save himself from a burgeoning backlog of polluter visits would be to stop issuing abatement orders altogether.

The 1970 Muskie bill rewarded citizen participation in the pollution control process by giving persons who provide information leading to a penalty fine against polluters one-tenth the amount collected. The Senator need no longer be concerned that he might be accused of being soft on citizens. That provision was deleted in the 1971 Muskie proposal.

Several valuable provisions from the 1970 Muskie proposal did manage to persist and make it into his 1971 draft bill in recognizable form. The Muskie bill would require all newly constructed industrial facilities to install the latest available pollution control techniques. (The 1971 draft goes on to specify that if closed-cycle systems are available, no discharge whatsoever will be permitted.) The Muskie proposal would protect workers who report pollution violations from reprisals by their employers. Muskie's bill would require states to hold public hearings on the water quality standards at least once every five years, to consider revising them upward. Lowering them would be prohibited. Federal agencies would not be allowed, under the Muskie bill, to purchase goods manufactured or services performed at commercial facilities that are undergoing Federal enforcement proceedings as violators of water quality standards. The President would be required to promulgate an order to insure that Federal agencies "effectuate the purpose . . . of this Act" in their loans, grants, and contracts. (This would presumably mean that, to the extent feasible, they would refuse to deal financially with polluters.) The Senator's proposal would eliminate a present obstacle in the law by permitting the Administrator to bring suit in Federal court on his own, rather than forcing him to depend on the Attorney General.

The Muskie citizen suit provision, while it does not provide financial incentives sufficient to stimulate broad citizen participation (only "costs of litigation" if the court deems it "appropriate"), is at least free of the wearisome qualifications tacked onto the Nixon citizen suit proposal. The clause that permits a citizen to sue the Administrator to force him to perform any "duty" suffers, unfortunately, from the same flaw as does the comparable clause in the Nixon bill. The Administrator has no legal duties to speak of in the enforcement area, but only discretionary authority.

Not all of the changes from the 1970 Muskie draft legislation are ones that environmentalists need look upon with suspicion. A 1970 requirement that the "technological feasibility" of abatement be considered in issuing cleanup orders disappeared in 1971. This is an extremely valuable improvement because it shifts the legal burden of developing control technology from the government back to private industry where it belongs. The latest Muskie proposal establishes a deadline for all polluters to be in compliance with Federal water quality standards three years after the standards are adopted by their state. (That helps, but it would be better yet to set an absolute deadline too. Unless that is done, the states can still delay cleanup by delaying standards adoption.) The Muskie bill now includes a $10,000-per-day penalty for "negligent" violations. The penalty is, unfortunately, discretionary, like everything else in the bill. And it is, of course, not as great as President Nixon's fines. Under the Nixon bill, any polluter—including "negligent" ones—can be fined $25,000 for first offense and $50,000 for second. (Except where there are so-called "hazardous substances" involved, the Muskie bill requires proof that the violation was committed "knowingly" to assess those larger fines. Muskie provides a jail sentence for the knowing offense, however, while Nixon does not.)

It now appears likely that the Muskie Subcommittee will, sometime in 1971, take the Nixon draft and Muskie's own proposal into committee executive session and eventually report out a bill to be voted on by the full Senate. There is at least a possibility that the final product will be somewhat better than either of its defective initial ingredients. A hopeful sign in this regard is the fact that Muskie's aides have conceded, privately, to insiders on Capitol Hill that the Senator's bill is not as strong as it could or ultimately should be. The strategy has apparently been to put out a modest initial effort, wait until the Nixon bill is on the table, and then go the President one—or hopefully two, three, or several—better by making additions in committee. Putting the Senator's strategy in its most generous light, the ostensible rationale behind it is that industrial lobbyists may be unprepared to attack what they cannot see in advance.

If this is what Senator Muskie plans, the public can only hope that he goes on from steps one and two to the final phase of improving his bill, at least to the point of fully recouping the heavy losses from his previous year's draft legislation. The Senator's waiting game, however, has serious hazards. Having been caught napping when new air pollution legislation passed in 1970, industry is, by all accounts, determined not to repeat its mistake. The fact that 1971, unlike 1970, is not an election year reduces the pressure on most legislators to take a strong stand on the environment and thereby increases the odds in industry's favor this time around. Whatever plans for improving the proposed bills the Senate Public Works Committee may now entertain could easily die a silent death under industrial pressure. The danger in Muskie's holding his best cards in reserve is that, having seen how easy Nixon's hand is to beat, the Senator may be tempted to use them very sparingly. The public deserves to see the strongest suits on the table from the beginning where the environment is at stake.

In the final analysis, the key to the pollution control program's legislative future does not rest with either of the two men press and public now usually look to for environmental leadership—Richard Nixon and Edmund Muskie—but rather with a less visible figure on the House side of Capitol Hill: Democratic Congressman John Blatnik of Minnesota. Throughout the 1960's, water pollution control measures pushed by Muskie through the Senate relatively unscathed have been dismembered in the industry-dominated House of Representatives, where they must pass before Blatnik's House Public Works Committee.

In a much earlier day Blatnik was known as a crusader for pollution reform. As Chairman of the Public Works Committee's Rivers and Harbors Subcommittee, he sponsored the Federal Water Pollution Control Act in 1956 and steered it skillfully through a hostile Congress. From 1956 through 1961 Congressman Blatnik was the principal driving force behind Congressional efforts to strengthen the Federal program. Historical accounts of the clean-water movement during the late 1950's are typically effusive in their praise for the Minnesota Congressman. Pollution control old-timers refer to him with reverence as the "father of the Federal water pollution control program."

But sometime in the early 1960's—no one can pinpoint exactly when—Father Blatnik dropped out as a vigorous force in the clean-water effort, and his Federal water pollution control family has been languishing for lack of effective House support ever since. Others in the House—Henry Reuss (D.-Wisc.), John Dingell (D.-Mich.), and Paul Rogers (D.-Fla.), for example—and Edmund Muskie in the Senate have taken the public spotlight from Blatnik on pollution control issues. What was left of his environmental image by the time 1969 rolled around was tarnished by public charges that he had attempted to suppress criticism and/or abatement efforts directed at Lake Superior's largest polluter and a large employer in his home district, the Reserve Mining Company. (See Chapter 7.) Congressman Blatnik's old Federal Water Pollution Control Act—once a badge of limited but legitimate accomplishment—has long since become an excuse for inaction. Persons close to the Congressman are conditioned to react defensively to suggestions that drastic changes are needed in Blatnik's law, as though its feeble scheme of non-regulation were somewhere very near the ultimate in legislative achievement.

The Task Force asked many Congressional observers and former Blatnik associates to explain the Congressman's untimely departure from the environmental front lines, and they suggested a number of possible reasons ranging from general fatigue, poor staff support, and lack of sufficient personal knowledge about pollution control to know the difference, all the way to diminished influence within the full Public Works Committee during the late 1960's. (When Maryland Congressman George Fallon took over the full committee chairmanship from New York's Charles Buckley in 1965, he reportedly allowed his subcommittee chairmen, including Blatnik, less freedom of maneuver than they had enjoyed before. Fallon—who lost his seat in the fall 1970 elections— was a fanatical devotee of the highway construction lobby, whose lack of concern about conservation spilled over into the water pollution field.) But whatever the reasons for Blatnik's diminished effectiveness, most observers familiar with the Congressman's record are essentially agreed on one conclusion: Blatnik's leadership on pollution control matters within his strongly pro-industry committee and within Congress generally over most of the past decade has been extremely weak.

Whatever Blatnik may have lacked in influence with his committee colleagues and fellow Congressmen prior to 1971, he now has. With the opening of the 92nd Congress in January 1971 the Minnesotan stepped up to the spot vacated by George Fallon as Chairman of the full Public Works Committee. With near-dictatorial powers over the committee that dispenses prized public works construction projects to Congressional districts throughout the land, Blatnik's stature in Congress is considerable. If he gives his full backing to strong pollution control legislation, it will almost certainly pass the House and will be a challenge that the Senate cannot ignore; if he does not, the likelihood of getting a good bill through is extremely low.

As John Blatnik steps into this new powerful position, the Task Force would urge that he rekindle the interest and aggressive drive that he showed as an environmental proponent some 10 years earlier, or, barring that, that he formally relinquish the water pollution baton— the jurisdiction over water pollution matters which now falls by right to his committee—to any one of a number of his House colleagues who would welcome the chance to run with it. He can be, if he chooses to, the main propelling force behind the dramatic changes that are needed if clean water is ever to be a reality. But if Con-

gressman Blatnik continues to sit on our most precious
public possession, the environment, as he has over much
of the past decade, we may wake up to discover—too
late—that there is precious little left for him or anyone
else to sit on.

Appendix A

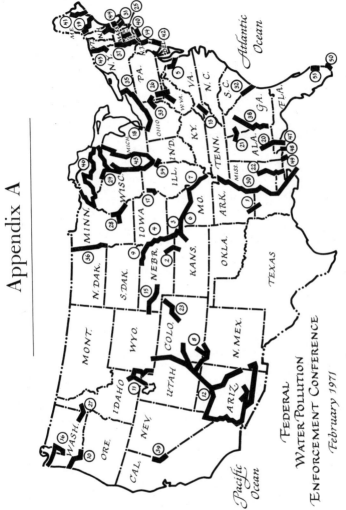

FEDERAL
WATER POLLUTION
ENFORCEMENT CONFERENCE
February 1971

FEDERAL WATER POLLUTION ENFORCEMENT CONFERENCES *
January 1957–February 1971

1. *Corney Creek Drainage System* (Ark.-La.)
   Hearing: January 16–17, 1957
   Initiated by Surgeon General, Public Health Service
2. *Big Blue River* (Neb.-Kan.)
   Conference: May 3, 1957
   Initiated by Surgeon General, Public Health Service
3. *Missouri River–St. Joseph, Missouri Area* (Mo.-Kan.)
   Conference: June 11, 1957
   Hearing: July 27–30, 1959
   Suit Filed: September 29, 1960
   Court Order: October 31, 1961
   Initiated by Surgeon General, Public Health Service
4. *Missouri River–Omaha, Nebraska Area* (Neb.-Kan.-Mo.-Ia.)
   Conference: (Session 1) June 14, 1957
              (Session 2) July 21, 1964
   Initiated by Surgeon General, Public Health Service
5. *Potomac River–Washington Metropolitan Area* (D.C.-Md.-Va.)
   Conference: (Session 1) August 22, 1957
              (Session 2) February 13, 1958
              (Session 3) April 2–4, May 8, 1969
              (Session 3 reconvened) May 21–22, 1970, October 13, 1970
              Progress meeting: December 8–9, 1970
   Initiated by Surgeon General, Public Health Service
6. *Missouri River–Kansas Cities Metropolitan Area* (Kan.-Mo.)
   Conference: December 3, 1967
   Hearing: June 13–17, 1960
   Initiated by Surgeon General, Public Health Service
7. *Mississippi River–St. Louis Metropolitan Area* (Mo.-Ill.)
   Conference: March 4, 1958
   Initiated by Missouri Health Division; Illinois Sanitary Water Board; Bi-State Development Agency
8. *Animas River* (Colo.-N.M.)
   Conference: (Session 1) April 29, 1958
              (Session 2) June 24, 1959
   Initiated by New Mexico Department of Public Health
9. *Missouri River–Sioux City Area* (S.D.-Ia.-Neb.-Kan.-Mo.)
   Conference: July 24, 1958

---

* Conference proceedings and reports are excellent sources of information on polluters and pollution and can be obtained from EPA.

Hearing: March 23–27, 1959
Initiated by Iowa Commissioner of Public Health

10. *Lower Columbia River* (Wash.-Ore.)
    Conference: (Session 1) September 10–11, 1958
                (Session 2) September 3–4, 1959
                (Session 3) September 8–9, 1965
    Initiated by Surgeon General, Public Health Service

11. *Bear River* (Ida.-Wyo.-Utah)
    Conference: (Session 1) October 8, 1958
                (Session 2) July 19, 1960
    Initiated by Utah Water Pollution Control Board

12. *Colorado River and All Tributaries* (Colo.-Utah-Ariz.-
    Nev.-Calif.-N.M.-Wyo.)
    Conference: (Session 1) January 13, 1960
                (Session 2) May 11, 1961
                (Session 3) May 9–10, 1962
                (Session 4) May 27–28, 1963
                (Session 5) May 26, 1964
                (Session 6) July 26, 1967
    Initiated by New Mexico Department of Public Health;
      Arizona State Department of Health; Nevada State
      Board of Health; Colorado Department of Public
      Health; Utah Water Pollution Control Board; Cali-
      fornia State Water Pollution Control Board

13. *North Fork of the Holston River* (Tenn.-Va.)
    Conference: (Session 1) September 28, 1960
                (Session 2) June 19, 1962
                (Session 3) Called and postponed
    Initiated by Tennessee Stream Pollution Control Board

14. *Raritan Bay* (N.J.-N.Y.)
    Conference: (Session 1) August 22, 1961
                (Session 2) May 9, 1963
                (Session 3) June 13–14, 1967
    Initiated by Surgeon General, Public Health Service

15. *North Platte River* (Neb.-Wyo.)
    Conference: (Session 1) September 21, 1961
                (Session 2) March 21, 1962
                (Session 3) November 20, 1963
    Initiated by Nebraska Department of Health

16. *Puget Sound* (Wash.)
    Conference: (Session 1) January 16–17, 1962
                (Session 2) September 6–7, October 6, 1963
    Initiated by Governor of Washington

17. *Mississippi River-Clinton, Iowa Area* (Ill.-Ia.)
    Conference: March 8, 1962
    Initiated by Secretary of Health, Education and Welfare

18. *Detroit River* (Mich.)
    Conference: (Session 1) March 27–28, 1962

(Session 2) June 15–18, 1965
Initiated by Governor of Michigan

19. *Androscoggin River* (N.H.-Me.)
Conference: (Session 1) September 24, 1962, February
6, 1963
(Session 2) October 21, 1969
Initiated by Secretary of Health, Education and Welfare

20. *Escambia River* (Ala.-Fla.)
Conference: October 24, 1962
Initiated by Florida State Board of Health

21. *Coosa River* (Ga.-Ala.)
Conference: (Session 1) August 27, 1963
(Session 2) April 11, 1968
Initiated by Secretary of Health, Education and Welfare

22. *Pearl River* (Miss.-La.)
Conference: (Session 1) October 22, 1963
(Session 2) November 7, 1968
Initiated by Secretary of Health, Education and Welfare

23. *South Platte River* (Colo.)
Conference: (Session 1) October 29, 1963
(Session 2) April 27–28, 1966
(Session 2 reconvened) November 10, 1966
Initiated by Governor of Colorado

24. *Menominee River* (Mich.-Wisc.)
Conference: November 6–8, 1963
Initiated by Secretary of Health, Education and Welfare

25. *Lower Connecticut River* (Mass.-Conn.)
Conference: (Session 1) December 2, 1963
(Session 2) September 27, 1967
Initiated by Secretary of Health, Education and Welfare

26. *Monongahela River* (W. Va.-Pa.-Md.)
Conference: December 17–18, 1963
Initiated by Secretary of Health, Education and Welfare

27. *Snake River–Lewiston, Idaho–Clarkston, Washington
Area* (Ida.-Wash.)
Conference: January 15, 1964
Initiated by Secretary of Health, Education and Welfare

28. *Upper Mississippi River* (Minn.-Wisc.)
Conference: (Session 1) February 7–8, 1964
(Session 2) February 28, March 1 and 20,
1967
Initiated by Secretary of Health, Education and Wel-
fare; Governors of Minnesota and Wisconsin

29. *Merrimack & Nashua Rivers* (N. H.-Mass.)
Conference: (Session 1) February 11, 1964
(Session 2) December 18, 1968
Workshops: October 20–21, 1970

Initiated by Secretary of Health, Education and Welfare; Governor of Massachusetts

30. *Lower Mississippi River* (Ark.-Tenn.-Miss.-La.)
Conference: May 5–6, 1964
Initiated by Secretary of Health, Education and Welfare

31. *Blackstone and Ten Mile Rivers* (Mass.-R. I.)
Conference: (Session 1) January 26, 1965
(Session 2) May 28, 1968
Initiated by Secretary of Health, Education and Welfare

32. *Lower Savannah River* (S.C.-Ga.)
Conference: (Session 1) February 2, 1965
(Session 2) October 29. 1969
Initiated by Secretary of Health, Education and Welfare

33. *Mahoning River* (Ohio-Pa.)
Conference: February 16–17, 1965
Initiated by Secretary of Health, Education and Welfare

34. *Grand Calumet River, Little Calumet River, Calumet River, Wolf Lake, Lake Michigan, and Their Tributaries* (Ill.-Ind.)
Conference: (Session 1) March 2–9, 1965
(Technical Session) January 4, 5, 31, February 1, 1966
(Reconvened) August 26, 1969
Initiated by Secretary of Health, Education and Welfare

35. *Lake Erie* (Mich.-Ind.-Ohio-Pa.-N.Y.)
(Session 1) August 3–5, 1965
(Session 2) August 10–12, 1965
(Session 3) March 22, 1967
(Session 4) October 4, 1968
(Session 5) June 3–4, 1970
Initiated by Secretary of Health, Education and Welfare; Governor of Ohio

36. *Red River of the North* (Minn.-N.D.)
Conference: September 14–15, 1965; January 18, March 4, 1966
Initiated by Secretary of Health, Education and Welfare

37. *Hudson River* (N.Y.-N.J.)
Conference: (Session 1) September 28–30, 1965
(Session 2) September 20–21, 1967
(Session 3) June 18–19, 1969
(Session 3 reconvened) November 25, 1969
Initiated by Secretary of Health, Education and Welfare; Governors of New York and New Jersey

38. *Chattahoochee River and Its Tributaries* (Ga.-Ala.)
Conference: (Session 1) July 14–15, 1966
(Session 2) February 17, 1970
Initiated by Secretary of Interior

39. *Lake Tahoe* (Calif.-Nev.)
    Conference: July 18–20, 1966
    Initiated by Secretary of Interior

40. *Moriches Bay and Eastern Section of Great South Bay and Their Tributaries* (N.Y.)
    Conference: (Session 1) September 20–21, 1966
    (Session 2) June 21, 1967
    Initiated by Secretary of Interior

41. *Penobscot River and Upper Penobscot Bay and Their Tributaries* (Me.)
    Conference: April 20, 1967
    Initiated by Secretary of Interior

42. *Eastern New Jersey Shore—from Shark River to Cape May* (N.J.)
    Conference: November 1, 1967
    Initiated by Secretary of Interior

43. *Lake Michigan* (Mich.-Ind.-Ill.-Wisc.)
    Conference: (Session 1) January 31, February 1–2, 5–7, March 7–8, 12, 1968
    (Session 2) February 25, 1969
    (Session 3) March 31, April 1, May 7, 1970
    Workshops held September 28–October 2, 1970
    Conference is scheduled to reconvene March 23–24, 1971
    Initiated by Governor of Illinois; Secretary of Interior

44. *Boston Harbor* (Mass.)
    Conference: (Session 1) May 20, 1968
    (Session 2) April 30, 1969
    Initiated by Secretary of Interior

45. *Lake Champlain* (N.Y.-Vt.)
    Conference: (Session 1) November 13, December 19–29, 1968
    (Session 2) June 25, 1970
    Initiated by Secretary of Interior; Vermont Department of Water Resources

46. *Lake Superior and Its Tributary Basin* (Wisc.-Minn.-Mich.)
    Conference: (Session 1) May 13–15, September 30, October 1, 1969
    (Session 2) April 29–30, August 12–13, 1970
    (Session 2 reconvened) January 14–15, 1971
    Initiated by Secretary of Interior

47. *Escambia River Basin* (Ala.-Fla.)
    Conference: January 20–21, 1970
    Initiated by Governor of Florida

48. *Perdido Bay* (Fla.-Ala.)
    Conference: January 22, 1970
    Initiated by Governor of Alabama
49. *Mobile Bay* (Ala.)
    Conference: January 27–28, 1970
    Initiated by Secretary of Interior
50. *Biscayne Bay* (Fla.)
    Conference: February 24–26, 1970
    Initiated by Governor of Florida
51. *Dade County* (Fla.)
    Conference: October 20–21, 1970
    Initiated by Governor of Florida

Appendix B

1963 LIST OF POTENTIAL ENFORCEMENT ACTIONS

[From testimony of Murray Stein before the U.S. Senate Special Subcommittee on Air and Water Pollution of the Committee on Public Works, June 17, 1963]

   1. Allegheny River (New York, Pennsylvania)
   2. Applegate River–Eliot River (California, Oregon)
   3. Arkansas River, area I (Colorado, Kansas)
   4. Arkansas River, area II (Kansas, Oklahoma)
   5. Arkansas River, area III (Oklahoma, Arkansas)
   6. Batten Kill (Vermont, New York)
   7. Green River (Kentucky, Tennessee)
   8. Big Sandy River (Kentucky, West Virginia, Virginia)
   9. Big Horn River (Wyoming, Montana)
  10. Big Sioux River (South Dakota, Iowa)
*11. Blackstone River (Massachusetts, Rhode Island) [31] †
 12. Bodcau River (Arkansas, Louisiana)
 13. Buntings Branch (Delaware, Maryland)
‡14. Byram River (Connecticut, New York)
 15. Connecticut River, upper (New Hampshire, Vermont)
*16. Connecticut River, lower (Massachusetts, Connecticut) [25]
 17. Catawba and Wateree Rivers (North Carolina, South Carolina)
*18. Chattahoochee, upper (Alabama, Georgia) [38]
 19. Chattahoochee, lower (Alabama, Georgia, Florida)
*20. Coosa River (Alabama, Georgia) [21]
 21. Delaware River (New York, Pennsylvania, New Jersey, Delaware)
 22. Des Moines River (Minnesota, Iowa, Missouri)
 23. Delores River (Colorado, Utah)
 24. French River (Massachusetts, Connecticut)
 25. French Broad River (Tennessee, North Carolina)
*26. Grand Calumet and Little Calumet Rivers (Indiana, Illinois) [34]
 27. Grand (Neosho) River (Kansas, Oklahoma)

* = Enforcement action has taken place (as of February 1971).
† Numbers in brackets refer to Enforcement Conferences.
‡ New Conference officially called by the Administrator. Not yet held.

28. Green River (Wyoming, Utah)
29. Hoosic River (Vermont, Massachusetts, New York)
30. Kanab Creek (Utah, Arizona)
31. Kanawha River, tributary to Ohio River (West Virginia, Ohio)
32. Klamath River (Oregon, California)
33. Leviathan Creek (California, Nevada)
34. Little Blue River (Nebraska, Kansas)
35. Lost River (Oregon, California)
*36. Mahoning River (Ohio, Pennsylvania) [33]
37. Malad Rivers (Idaho, Utah)
38. Marais Des Cygnes River (Kansas, Missouri)
39. McElmo Creek (Colorado, Utah)
*40. Menominee River (Wisconsin, Michigan) [24]
*41. Merrimack River (New Hampshire, Massachusetts) [29]
*42. Mississippi River, area I (Minnesota, Wisconsin) [28]
*43. Mississippi River, area IX (Memphis, Tenn.–Vicksburg, Miss.) (Tennessee, Arkansas, Mississippi, Louisiana) [30]
*44. Mississippi River, area X (Vicksburg, Miss.–mouth) (Mississippi, Louisiana) [30]
45. Missouri River, including lower Yellowstone River (Montana, North Dakota, South Dakota)
*46. Monongahela River (West Virginia, Maryland, Pennsylvania) [26]
47. Montreal River (Michigan, Wisconsin)
*48. Nashua River (Massachusetts, New Hampshire) [29]
49. Nolichucky River (North Carolina, Tennessee)
50. Ochlockonee River (Georgia, Florida)
51. Ohio River, I–Pittsburgh, Pa.–Pennsylvania State Line (Ohio, Pennsylvania)
52. Ohio River, II–Pennsylvania State Line–Huntington, W.Va. (West Virginia, Ohio)
53. Ohio River, III–Huntington to above Cincinnati (Kentucky, West Virginia, Ohio)
54. Ohio River, IV (Indiana, Kentucky, Ohio)
55. Ohio River, V (Illinois, Kentucky, Indiana)
56. Ouachita River (Arkansas, Louisiana)
57. Pigeon River (North Carolina, Tennessee)
58. Pawcatuck River (Rhode Island, Connecticut)
59. Pea and Choctawhatchee Rivers (Alabama, Florida)
*60. Pearl River (Louisiana, Mississippi) [22]
61. Piscataqua River (New Hampshire, Maine)
62. Potomac River, Luke–Cumberland area (Maryland, West Virginia)
63. Quinebaug River (Massachusetts, Connecticut)
64. Red River, upper (Arkansas, Oklahoma, Texas)

65. Red River, lower (Arkansas, Louisiana)
*66. Red River of the North (Minnesota, North Dakota, South Dakota) [36]
67. Rio Grande River (Texas, New Mexico)
68. Roanoke River (Virginia, North Carolina)
69. Rock River (Illinois, Wisconsin)
70. Saco River (New Hampshire, Maine)
71. St. Croix River (Minnesota, Wisconsin)
*72. Hudson and East Rivers (tributary to Raritan Bay enforcement area) [37]
73. Snake River, area II (Idaho, Oregon)
*74. Snake River, area III (Washington, Idaho) [27]
75. St. Louis River (Wisconsin, Minnesota)
76. St. Mary's River (Georgia, Florida)
77. Chattoga, Tugaloo, Seneca and upper Savannah Rivers (South Carolina, Georgia)
*78. Savannah River, lower (South Carolina, Georgia) [32]
*79. Savannah River, mouth (South Carolina, Georgia) [32]
*80. South Platte (Colorado, Nebraska) [23]
81. Suwannee River (Georgia, Florida)
82. Susquehanna River (north branch) (Pennsylvania, New York)
*83. Ten Mile River (Rhode Island, Massachusetts) [31]
84. Tennessee River (Georgia, Tennessee, Alabama)
85. Verdigris River (Kansas, Oklahoma)
86. Virgin River (Utah, Arizona, Nevada)
87. Wabash River (Illinois, Indiana)
88. Warm Springs Run (West Virginia, Maryland)
89. Yellowstone River, lower (Montana, North Dakota)
*90. Youghiogheny River (West Virginia, Maryland, Pennsylvania) [26]

Appendix C

COMPOSITION OF STATE
POLLUTION BOARDS †

Key: * Means state pollution board contains representatives
of basic pollution sources (industry, agriculture, county
and city governments).
° Means state board is free of such representation.
"No Boards" means air and water pollution regulation
statewide is handled by a full-time state agency.

	Water Board	Combination Air-Water Board
Alabama	*	
Alaska		No Boards
Arizona		No Boards
Arkansas		*
(1) California	*	
Colorado	*	
Connecticut	*	
Delaware		*
Florida		°
Georgia	*	
Hawaii		°
Idaho	*	
Illinois		No Boards
Indiana	*	
Iowa	*	
Kansas		°
Kentucky	*	
Louisiana	*	
Maine		*
Maryland		No Boards
Massachusetts		°
Michigan	*	
Minnesota		*
Mississippi	*	
Missouri	*	

† Adapted from a chart in the article "Polluters Sit on Anti-Pollution Boards," by Gladwin Hill, *The New York Times,* December 7, 1970. © 1970, *The New York Times,* Reprinted by permission of the publisher.

	Water Board	Combination Air-Water Board	
Montana	*		
Nebraska	*		
Nevada			(2)
New Hampshire	*		
New Jersey	No Boards		
New Mexico		o	
New York	No Boards		(3)
North Carolina		*	
North Dakota	*		
Ohio	*		
Oklahoma		*	
Oregon		*	
Pennsylvania	*		
Rhode Island	No Boards		
South Carolina		*	
South Dakota	*		
Tennessee	*		
Texas	*		
Utah	*		
Vermont	o		
Virginia	o		
Washington	No Boards		(4)
West Virginia			
Wisconsin		*	(5)
Wyoming			(6)

(1) Pollution sources represented in regional branches of State Water Board. (2) Water under State Board of Health. (3) State Environmental Board is advisory. (4) Interest conflicts banned by law. (5) Water under State Division of Water Resources. (6) Water Pollution Control Council is advisory.

Appendix D

NATIONAL INDUSTRIAL POLLUTION
CONTROL COUNCIL

Members as of February 17, 1971

1. *Chairman:* Bert S. Cross, Chairman of Board, Chief Executive Officer, 3–M Company
2. *Vice-Chairman:* Willard F. Rockwell, Jr., Chairman of Board, President, North American Rockwell Corporation
3. Birny Mason, Jr., Chairman of Board, Union Carbide Corporation
4. Charles H. Sommer, Chairman of Board, Monsanto Company
5. Clifford D. Siverd, President, Chief Executive Officer, American Cyanamid Company
6. Herbert F. Tomasek, President, Chemagro Corporation
7. Howard J. Morgens, President, Procter and Gamble Company
8. Milton C. Mumford, Chairman of Board, Lever Brothers Company
9. C. W. Cook, Chairman of Board, Chief Executive Officer, General Foods Corporation
10. Howard C. Harder, Chairman of Board, CPC International, Inc.
11. Robert W. Reneker, President, Chief Executive Officer, Swift and Company
12. Charles R. Orem, President, Armour and Company
13. James P. McFarland, President, Chief Executive Officer, General Mills
14. Robert J. Keith, Chairman of Board, Chief Executive Officer, The Pillsbury Company
15. Donald M. Kendall, Chairman of Board, Chief Executive Officer, Pepsico, Inc.
16. William F. May, Chairman of Board, President, American Can Company
17. Ellison L. Hazard, Chairman of Board, President, Continental Can Company, Inc.
18. Edwin D. Dodd, President, Chief Executive Officer, Owens-Illinois, Inc.

19. John L. Gushman, President, Anchor Hocking Corporation

20. Leo H. Schoenhofen, Chairman, Chief Executive Officer, Container Corporation of America

21. C. Raymond Dahl, Chairman of Board, Chief Executive Officer, Crown-Zellerbach

22. Edmund F. Martin, (Retired) Chairman of Board, Chief Executive Officer, Bethlehem Steel

23. Thomas F. Patton, Chairman of Board, Republic Steel Company

24. J. F. Jamieson, President, Standard Oil Company of New Jersey

25. Robert O. Anderson, Chairman of Board, Chief Executive Officer, Atlantic-Richfield Corporation

26. Frank R. Milliken, Chairman of Board, Kennecott Copper Corporation

27. Gilbert W. Humphrey, Chairman of Board, Hanna Mining Company

28. Thomas C. Mullins, President, Peabody Coal Company

29. Russell DeYoung, Chairman of Board, Goodyear Tire and Rubber Company

30. J. Ward Keener, Chairman of Board, B. F. Goodrich Company

31. Karl R. Bendetsen, Chairman of Board, Chief Executive Officer, U.S. Plywood-Champion Papers, Inc.

32. Norton Clapp, Chairman of Board, Weyerhaeuser Company

33. Cris Dobbins, Chairman of Board, President, Ideal Basic Industries, Inc.

34. Robinson F. Barker, Chairman of Board, PPG Industries, Inc.

35. Edwin N. Cole, President, General Motors

36. L. Anthony Iacocca, President, Ford Motor Company

37. Benjamin F. Biaggini, Jr., President, Southern Pacific Company

38. John M. Budd, President, Burlington Northern, Inc.

39. Charles C. Tillinghast, Jr., Chairman of Board, Trans-World Airlines, Inc.

40. Frank A. Nemec, President, Chief Operating Officer, Lykes-Youngstown Corporation

41. Schermer L. Sibley, President, Chief Executive Officer, Pacific Gas and Electric Company

42. Lelan F. Sillin, Jr., President, Northeast Utilities

43. Fred J. Borch, Chairman of Board, Chief Executive Officer, General Electric Corporation

44. Donald C. Burnham, Chairman, Westinghouse Electric Corporation

45. Paul L. Davies, Senior Director, FMC Corporation

46. Arthur J. Santry, Jr., President, Combustion Engineering, Inc.
47. Stephen D. Bechtel, Jr., President, Bechtel Corporation
48. Ralph Evinrude, Chairman of Board, Outdoor Marine
49. Rodney C. Gott, Chairman, President, American Machine and Foundry, Inc.
50. Arch N. Booth, President, U.S. Chamber of Commerce
51. William P. Gullander, President, National Association of Manufacturers
52. W. K. Coors, President, Adolph Coors Brewers
53. Bertrand L. Perkins, Chairman of Board, President, Morrison-Knudsen Company, Inc.
54. J. Simon Fluor, Honorary Chairman of Board, Fluor Corporation
55. Adolph B. Kurz, Vice-President, Keystone Shipping Company
56. George G. Zipf, President, Chief Executive Officer, Babcock and Wilcox Company
57. Robert V. Hansberger, Chairman, Boise Cascade
58. W. P. Gwinn, Chairman of Board, Chief Executive Officer, United Aircraft Corporation
59. John Corcoran, President, Director, Consolidation Coal Company, Inc.
60. Alexander H. Galloway, Chairman of Board, President, R. J. Reynolds Industries, Inc.
61. Alexander B. Trowbridge, Jr., President, The Conference Board

Appendix E

Division of Water Quality Standards
OFFICE OF ENFORCEMENT AND STANDARDS COMPLIANCE

WATER QUALITY STANDARDS APPROVED UNDER THE
FEDERAL WATER POLLUTION CONTROL ACT, AS AMENDED

February 1971

Jurisdiction	Status	Approval Dates	Jurisdiction	Status	Approval Dates
Alabama	P	2/15/68	New Jersey	PA	3/13/68
Alaska	P	2/20/68	New Mexico	FA	7/09/68
Arizona	FA	9/27/68			11/19/68
		1/17/69			8/21/69
Arkansas	FA	8/07/67	New York	F	7/18/67
		11/19/69			8/07/67
California	PA	1/09/69	North Carolina	FA	5/16/68
Colorado	P	10/21/68			1/20/71
		10/09/69	North Dakota	FA	8/07/67
Connecticut	FA	2/15/68			5/22/70
		4/21/70	Ohio	PA	3/04/68
Delaware	P	3/13/68			9/13/68
Florida	FA	1/17/69			1/02/69
Georgia	F	7/18/67	Oklahoma	FA	2/28/68

State	Code	Date
Hawaii	F	3/13/68
Idaho	F	8/07/67
		2/15/68
Illinois	PA	1/27/68
Indiana	FA	7/18/67
		1/20/71
Iowa	P	1/16/69
Kansas	PA	4/25/69
Kentucky	P	3/20/69
Louisiana	P	2/12/68
		12/17/69
Maine	P	6/03/68
		12/12/68
Maryland	F	8/07/67
Massachusetts	FA	8/07/67
		11/25/70
Michigan	PA	4/17/68
Minnesota	FA	6/18/68
		11/26/69
Mississippi	P	5/06/68
Missouri	FA	2/08/68

State	Code	Date
Oregon	FA	2/17/70
		7/18/67
		9/13/68
		12/17/68
Pennsylvania	F	1/16/70
Rhode Island	FA	5/21/68
		1/30/68
		11/19/68
South Carolina	PA	1/20/71
		10/21/68
		4/21/69
South Dakota	F	8/07/67
Tennessee	P	2/28/68
Texas	FA	1/27/68
		5/02/69
Utah	FA	12/31/68
Vermont	P	6/27/68
Virginia	P	1/17/69
Washington	FA	1/22/68
West Virginia	P	5/21/68
Wisconsin	FA	1/24/68

Jurisdiction	Status	Approval Dates	Jurisdiction	Status	Approval Dates
		6/26/68	Wyoming	FA	11/27/68
Montana	FA	7/30/70	District of		4/17/68
		2/29/68	Columbia	FA	1/17/69
		1/17/69	Guam	FA	6/12/68
Nebraska	FA	12/19/68	Puerto Rico	FA	12/30/68
Nevada	FA	6/27/68	Virgin Islands	FA	2/28/68
New Hampshire	FA	8/16/68			4/28/70
		12/01/70			
				(27)	
				(7)	
				(7)	
				(13)	

Legend:
FA—Fully approved water quality standards—anti-degradation statement included
F —Approved water quality standards—anti-degradation statement not included
PA—Partially approved water quality standards—anti-degradation statement included
P —Partially approved water quality standards—anti-degradation statement not included

STATUS SUMMARY

	States
Fully approved with anti-degradation statement (FA)	27
Standards approved wholly or in part	54
Approved anti-degradation statement	34

Appendix F

ARSENIC CONTENT IN CLEANING PRODUCTS

Detergent	Arsenic Content ug/g (ppm)	Detergent	Arsenic Content ug/g (ppm)
Biz	35.8	Amway Dish Drops (liquid)	<1.0
Axion	13.3	Amway Industroclean solvent detergent (liquid)	<1.0
Dash	10.0	White King Detergent (with Borax)	5.0
Salvo	3.8	Borateem	10.0
Drive	<1.0	Spic and Span	<1.0
Oxydol	31.2	White King Soap	<1.0
Tide	8.5	Pine Sol (liquid)	<1.0
Ajax	8.6	Ajax All Purpose (liquid)	2.0
Bold	11.4	Lux Liquid (liquid)	<1.0
Cold Water All	<1.0	Downy	<1.0

Detergent	Arsenic Content ug/g (ppm)	Detergent	Arsenic Content ug/g (ppm)
Dreft	<1.0	Mr. Clean (liquid)	2.0
Gain	2.0	Whistle (liquid)	<1.0
Punch	6.4	Calgon	<1.0
Duz	25.1	Miracle White (liquid)	<1.0
Bonus	6.9	Instant Fels Soap	<1.0
Fab	13.8	Addit (liquid)	<1.0
Cheer	6.2	Duz Heavy Duty Detergent	22.0
Breeze	3.5	Cold Water Surf	<1.0
Cold Power	<2.0	Enzyme Brion	31.0
Diaper Pure	<1.0	Arm & Hammer Sal Soda	<1.0
Wisk	<1.0	Rinso Detergent with chlorine bleach	<1.0
Trend	4.2	Blue Rain Drops	16.0
Amway Water Softener	<1.0	Dash	24.0
Amway SA-S	<1.0	Snowy Bleach (liquid)	<1.0
Amway Trizyme	<1.0	Ivory Snow	<1.0
Amway Automatic Dishwasher	2.0		

Source: Federal Water Quality Administration, September 1970.

Appendix G

Purpose		Manufacturer	Percentage Phosphate as STPP*
Pre-soaks	Biz	Procter & Gamble	73.9
	Enzyme Brion	Purex	71.4
	Amway Trizyme		71.2
	Axion	Colgate-Palmolive	63.2
Laundry Detergents	Blue Rain Drops		63.2
	Salvo	Procter & Gamble	56.6
	Tide	Procter & Gamble	49.8
	Amway SA-8		49.3
	Coldwater Surf	Lever Brothers	48.2
	Drive	Lever Brothers	47.4
	Oxydol	Procter & Gamble	46.6
	Bold	Procter & Gamble	45.4
	Cold Water All (powder)	Lever Brothers	45.4
	Ajax Laundry	Colgate-Palmolive	44.6
	Cold Power	Colgate-Palmolive	44.6
	Punch	Colgate-Palmolive	44.2
	Dreft	Procter & Gamble	41.9
	Rinso with Chlorine Bleach	Lever Brothers	41.0
	Gain	Procter & Gamble	39.5
	Duz	Procter & Gamble	38.3
	Bestline B-7	Bestline Products	38.0
	Bonus	Procter & Gamble	37.5
	Breeze	Lever Brothers	37.2
	Cheer	Procter & Gamble	36.3
	Fab	Colgate-Palmolive	34.8
	White King (with Borax)	White King Soap Co.	34.7

* Phosphate contents are reported as percentage STPP (sodium tripolyphosphate), the chemical form of phosphate most widely used in detergent formulations, although some of the products contain phosphate in other forms. The percentage of STPP in a product is 3.9542 times the percentage of elemental phosphorus.

Chicago, Ill., has had a ban on the sale of laundry detergents containing more than 8.7% phosphorus (i.e., 34.4% phosphate) in effect since February 1, 1971. Canada has had a ban on the manufacture of laundry detergents containing more than 20% phosphate in effect since August 1, 1970.

Some percentages may have changed since the Federal Water Quality Administration released this list in September 1970.

Source: Federal Water Quality Administration, September 6, 1970.

PERCENTAGE OF PHOSPHATES IN DETERGENTS
AND SOAPS—(*Continued*)

Purpose		Manufacturer	Percentage Phosphate as STPP*
	Royalite		21.7
	Instant Fels Soap	Purex	16.6
	Wisk (liquid)	Lever Brothers	14.2
	Par Plus	Par Chemical	4.3
	Addit (liquid)		2.2
	Ivory Liquid	Procter & Gamble	1.9
	Lux Liquid	Lever Brothers	1.9
	White King Soap	White King Soap Co.	<1.0
	Cold Water All (liquid)	Lever Brothers	<1.0
Automatic Dishwasher Detergents	Amway		60.0
	Cascade	Procter & Gamble	54.5
	All	Lever Brothers	54.0
	Calgonite	Calgon	49.4
	Electrosol	Economics Lab., Inc.	34.8
Household Cleaners	Ajax All Purpose	Colgate-Palmolive	28.5
	Mr. Clean	Procter & Gamble	27.0
	Whistle	Drackette	3.1
	Pinesol	American Cyanamid	<1.0
Miscellaneous	Snowy Bleach	Gold Seal	36.4
	Borateem	U.S. Borax	<1.0
	Downy	Procter & Gamble	<1.0
	Amway Dish Drops		<1.0

Appendix H

Amway SA-8	Procter & Gamble
Cheer	Procter & Gamble
Gain	
H.L.D.	
K-50	
Laundri-Maid	
Liquid All	Lever Brothers
Loft	Sanford Chemical Co.
Phos-Free	C.P. Baker Co.
Roundy's	
Sav-Us	Hadco Chemical Co.
Ultra	Burford Division of Solar Chemical Co.
Valley-Dew	Fremont Industries, Inc.

* Nitrilotriacetic Acid. List is not comprehensive.

Source: Environmental Protection Agency, December 1970.

Appendix I

Aftermath

Following the April 12, 1971, release of this report, Senator Muskie's office issued the following statement:

I have reviewed the Nader Task Force Report on Water Pollution with great interest and anticipation. It is both revealing and disturbing—revealing because it provides an excellent insight into specific water pollution situations and disturbing because of the inadequacies in the Federal-State water pollution control effort which it points out.

David Zwick and his associates have provided the Congress a frank, open and important challenge. Their criticism of the existing program and pending legislation, including my own bill, will be a useful supplement to the testimony which Mr. Zwick presented to the Subcommittee last month.

I sincerely hope my colleagues and the Administration will carefully review the recommendations of this report. From this material and the testimony the Subcommittee on Air and Water Pollution has received this year, tough and effective legislation should result.

On April 13, Environmental Protection Agency Administrator William Ruckleshaus issued this statement on *Water Wasteland:*

Many of Mr. Nader's conclusions are our conclusions. That is why, over two years ago, we began changing the emphasis and force of the national cleanup of our waterways. The most important institutional change for combatting water pollution—indeed for battling pollution in all its insidious forms—was creation of the Environmental Protection Agency. When I took the job of heading this agency, I did so without illusions. I realized that we were dealing with decades of neglect at a time when environmental concern was cresting, when our citizens were demanding instant solutions. That is their right. America the beautiful has been compromised and dirtied, sometimes beyond belief. The full realization of this neglect came with Earth Day a year ago. There is nothing more vile than a filthy stream or river or lake or seaside. Our agency is dedicated to nothing less than the restoration of those waters that have been subjected to the abominable abuse that is chronicled in the Nader report.

We recognize with Mr. Nader that enforcement is

an important key to an effective cleanup of our waterways. That is why we are asking for radical changes in the law. We want to scrap the cumbersome practices of the past and replace them with swift public hearings to crack down on polluters.

We have proposed legislation in which private citizens can take legal action against water quality standards' violations. We have asked for a tripling of enforcement monies for the States—from $10 million to $30 million over the next four years.

We will shortly, in concert with the Corps of Engineers and the Council on Environmental Quality, begin to require industries to apply for a permit before they are allowed to discharge wastes to our waterways. States must certify that the discharges do not violate water quality standards. Where State Water Quality Standards are inadequate or so weak as to be inconsistent with the purposes of the Federal Water Pollution Control Act, EPA will override State recommendations and apply appropriate water quality standards.

We can cite some successes in the enforcement field. We believe that we have set a precedent by requiring all new power plants to be situated along Lake Michigan to have cooling towers in order to avoid the potential disastrous effects of thermal pollution. This policy could well become a national policy. We have taken legal action against three of the Nation's largest cities for violation of water quality standards. We intend to move against others which ignore these requirements. Since the EPA came into existence four months ago, the agency has requested the Justice Department to file six suits against industries under the 1899 Refuse Act for water pollution violations. Last July, cases involving 10 mercury polluters were referred to the Justice Department for legal action. In addition, forty other industrial mercury polluters were identified. All of these firms agreed to a voluntary compliance schedule to reduce mercury discharges to below detectable limits. To date, a reduction of 94 percent in mercury discharges has resulted. The Justice Department has taken legal action to avoid thermal pollution of Biscayne Bay in Florida: damages have been collected from a shipping firm for oil pollution of Cook Inlet in Alaska; a Federal-State enforcement conference that began this week in New Haven, Connecticut, will require many industries discharging wastes to Long Island Sound to provide the most up-to-date technology for purifying their effluents. Granted, this is only a start.

Mr. Nader and his able assistants are right: we are in danger of creating a 'water wasteland' if we permit to happen in the future what has in the past. But I think everyone now realizes the consequences of indecision and inaction. The country has experienced an environmental awakening. It no longer has to be convinced that we must act.

The report that Mr. Nader and his associates have put together is a commendable, thorough and far-reaching examination of where we have been: of documented apathy, ignorance, arrogance, and a laissez faire ethic that no longer fits the times. My responsibility is to chart where we are going. I have called upon industry to develop a new environmental ethic. The 'Water Wasteland' is an important report. Because of its importance, I have directed our Water Quality Office to analyze its contents in depth. Once that review is completed, we intend to convey our thoughts on its substance to Mr. Nader and his associates. Our agency received his report last week. It is a report that has been long in preparation, and one which requires more than an idle perfunctory response.

One last comment. We welcome the advice and criticism of all public interest groups. I personally think an adversary relationship between government and its citizens is healthy and vital. But I also think—when it comes to the environment we all must share—that there is a commonalty of interest so compelling that all of us must join this cause to save that upon which we are all dependent—our air, our water, our land. I would hope that we will be able to call upon the resources of the Center for Study of Responsive Law and all other public interest groups, upon industry, municipalities, other branches of government, our citizens to help and aid us in seeing that we can avoid that ultimate catastrophe implied by the title of Mr. Nader's report.

On April 14—two days after public release of the report and six days after EPA had received advance copies—Ruckleshaus called a Federal Enforcement Conference on the Houston Ship Channel, to be held in June 1971. (See Chapter VI.) Local authorities publicly attributed the calling of the conference to the Nader Report. EPA regional officials later agreed, privately.

On April 23, 1971, at a reconvening of the Lake Superior Enforcement Conference, the Reserve Mining Company learned that it would soon receive a notice, under the Water Quality Act of 1965, giving it 180 days to come up with a plan for halting its dumping of taconite tailings or face Federal court action. Federal officials at the conference declared for the first time that the company's plan for disposing of its wastes in the lake was unacceptable. (See Chapter VII.)

On May 17, Muskie's Senate Subcommittee on Air and Water Pollution convened hearings on the corporate practice of making a community and its labor force choose between environmental protection and jobs. Ralph Nader, who had made a public request in January 1971 that hearings be held on this topic, was the leadoff witness. Mr. Nader's recommendations to the Subcommittee were based, in large part, on the recommendations made in *Water Wasteland*. (See Chapter XX.)

The Subcommittee went into closed executive session in early June 1971 to begin work on reporting a water pollution bill. The House Public Works Committee held oversight hearings on the Federal water pollution control program in June, and had scheduled initial hearings on proposed water pollution legislation for mid-July.

Notes

Chapter I

1. Bethamy Probst, "If Fish Are Dying from Pollution, the People Can't Be Far Behind," *Tampa Tribune,* April 23, 1970.
2. Barry Commoner, *Science and Survival* (1966), p. 28.
3. "Community Water Supply Study: Significance of National Findings," U.S. Department of Health, Education and Welfare, Public Health Service, July 1970, p. 7.
4. Henry J. Graeser, Director, Water Utilities Department, Dallas, Texas, commenting on a paper by Robert R. Harris, former Director of the Bureau of Water Hygiene, entitled "The Water Activities of the Bureau of Water Hygiene."
5. James H. McDermott, Director, Bureau of Water Hygiene, "Future Program of the Bureau of Water Hygiene," paper presented at 50th Anniversary Meeting of the Conference of State Sanitary Engineers, Minneapolis, Minnesota, May 26, 1970, p. 1.
6. "Community Water Supply Study: Significance of National Findings," *op. cit.,* p. 10.
7. *Ibid.*
8. W. C. Hueper and W. D. Conway, *Chemical Carcinogenesis and Cancers,* (Springfield, Illinois: Charles C. Thomas, Publisher, 1964), p. 692.
9. Hueper and Conway, *op. cit.,* p. 696, citing J. C. Diehl and S. W. Tromp, *Sticht. ter Bevoerd. van de Psych. Phys. Leiden,* 1953, p. 120.
10. "Birth Defects and Their Environmental Causes," *Medical World News,* January 22, 1971, p. 50
11. Howard J. Sanders, "Chemical Mutagens: The Road to Genetic Disaster?" *Chemical and Engineering News,* May 19, 1969, p. 56; and "Birth Defects and Their Environmental Causes," *op. cit.,* p. 54.
12. "Birth Defects and Their Environmental Causes," *op. cit.,* p. 57.
13. Sanders, *op. cit.*
14. *Report of the Secretary's Commission on Pesticides,* U.S. Department of Health, Education and Welfare, December 1969, p. 571.

15. James F. Crow, "Chemical Risk to Future Generations," *Scientist and Citizen*, June–July, 1968, p. 113.

16. Testimony of Henry A. Schroeder at Hearings before the Subcommittee on Energy, Natural Resources, and the Environment of the Committee on Commerce, U.S. Senate, "Effects of Mercury on Man and the Environment," August 26 and 27, 1970, Pt. 3, p. 692.

17. B. Aberg, "Metabolism of Methyl Mercury (Mercury 203) Compounds in Man," *Archives of Environmental Health*, Vol. 19, October 19, 1969, pp. 478–484.

18. Richard R. Leger, "Mounting Peril: Mercury Poisoning of Food Supply Alarms Officials," *Wall Street Journal*, April 28, 1970.

19. "Mercury Stirs More Pollution Concern," *Chemical and Engineering News*, June 22, 1970, p. 37.

20. *Wall Street Journal*, January 15, 1971.

21. "Mercury Stirs More Pollution Concern," *op. cit.*

22. Victor Cohn, "Mercury Contamination May Require Curb on U.S. Fish Eating," *Washington Post*, July 26, 1970.

23. "Survey of Metal Carcinogenesis," *Progress in Experimental Tumor Research*, Vol. 12 (1969), pp. 102–133. Lead acetate, nickel, and arsenic were found to be carcinogenic (although possibly it is the selenium which occurs commonly with arsenic which is carcinogenic when ingested). Other substances for which carcinogenicity is disputed are cadmium, zinc, chromium, and cobalt.

24. *Drinking Water Standards:* 1962, U.S. Public Health Service (Washington: U.S. Government Printing Office, 1969), p. 25.

25. Victor Cohn, "Hill Unit Told Arsenic, Lead Found in Some Drinking Water," *Washington Post*, July 31, 1970.

26. *Drinking Water Standards, op. cit.*, p. 25.

27. E. L. Goldblatt, *et al.*, "The Unusual and Widespread Occurrence of Arsenic in Well Waters of Lane County, Oregon," paper describing a study by the Lane County, Oregon, Health Department, 1963.

28. Henry A. Schroeder, "The Water Factor," *New England Journal of Medicine*, April 10, 1969, pp. 836–837; "Relation between Mortality from Cardiovascular Disease and Treated Water Supplies," *Journal of the American Medical Association*, April 23, 1960, pp. 1902–08; and "Municipal Drinking Water and Cardiovascular Death Rates," *Journal of the American Medical Association*, January 10, 1966, pp. 125–129.

29. Henry A. Schroeder, Hearings before the Subcommittee

on Energy, Natural Resources, and the Environment, *op. cit.,* p. 693.

30. *Ibid.,* p. 696.

31. "Nitrates, Nitrites, and Methemoglobinemia," National Institute of Environmental Health Sciences, National Institutes of Health, Research Triangle Park, North Carolina, May 1970. Figure 3 in the Appendix shows that 500,000 tons were used in 1945, whereas by 1970 use had increased to 7 million tons.

32. Robert H. Harmeson, *et al.* "The Nitrate Situation in Illinois," paper presented at the 90th AWWA [American Water Works Association] Conference, June 23, 1970, p. 8. In Illinois the growth in the use of nitrate fertilizers between 1960 and 1970 was from 120,000 tons to 600,000 tons.

33. Elliott F. Winton, "Public Health Aspects of Nitrate in Drinking Water: Current Research and Recommendations," paper presented at the 90th AWWA Conference, June 23, 1970, p. 1.

34. Samuel S. Epstein, "Toxicological and Environmental Implications on the Use of Nitrilotriacetic Acid as a Detergent Builder," Staff Report for U.S. Senate Committee on Public Works, December 1970, p. 8, citing N. Gruener and H. P. Shuval, "Health Aspects of Nitrates in Drinking Water," Developments in Water Quality Research, Proceedings of the Jerusalem International Conference on Water Quality and Pollution Research, Ann Arbor, Michigan (London: Humphrey Science Publishers, June 1969).

35. *Ibid.*

36. *Ibid.,* p. 8, citing F. N. Subbotin, "The Nitrates of Drinking Water and Their Effect on the Formation of Methaemoglobin," *Gigiena I. Sanitariia 2,* 1961, p. 13.

37. *Ibid.,* citing A. Gelperin, *Medical World News,* July 17, 1970, p. 4.

38. *Ibid.,* citing G. E. Hein, "Reaction of Tertiary Amines with Nitrous Acid," *Journal of Chemical Education,* Vol. 40, 1963, p. 181; and S. S. Epstein and W. Lijinsky, "Nitrosamines as Environmental Carcinogens," *Nature,* Vol. 225, 1970, p. 21.

39. Barry Commoner, "Threats to the Integrity of the Nitrogen Cycle: Nitrogen Compounds in Soil, Water, Atmosphere and Precipitation," paper presented at the Annual Meeting of the American Association for the Advancement of Science, Dallas, Texas, December 26, 1968.

40. Public statement of the City Council, Delano, California, September 18, 1969.

41. "First Quarterly Progress Report—Delano Nitrate Study," California Department of Public Health, April 10, 1970, p. 3.

42. "Birth Defects and Their Environmental Causes," *op. cit.*, p. 51.

43. "Suit Asks a Halt to DDT Discharge," *The New York Times,* June 6, 1970, p. 21.

44. *Report of the Secretary's Commission on Pesticides and Their Relationship to Environmental Health,* U.S. Department of Health, Education and Welfare, December 1969, pp. 665–666.

45. Dioxin is roughly 100,000 to 1 million times as teratogenic as thalidomide in certain species studied, such as the chick and the rat. (Dr. Jacqueline Verrett, Hearings before the Subcommittee on Energy, Natural Resources, and the Environment of the Committee on Commerce [91st Congress, 2nd Session], "Effects of 2,4,5-T on Man and the Environment," April 7, 15, 1970, p. 202.)

   The lowest effective teratogenic dose of thalidomide for human beings, however, is 0.5 mg/kg/day, whereas for rats it is 50 mg/kg/day. ("Birth Defects and Their Environmental Causes," *op. cit.*, p. 49). Since human beings are, therefore, about 100 times more sensitive to thalidomide than are rats, the first figures have been divided by 100 to estimate the potential effect of dioxins on humans.

46. See *Federal Register,* Vol. 30, February 18, 1965, p. 2315, for history of FDA actions concerning the banning of DES in chicken feed.

47. "Animal Feed Adjuncts," Note prepared for Joint FAO-WHO Food Standards Program, Codet Committee on Food Administration, 4th Session, September 11–15, 1967, Ref. CCFA-67-21, pp. 10–12.

48. *Bell v. Goddard,* 366 F. 2d 177 at 182 (7th Cir. 1966).

49. James Schwartz, "It Tastes Aged, But Not Vintage: If You Can Drink It, That Funny-Flavored Water Is Harmless," *Louisville Times,* May 22, 1970.

50. Kenneth M. Mackenthun and Lowell E. Keup, "Biological Problems Encountered in Water Supplies," paper presented at the 90th Annual AWWA Conference, June 23, 1970, pp. 17, 31.

51. Lewis Herber, *Crisis in Our Cities* (Englewood Cliffs, New Jersey: Prentice-Hall, Inc.), p. 98.

52. Mackenthun and Keup, *op. cit.*, p. 31. See Table 1: Organisms Creating Problems in Water Supplies.

53. *Ibid.*, p. 10.

54. W. M. Ingram and K. M. Mackenthun, "Animal In-

festations in Distribution Systems," *Proceedings—Fifth Sanitary Engineering Conference,* January 29–30, 1963, p. 2.

55. Shih L. Chang and Leland J. McCabe, "Health Aspects of Wastewater Reuse," Public Health Service, Bureau of Water Hygient, Cincinnati Office, undated.

56. *Congressional Record* (91st Congress, 2nd Session), December 10, 1970, pp. H 11529–30.

57. Herber, *op. cit.,* 96.

58. Kenneth E. Boulding, "The Economics of the Coming Spaceship Earth," in *The Environmental Handbook,* Garrett de Bell, editor (New York: Ballantine/Friends of the Earth, 1970) p. 96.

59. Rachel Carson, *Silent Spring* (Greenwich, Connecticut: Fawcett Publications, 1962), pp. 50–52.

60. Roy Reed, "Pesticides Make Cotton Prosper But Endanger Life, " *The New York Times,* December 16, 1969. The *Times* quotes Dr. Ferguson as asserting that resistance to pesticides constitutes a genetic change in the newly resistant species. It thus raises questions about future biological potentials of the species, such as "What other characteristics are being selected along with resistance?"

61. *The New York Times,* October 30, 1970, p. 1.

62. Gladwin Hill, "Mercury Hazard Found Nationwide," *The New York Times,* September 11, 1970.

63. *HEW News,* U.S. Department of Health, Education and Welfare, Food and Drug Administration News Release #71-5, February 5, 1971.

64. David Bird, "Canadians Wary on Mercury Rate," *The New York Times,* February 17, 1971.

65. Victor Cohn, "Mercury Contamination May Require Curb on U.S. Fish Eating," *Washington Post,* July 26, 1970.

66. *FDA Fact Sheet,* "Mercury Residues in Canned Tuna," Food and Drug Administration, December 15, 1970.

67. David Bird, "Fisheaters Found to Have a High Level of Mercury," *The New York Times,* January 14, 1971.

68. Dr. Charles F. Wurster, "It's Polluting All the World," *Washington Post,* May 4, 1969.

69. "Fish, Wildlife, and Pesticides," U.S. Department of the Interior, Fish and Wildlife Service, 1966.

70. Wurster, *ibid.*

71. Henderson, Johnson, and Inglis, "Organochlorine Insecticide Residues in Fish," *Pesticides Monitoring Journal,* December 1969, pp. 145–171.

72. *Congressional Record* (91st Congress, 1st Session), April 4, 1969, p. 2.

73. *The Nuclear Industry 1970,* U.S. Atomic Energy Commission, 1970, p. 153.

74. *Environmental Quality,* The First Annual Report of the Council on Environmental Quality, August 1970, p. 143.

75. *Resources and Man,* A Study and Recommendations by the Committee on Resources and Man of the Division of Earth Sciences, National Academy of Sciences, National Research Council (San Francisco: W. H. Freeman and Company), p. 235.

76. *Ibid.,* p. 233.

77. Charles H. Fox, "Radioactive Wastes," U.S. Atomic Energy Commission, Division of Technical Information, 1969 (revised), p. 4.

78. *Resources and Man, op. cit.,* p. 234.

79. Dennis Farney, "Atom-Age Trash: Finding Places to Put Nuclear Waste Proves a Frightful Problem," *Wall Street Journal,* January 25, 1971, p. 1. (The 99% figure was confirmed by the Atomic Energy Commission.)

80. Two interesting general discussions of the radioactive waste disposal problem, in addition to Farney's *Wall Street Journal* article, are Roger Rapoport, "Caution: This Garbage May Be Hazardous to Your Health," *Los Angeles Times,* October 25, 1970; and Donald M. Rothberg, "Under the Mushroom Cloud" (four-part series), *Washington Evening Star,* reprinted in *Congressional Record,* (91st Congress, 2nd Session), July 30, 1970, pp. S12451–55.

81. Bob Smith, "A.E.C. Scored on Storing Waste," *The New York Times,* March 7, 1970.

82. "Radioactive Wastes," *op. cit.* pp. 25–26.

83. *Ibid.,* p. 26; and *Progress and Problems in Programs for Managing High-Level Radioactive Wastes,* Report to the Joint Committee on Atomic Energy, Congress of the United States, by the Comptroller General of the United States, January 29, 1971, p. 22.

84. *Ibid.,* pp. 20–21.

85. *Ibid.,* p. 15.

86. Committee on Geologic Aspects of Radioactive Waste Disposal, of the National Academy of Sciences-National Research Council, Division of Earth Sciences, *Report to the Division of Reactor Development and Technology, United States Atomic Energy Commission,* May 1966, p. 13.

87. *Ibid.,* p. 21.

88. Committee on Geologic Aspects of Radioactive Waste Disposal, *op. cit.,* p. 11.

89. Dennis Farney, *op. cit.,* p. 10.

90. *Resources and Man, op. cit.,* 235.

91. *Progress and Problems in Programs for Managing High-Level Radioactive Wastes, op. cit.,* p. 16.

92. Committee on Geologic Aspects of Radioactive Waste Disposal, *op. cit.,* p. 76.

93. *Progress and Problems in Programs for Managing High-Level Radioactive Wastes, op. cit.,* p. 34.

94. *Ibid.,* p. 43.

95. *Ibid.,* p. 74.

96. *Ibid.,* pp. 73–75.

97. *Progress and Problems in Programs for Managing High-Level Radioactive Wastes, op. cit.,* p. 39.

98. *Ibid.,* p. 44.

99. Committee on Geologic Aspects of Radioactive Waste Disposal, *op. cit.,* p. 70.

100. *Examination of the Waste Treatment and Disposal Operations at the National Reactor Testing Station, Idaho Falls, Idaho,* Federal Water Quality Administration, Northwest Region, Portland, Oregon, April 1970, p. 19.

101. Dennis Farney, *op. cit.,* p. 1.

102. *Examination of the Waste Treatment and Disposal Operations at the National Reactor Testing Station, Idaho Falls, Idaho, op. cit.,* p. 3.

103. *The Nuclear Industry, op. cit.,* p. 293.

104. *Progress and Problems in Programs for Managing High-Level Radioactive Wastes, op. cit.,* p. 49.

105. Arthur R. Tamplin and John W. Gofman, *"Population Control" Through Nuclear Pollution* (Chicago: Nelson-Hall Company, 1970), p. 22.

106. *Ibid.,* p. 27.

107. The maximum permissible concentrations are listed in Title 10, Part 20, of the *Code of Federal Regulations,* "Standards for Protection Against Radiation."

108. John W. Gofman and Arthur R. Tamplin, "Low Dose Radiation, Chromosomes, and Cancer," paper presented at the 1969 IEEE Nuclear Science Symposium, San Francisco, California, October, 29, 1969, p. 7.

109. Arthur R. Tamplin, Testimony Presented Before the Environmental Quality Study Council, State of California, January 21, 1971, p. 5.

110. Gofman and Tamplin, "Low Dose Radiation, Chromosomes, and Cancer," *op. cit.,* p. 7.

111. *Environmental Quality,* The First Annual Report of the Council on Environmental Quality, *op. cit.,* p. 27.

112. Arthur R. Tamplin, "The Regulation of Man-made Radiation in the Biosphere," in Harry Foreman, editor,

Nuclear Power and the Public (Minneapolis: University of Minnesota Press), p. 46.

113. "Atomic Radiation Safety Standards," in "Atomic Power: Paradise Lost or Found?" *Cincinnati Enquirer,* 1971, p. 8. Emphasis added.

114. John C. Devlin, "Connecticut Said to Keep Polluted Beaches Open to Avert Slum Violence," *The New York Times,* July 9, 1970.

115. A. Gene Gazley, "Use of DDT on Way Out, But Effects May Last Years," article quoted in *Cincinnati Enquirer,* November 30, 1969. Gazley is with the Michigan Department of Natural Resources.

116. Leroy S. Houser and Frank J. Silva, "National Register of Shellfish Production Areas," Public Health Service, Department of Health, Education and Welfare, 1966.

117. "The National Estuarine Pollution Study," Department of the Interior, 1970, p. 292.

118. The *Jackson* [Mississippi] *Daily News,* June 29, 1970, citing biologist Walter Thomsen of Folsom, California.

119. "Pollution of Tiny Harbor Shows National Problem," *Tallahassee Democrat,* March 1, 1970.

120. Don. E. Weaver, "Pollution Threatens Earth's Livability," *Cincinnati Post and Times-Star,* January 1, 1970.

121. Chuck Anderson, "Marine Life Sanctuary 'Runs Sacred,'" *Cincinnati Post and Times-Star,* January 1, 1970.

122. Hearings on "Migrant and Seasonal Farmworkers' Powerlessness" (Pesticides and the Farmworker), Senate Committee on Labor and Public Welfare Subcommittee on Migratory Labor Part 6C, p. 3888, September, 30, 1969.

123. M. W. Oberle, "Lead Poisoning: A Preventable Childhood Disease of the Slums," *Science,* September 5, 1969, p. 922. Oberle's figures are for the costs to society of any child born severely mentally retarded.

124. Albert E. Forster, "A Matter of Survival," *Proceedings of the National Conference on Water Pollution,* U.S. Department of Health, Education and Welfare, Public Health Service, 1960, p. 15.

125. U.S. Department of Agriculture, Soil Conservation Service, *Water and the Land,* Pamphlet No. TP-147, July 1965, p. 3. One expert has predicted that national demand for water will reach 630 billion gallons per day by 1980 (Clarence Klassen, "Water Quality Management—A National Necessity," *Proceedings of the National Conference on Water Pollution,* p. 140).

126. *Ibid.*

Chapter 2

1. Colonel Cliff Atkinson (U.S. Marines, Ret.), Director of Public Affairs of the American Water Works Association, estimates that personal use ranges between 50 and 100 gallons per day; a 40–65 gpd estimate is in *The Economics of Clean Water* (U.S. Department of Interior, FWQA, 1970, Vol. I, pp. 122, 135). The breakdown on specific home uses comes from Luna B. Leopold and W. B. Langbein, *A Primer on Water*, U.S. Geological Survey, 1960, p. 32.

2. Murray Stein, "Problems and Programs in Water Pollution," *Natural Resources Journal*, December, 1962, p. 397.

3. U.S. Department of Interior, FWQA, *Municipal Waste Facilities in the United States: Statistical Summary, 1968 Inventory*, Publication No. CWT-6, 1970, p. 14.

4. Based on data from *Municipal Waste Facilities*, p. 14, and Stein, "Problems and Programs," p. 397.

5. FWQA, *Clean Water for the 1970's: A Status Report*, 1970, p. 4.

6. *Municipal Waste Facilities*, p. 4.

7. *Ibid.*

8. *Clean Water for the 1970's*, p. 4.

9. U.S. Environmental Protection Agency, *Sewage Facilities Construction, 1970*, agency compendium of computer data, unpaged.

10. FWQA, *The Cost of Clean Water* [1968], Vol. I—Summary Report, p. 3. This report was issued pursuant to a Congressional requirement that FWQA make an annual study of the costs of abating pollution.

11. FWQA, *The Economics of Clean Water* [1970], Summary Report, p. 8.

12. *Ibid.*

13. *Cost of Clean Water* [1968], Vol. I, p. 3.

14. *Economics of Clean Water* [1970], Vol. I, pp. 124, 126, 134.

15. "Sewer Services and Charges," *Urban Data Service*, February 1970, pp. 3–5, 6, 27–30.

16. *Cost of Clean Water* [1968], Vol. I, p. 21; and *The Economics of Clean Water* [1970], Vol. I, p. 136.

17. Stein, "Problems and Programs," p. 397.

18. *Cost of Clean Water* [1968], Vol. I, p. 17.

19. *Ibid.*, Vol. III, "Industrial Waste Profile No. 1, Blast Furnaces and Steel Mills."

20. *Ibid.*, "Industrial Waste Profile No. 4, Textile Mill Products."

21. Ben Edwards, "Crusade for Cleanliness," Charleston (W. Va.) *Sunday Gazette-Mail,* April 21, 1968, p. 4m.
22. *Ibid.,* p. 5m.
23. From figures for water use in different industries, given in *Cost of Clean Water* [1968], Vol. I, p. 17.
24. *Paper Profits: Pollution in the Pulp and Paper Industry* (Washington, D.C.: Council on Economic Priorities, 1970), Section II, pp. P-1 through P-5. This valuable study includes a plant-by-plant evaluation of air and water pollution control at 131 pulp mill locations of the 24 largest U.S. pulp and paper companies, their status of compliance with pollution laws, and cost of upgrading control technology to state-of-the-art standards.
25. *Ibid.,* p. L-2.
26. *Ibid.,* p. 8, citing "Pollution by Pulp Mills Stirs More Localities to Press for Curbs," *Wall Street Journal,* June 25, 1970.
27. *Ibid.,* Section II, p. L-22.
28. *Paper Profits,* Sec. II, p. T-6.
29. *Ibid.,* pp. L-29, L-31, L-25.
30. *Ibid.,* p. 3.
31. James M. Fallows, *The Water Lords,* Ralph Nader's Study Group Report on Industry and Environmental Crisis in Savannah, Georgia (New York: Grossman Publishers, 1971), p. 49.
32. *Paper Profits,* pp. 7–9.
33. *Ibid.,* p. 46.
34. *Business Week,* October 4, 1969, p. 118.
35. *Business Week,* April 11, 1970, p. 66.

Chapter 3

1. William Steif, "Water Pollution Unit Is Shaken Up," *Washington Daily News,* January 13, 1970.

Chapter 4

1. *Detergents: Sub-Council Report, October 1970,* National Industrial Pollution Control Council, p. 7. (Hereafter cited as NIPCC Report).
2. E. E. Angino *et al.,* "Arsenic in Detergents: Possible Danger and Pollution Hazard," *Science,* April 17, 1970, p. 389.
3. Testimony of Dr. Ernest Angino at Hearings before the Subcommittee on Air and Water Pollution of the Committee on Public Works, U.S. Senate, 91st Congress, 2nd Session, June 8, 1970, Part 4, p. 1385. (Hereafter, hearings before this subcommittee are cited as "1970 Senate Hearings.")

4. Statement of Hon. Gaylord Nelson at Hearings before a Special Subcommittee on Air and Water Pollution of the Committee on Public Works, U.S. Senate, 88th Congress, 1st Session, June 19, 1963, p. 344.

5. Hearings before the Conservation and Natural Resources Subcommittee of the Committee on Government Operations, U.S. House of Representatives, 91st Congress, 1st Session, December 15, 1969, p. 33. (Hereafter cited as "1969 House Hearings.")

6. *Ibid.*, p. 74.

7. *Ibid.*, p. 33.

8. *Ibid.*, p. 74.

9. "Biodegradibility of Detergents—A Story About Surfactants," *Chemical and Engineering News,* March 18, 1963, p. 111.

10. Announcement printed in Hearings before a Subcommittee of the Committee on Government Operations [Subcommittee on Natural Resources and Power], U.S. Senate, 88th Congress, 1st Session, June 10, 1963, p. 1102.

11. NIPCC Report, p. 14.

12. FWQA, *Municipal Waste Facilities,* pp. 5, 14.

13. Each adult contributes an average 1.4 pounds of phosphorus per year, and an average of 1.5–2.0 pounds more through the use of detergents, according to Assistant Secretary of Interior Carl Klein, 1969 House Hearings, p. 5. A report by the International Joint Commission, *Pollution of Lake Erie, Lake Ontario, and the International Section of the St. Lawrence River,* 1970, p. 82, found that 70% of the phosphorus in sewage in the U.S. portion of the basin and 50% in Canada came from detergents.

14. Committee on Government Operations, *Phosphates in Detergents and the Eutrophication of America's Waters,* House Report No. 91–1004, 1970, p. 12.

15. *Ibid.*, p. 4ff.

16. International Joint Commission Report, 1970, p. 37. In Lake Erie, for example, 70% of the total phosphorus loading is attributable to municipal and industrial sources, while only 30–40% of the nitrogen is. (Report to the International Joint Commission by the International Lake Erie and Lake Ontario–St. Lawrence River Water Pollution Boards, 1969, Vol. I—Summary, pp. 69–71).

17. FWPCA, *Lake Erie Report: A Plan for Pollution Control* (1968), p. 32.

18. U.S. Department of Health, Education and Welfare,

Report on Pollution of Lake Erie and its Tributaries, July 1965, Part I, p. 1.

19. 1969 House Hearings, pp. 54, 56.
20. *Ibid.,* p. 54ff.
21. *Ibid.,* p. 74.
22. *Ibid.,* p. 33.
23. *Ibid.,* pp. 57–59.
24. Testimony of Dr. A. F. Bartsch, FWQA, *Ibid.,* p. 14ff.
25. Letter from Dr. Stephan to E. Scott Pattison, *Ibid.,* pp. 60–1.
26. Statement of William H. Rodgers, Jr., before the Subcommittee on Intergovernmental Relations, Committee on Government Operations, U.S. Senate, October 6, 1970, p. 4, citing House Intergovernmental Relations Subcommittee, *Deficiencies in Administration of Federal Insecticide, Fungicide and Rodenticide Act,* House Report No. 91–637, 91st Congress, 1st Session, 1969, p. 7.
27. Michael C. Jensen, "Will Phosphates Hurt Procter & Gamble?" *New York Times,* February 28, 1971.
28. Soap and Detergent Association press release, January 25, 1971, p. 2.
29. FWQA, *Municipal Waste Facilities in the United States,* pp. 5, 14.
30. International Joint Commission Report, 1970, p. 125.
31. 1969 House Hearings, p. 28.
32. Report to the International Joint Commission, 1969, pp. 79, 80.
33. Testimony of Dr. Ernest Angino, 1970 Senate Hearings, Part 4, p. 1386.
34. Committee on Government Operations, *Phosphates in Detergents,* pp. 40–1 for table showing average hardness of water supply of 100 largest U.S. cities.
35. *Ibid.,* p. 39.
36. 1970 Senate Hearings, Vol. 3, p. 1105.
37. Staff Report, U.S. Senate Committee on Public Works, *Toxicological and Environmental Implications of the Use of Nitrilotriacetic Acid as a Detergent Builder,* December 1970, p. 6. (Hereafter cited as Senate staff report, *Toxicological Implications of NTA*).
38. 1969 House Hearings, p. 71.
39. Petition by William H. Rodgers, Jr., to Joseph W. Shea, Secretary, Federal Trade Commission, December 4, 1970, p. 4, citing Leading National Advertisers, Inc., National Advertising Investments 5 (1970).
40. Committee on Government Operations, *Phosphates in Detergents,* pp. 63–64.

41. Reprinted in 1970 Senate Hearings, Part 3, pp. 1195–6.
42. U.S. Department of Interior News Release, FWQA, "Additional Findings on Phosphates in Detergents Reported," September 6, 1970.
43. "Detergent Manufacturers to Label Phosphate Products," Soap and Detergent Association news release, November 9, 1970.
44. Federal Trade Commission, "Labelling and Advertising Requirements for Detergents; Notice of Public Hearing and Opportunity to Submit Data, Views or Arguments Regarding Proposed Trade Regulation Rule," January 25, 1971.
45. Rodgers' petition to FTC, p. 4.
46. "Comment by an SDA spokesman on the Proposed FTC Trade Regulation on Detergent Labelling," Soap and Detergent Association press release, January 25, 1971.
47. "Detergent Maker Sues on Labelling," *New York Times,* April 7, 1971; "Lever Brother Bid Denied in U.S. Court," *New York Times,* April 20, 1971.
48. "U.S. Aides Ask Delay on Phosphate Move," *New York Times,* April 27, 1971.
49. Richard D. Lyons, "Phosphates: A Sudden Boom for Detergents that Don't Pollute," *New York Times,* January 31, 1971.
50. Jay Walz, "Detergent Curb Begins in Canada," *New York Times,* August 2, 1970.
51. Edward Schreiber, "City Council Passes Ban on Use of Phosphorus for Detergents," *Chicago Tribune,* October 15, 1970.
52. "Curbs on Phosphate Voted in Akron Pollution Move," *New York Times,* December 13, 1970.
53. David Bird, "Akron Detergent Battle Seen as Vital," *New York Times,* January 31, 1971.
54. U.S. Department of State, Excerpts from Transcript of Press, Radio and Television News Briefing, January 14, 1971.
55. 1969 House Hearings, p. 45ff.
56. *Ibid.,* p. 299.
57. Rodgers' petition to FTC, p. 4, citing Leading National Advertisers, Inc., National Advertising Investments 5 (1970).
58. 1969 House Hearings, p. 85.
59. Rodgers petition to FTC, citing Colgate-Palmolive, 1969 Annual Report, p. 3.
60. 1969 House Hearings, p. 43.
61. *Ibid.,* p. 88.
62. *Detergents and the Environment,* pamphlet published

by Lever Brothers Company, reprinted in 1970 Senate Hearings, Part 3, p. 1156.

63. 1969 House Hearings, p. 91.
64. 1970 Senate Hearings, Part 3, p. 1101.
65. Senate staff report, *Toxicological Implications of NTA,* p. 38.
66. 1969 House Hearings, p. 69.
67. Soap and Detergent Association press release, March 26, 1971, p. 3.
68. 1970 Senate Hearings, Part 3, p. 1159.
69. Statement by the Soap and Detergent Association to Committee on the Environment, General Assembly, State of Connecticut, March 5, 1971, p. 8.
70. Edward Schreiber, "City Council Passes Ban on Use of Phosphates for Detergents," *Chicago Tribune,* October 15, 1970.
71. David Bird, *New York Times,* January 31, 1971.
72. 1969 House Hearings, pp. 110, 172.
73. *Ibid.,* p. 72.
74. 1970 Senate Hearings, Part 5, pp. 1988–9.
75. Samuel S. Epstein, "NTA," *Environment,* September 1970; Senate Staff Report, *Toxicological Implications of NTA.*
76. Paul Delaney, "NTA is Removed from Detergents," *New York Times,* December 19, 1970; Statement on NTA by EPA Administrator Ruckelshaus and Surgeon General Steinfeld, December 18, 1970.
77. Senate Staff Report, *Toxicological Implications of NTA,* p. 2.
78. Delaney, *New York Times,* December 19, 1970.
79. 1970 Senate Hearings, Part 4, p. 1381.
80. Senate Staff Report, *Toxicological Implications of NTA,* p. 49.

Chapter 5

1. Cecil H. Wadleigh, *Wastes in Relation to Agriculture and Forestry,* U.S. Department of Agriculture, Miscellaneous Publication No. 1065, 1968, p. 36.
2. *Ibid.*
3. *Ibid.,* p. 35, citing "Water Resources Activities in the United States: Water Requirements for Pollution Abatement," Committee Print No. 9 of the Senate Select Committee on National Water Resources, 86th Congress, 2nd Session, 1960.
4. Donald B. Carr, *Death of the Sweet Waters* (New York: W. W. Norton and Co., 1966), p. 128.
5. Report to the President, *Control of Agriculture-Related Pollution,* Submitted by the Secretary of Agriculture

and the Director of the Office of Science and Technology, January 1969 (out of print), p. 25.

6. The quote appears in Wadleigh, p. 42. Source given is Keith Fry, *Land Runoff: A Factor in Potomac Basin Pollution,* Interstate Commission on the Potomac River Basin, July 1966.

7. Report to the President, p. 25.

8. G. L. Hutchison and F. G. Viets, Jr., "Nitrogen Enrichment of Surface Water by Absorption of Ammonia Volatized from Cattle Feedlots," *Science,* October 24, 1969, pp. 514–15.

9. Report to the President, p. 26.

10. Estimates in this range were made in FWQA, *Nutrients in the Potomac River Basin,* an International Joint Commission, *Report on Project Hypo,* which concerned the Great Lakes.

11. *Environmental Quality,* First Annual Report, Council on Environmental Quality, August 1970, p. 131.

12. Morton Mintz and Nick Kotz, "Pesticide Firms Aided Whitten Book," *Washington Post,* March 14, 1971.

13. *Environmental Quality,* pp. 37–38.

14. William McAlister, "Foes of More Filling in San Francisco Bay Clash with Developers," *Wall Street Journal,* April 14, 1969.

15. *Environmental Quality,* p. 177.

16. U.S. Department of Interior, *National Estuarine Pollution Study,* 1970, p. 138.

17. Reg. No. 1145–2–303, 304 and Cir. No. 1145–2–18, *Federal Register,* May 28, 1970, p. 8280.

18. Edward J. Cleary, *The ORSANCO Story: Water Quality Management in the Ohio Valley Under an Interstate Compact* (Baltimore: John Hopkins Press, Published for Resources for the Future, 1967), p. 168.

19. *Ibid.*

20. U.S. Department of Interior, Bureau of Mines, *Environmental Effects of Underground Mining and Mineral Processing,* unpublished, 1969, p. 105. The report is on file in the Washington office of the Bureau of Mines.

21. *Ibid.,* p. 100.

22. Terri Aaronson, "Problems Underfoot," *Environment,* November 1960, p. 24.

23. U.S. Bureau of Mines, *Environmental Effects of Mining,* p. 19.

24. Aaronson, "Problems Underfoot," p. 24.

25. U.S. Bureau of Mines, *Environmental Effects of Mining,* p. 19.

26. Hearings before the Subcommittee on Air and Water Pollution of the Committee on Public Works, U.S.

Senate, 90th Congress, 1st Session, Part I, July 13, 1967, p. 323.

Chapter 6

1. Barry Commoner, "Lake Erie, Aging or Ill?" *Scientist and Citizen,* December 1968, pp. 255–7; U.S. Department of Health, Education and Welfare, *Proceedings, Conference in the Matter of Pollution of Lake Erie and its Tributaries,* August 3–6, 1965, pp. 126–130.
2. FWQA, *Clean Water for the 1970's: A Status Report,* 1970, p. 22.
3. Hearings before a Special Subcommittee on Air and Water Pollution, Committee on Public Works, U.S. Senate, 88th Congress, 1st Session, June 17, 1963, p. 52.
4. *Ibid.,* p. 53.
5. *Ibid.*
6. Leroy S. Houser and Frank J. Silva, *National Register of Shellfish Production Areas,* U.S. Department of Health, Education and Welfare, Public Health Service, Publication No. 1500, 1966, pp. 2–3.
7. Ronald G. Macomber, *Conditional Areas—A Time for Decision,* U.S. Department of Health, Education and Welfare, Public Health Service, January 1968, p. 10 and Appendix.
8. Elinore M. Barrett, *California Oyster Industry,* California Department of Fish and Game, Fish Bulletin No. 123, 1963, pp. 7, 19–21.
9. Russell Train, "The Challenge of the Estuary," *Proceedings of the National Shellfisheries Association, Vol. 59—June 1969* (Washington, D.C.: U.S. Department of Interior, Fish and Wildlife Service, 1969), p. 16.
10. FWQA, *Delaware Estuary Comprehensive Study: Preliminary Report and Findings,* 1966, p. i. (Hereafter cited as *Delaware Estuary Study*).
11. *Ibid.,* p. vii.
12. Carr, *Death of the Sweet Waters,* p. 169.
13. *Delaware Estuary Study,* p. v.
14. Ohio River Valley Sanitation Commission, 21st Yearbook, 1969, p. 22.
15. "Houston Ship Channel: A Port Set to Explode," by the Associated Press, *New York Times,* April 19, 1970.
16. Bob Eckhardt, "Death of Galveston Bay," *Transactions of the 33rd North American Wildlife and Natural Resources Conference* (Washington, D.C.: Wildlife Management Institute, 1968), p. 84.
17. L. J. Carter, "Galveston Bay: Test Case of an Estuary in Crisis," *Science,* February 20, 1970, p. 1102.
18. *Ibid.,* p. 1106.

19. Harold Scarlett, "LBJ Board Says Channel Pollution Overwhelming;" *Houston Post,* September 13, 1967.
20. Hal Wimberly, "LBJ's Pollution Panel on Critical Tour," *Houston Post,* June 29, 1967.
21. Eckhardt, "Death of Galveston Bay," p. 84.

Chapter 7

1. U.S. Department of Interior, Federal Water Pollution Control Administration (FWPCA), *An Appraisal of Water Pollution in the Lake Superior Basin,* April 1969, p. 21. (Hereafter referred to as *Appraisal.*)
2. *Ibid.*
3. *Ibid.*
4. FWPCA, *Lake Erie Report: A Plan for Pollution Control,* 1968, p. 32.
5. Robert H. Rainey, "Natural Displacement of Pollution from the Great Lakes," *Science,* March 10, 1967, p. 1242–3. Rainey's calculations are based on a theoretical model. He suggests that the drainage rates for Lake Superior and Michigan are actually even slower.
6. Bernie Shellum, "LeVander and Pollution: The Two-Year Making of a Militant," *Minneapolis Tribune,* October 5, 1969.
7. The report—"Environmental Impacts of Taconite Waste Disposal in Lake Superior"—has not been published. It is on file in the Chicago and Arlington, Va., offices of EPA.
8. Eric Jones, "Investment Opportunities in the Taconite Region," *The Commercial and Financial Chronicle,* September 8, 1966.
9. Ron Way, "U.S. Study Finds Taconite Tailings Pollute Superior," *Minneapolis Tribune,* January 16, 1969.
10. Edwards is quoted by *New York Times* reporter Gladwin Hill in Hill's article, "Lake Superior, Private Dump," *The Nation,* June 23, 1969, p. 796.
11. Albert Eisele, "Interior Official Defends Superior Pollution Report," *Duluth News-Tribune,* January 30, 1969.
12. "LeVander Backs Parley on Pollution," *Duluth News-Tribune,* January 17, 1969.
13. FWPCA, *Proceeding, Conference—Pollution of Lake Superior and its Tributary Basin, Minnesota–Wisconsin–Michigan,* May 13–15, 1969, p. 33.
14. *Ibid.,* p. 188.
15. *Ibid.,* p. 674.
16. *Appraisal,* pp. 55–79.
17. *Ibid.,* p. 24.
18. Hearings before a special subcommittee on Air and

Water Pollution of the Committee on Public Works, U.S. Senate, 88th Congress, 1st Session, June 17, 1963, p. 55.

Chapter 8

1. U.S. Congress, Senate, *Congressional Record,* 91st Congress, 1st Session, Vol. 115, Part 7; April 18, 1969, p. 9674.
2. Carr, *Death of the Sweet Waters,* p. 152.
3. U.S. Department of Health, Education and Welfare, *Proceedings, Conference in the Matter of Pollution of the Interstate Waters of the Mahoning River and its Tributaries,* February 16–17, 1965, Vol. I, pp. 62–65.
4. *Ibid.,* Vol. II, p. 463.
5. *Ibid.,* pp. 520–21.
6. *Ibid.,* Vol. I, p. 163.
7. *Ibid.,* pp. 163–4.
8. *Ibid.,* p. 127.
9. *Ibid.,* pp. 137–40.
10. *Ibid.,* p. 138.
11. *Ibid.,* pp. 13–14.
12. *Ibid.,* p. 13.
13. Hearings before a Special Subcommittee on Air and Water Pollution of the Committee on Public Works, U.S. Senate, 88th Congress, 1st Session, June 17, 1963, p. 79.
14. *Proceedings* [Mahoning River Conference], Vol. I, p. 20.
15. *Ibid.,* p. 22.
16. James Schwartz, "Mercury Throws Sparks: Kentucky, U.S. Officials Deadlock over Fish Contamination," Louisville (Ky.) *Courier-Journal,* July 10, 1970.
17. Thomas Edwards, "Beautiful Ohio isn't: Polluters fouling our river," *Cincinnati Post & Times-Star,* April 10, 1970.
18. *Ibid.*

Chapter 9

1. Hearings before the Subcommittee on Public Works of the Committee on Appropriations, U.S. House of Representatives, 91st Congress, 1st Session, Part 3, May 1, 1969, p. 1108.
2. Betty Klaric, "Report on Pollution Expected This Month," *Cleveland Press,* January 9, 1965.
3. U.S. Department of Health, Education and Welfare, *Proceedings, Conference in the Matter of Pollution of Lake Erie and its Tributaries* [Cleveland], August 3–6, 1965, pp. 25–26.

4. *Ibid.,* pp. 27–30.
5. John Poland, "Eat, Drink and Be Sick," *Medical World News,* September 26, 1969, p. 30.

Chapter 10

1. Murray Stein, "Problems and Programs in Water Pollution," *Natural Resources Journal,* December 1962, p. 409.
2. Hearings before the Subcommittee on Air and Water Pollution, Committee on Public Works, U.S. Senate, 89th Congress, 2nd Session, May 11, 1966, p. 452.
3. Murray Stein, "Problems and Programs," p. 396.
4. Gladwin Hill, "Reclaimed Water Gets Top Priority," *New York Times,* May 25, 1969.
5. The conferences were: Eastern New Jersey, 1967; Potomac River, 1970; Escambia Bay, 1970.
6. U.S. Department of Interior, Federal Water Pollution Control Administration (FWPCA), *Proceedings, Conference in the Matter of Pollution of the Interstate Waters of the Hudson River and its Tributaries—New York and New Jersey,* Second Session, September 20–21, 1967, pp. 45–46.
7. For example, the requirements set for Animas River in 1959; Moriches Bay in 1967; Penobscot River in 1967; and Escambia River Basin in 1970.
8. FWPCA, *Proceedings, Conference in the Matter of Pollution of the Interstate Waters of the Potomac River,* April–May, 1969, pp. 198–200.
9. They were: Coosa River, 1968; Red River of the North, 1965; and Upper Mississippi, 1967.
10. FWPCA, *Potomac River Water Quality: Washington, D.C. Metropolitan Area,* 1969, p. 11.
11. FWPCA, *Proceedings* [Detroit River and Lake Erie Conference], 1962, p. 403.
12. *Proceedings,* [Potomac River Conference], April, 1969, p. 273.
13. FWPCA, *Report of Water Quality Sub-Task Force on Project Potomac,* February 1967, p. 27.
14. *Potomac River Water Quality,* p. 7.
15. *Ibid.,* p. 76.
16. *Proceedings* [Potomac River Conference], April–May, 1969, p. 43.
17. FWPCA, *Proceedings, Progress Evaluation Meeting, Conference in the Matter of Pollution of Lake Erie and its Tributaries—Indiana, Michigan, New York, Ohio, and Pennsylvania,* June 4, 1968, Vol. II, pp. 446, 534.
18. *Saturday Review,* November 22, 1969, p. 55.
19. *Grand Calumet River, Little Calumet River, Calumet*

River, Wolf Lake, Lake Michigan, and their Tributaries: Conference Proceedings,* March 1965, p. 84.

20. *Conference Proceedings,* [Grand Calumet River], December 1968–January 1969, p. 348.
21. *Ibid.,* p. 401.
22. *Ibid.,* pp. 690–691.
23. U.S. Steel, *Annual Report,* 1969, p. 21.
24. *Air and Water News,* May 5, 1969, p. 7.
25. Schmetterer gave this account in a speech he made in Chicago, reported in Max Sonderby, "Interior Dept. Blamed For Lull in Pollution War," *Chicago Sun-Times,* April 23, 1970, p. 4.
26. Gladwin Hill, "Industrialists Back Clean-Water Fight but Call Profits Essential," *New York Times,* October 24, 1969.
27. Joseph Kraft, "Guidelines are Needed to Separate the Polluters from the Good Ones," *Philadelphia Evening Bulletin,* January 28, 1970.
28. Casey Bukro, "Detroit's Tale of Two Rivers: One Improves While Other Worsens," *Chicago Tribune,* October 15, 1970.
29. Mike Royko, "Steel yourself: 'Fox' is back!" *Chicago Daily News,* December 2, 1970.
30. *Ibid.*
31. *Ibid.*

Chapter 11

1. Hearings before a Special Subcommittee on Air and Water Pollution, Committee on Public Works, U.S. Senate, 88th Congress, 1st Session, June 17, 1963, pp. 58–59.
2. *Ibid.,* p. 56.

Chapter 12

1. Ridgeway, *The Politics of Ecology,* p. 104.
2. Transcript of 180-day informal hearing on Republic Steel, October 1969, Cleveland, on file in Arlington, Va., office of enforcement records.
3. Hearings before the Subcommittee on Air and Water Pollution, Committee on Public Works, U.S. Senate, 89th Congress, 2nd Session, May 11, 1966, p. 449.
4. *Ibid.*
5. Committee on Government Operations, *The Establishment of a National Industrial Wastes Inventory,* House Report No. 91–1717, 1970.
6. Reinemer, "Do Advisory Panels Have Bias?" p. 39.
7. E. W. Kenworthy, "U.S. Pollution Control Panel Bars Environmental and Consumer Observers," *New York Times,* October 15, 1970.

8. Committee on Government Operations, *Establishment of National Wastes Inventory,* p. 31.

Chapter 13

1. Dr. Crumbine's complaint is mentioned in a letter from Dwight F. Metzler to Frederick W. Slater, of the *St. Joseph News-Press,* December 7, 1970, on file in the *News-Press* office.
2. *Missouri River—St. Joseph, Missouri Area: Conference Proceedings,* 1957.
3. "In the Matter of the Pollution of the Interstate Waters of the St. Joseph, Missouri Area: Findings, Conclusions, and Recommendations of the Hearing Board," (mimeo), July 30, 1959, p. 5.
4. *Ibid.,* p. 7.
5. Letter from HEW Secretary Fleming to Attorney General William Rogers, August 1, 1960.
6. *United States v. City of St. Joseph, Missouri,* Docket No. 1077 in the U.S. District Court for the Western District of Missouri, St. Joseph Division. Order: October 31, 1961.
7. Letter from Joseph Wood, St. Joseph City Attorney, to Herbert Clare, U.S. Public Health Service, May 21, 1962.
8. Transcript of Hearing Proceedings, *United States v. City of St. Joseph, Missouri,* March 27, 1967.
9. U.S. Department of Interior, Federal Water Pollution Control Administration, "St. Joseph, Missouri, to Provide Secondary Treatment," *Water Pollution News,* October 29, 1969.
10. Hearing Proceedings, *United States v. City of St. Joseph, Missouri,* March 27, 1967.
11. *Ibid.*
12. Hearings before the Subcommittee on Air and Water Pollution, Committee on Public Works, U.S. Senate, 89th Congress, 2nd Session, April 27, 1966, p. 302.
13. *Ibid.,* p. 546.

Chapter 14

1. James Quigley, "The Solution to Water Pollution," *Consulting Engineer,* March 1967, p. 135.
2. Hearings before the Subcommittee on Air and Water Pollution, Committee on Public Works, U.S. Senate, 89th Congress, 2nd Session, May 11, 1966, p. 431.
3. Illinois Sanitation Water Board, *Water Quality Standards,* SWB–7–1.01.
4. *Ibid.*
5. *Ibid.,* SWB–8–1.03b.

6. U.S. Department of Interior news release, "Water Quality Degradation Issue Resolved," February 8, 1968.
7. Illinois Board, *Water Quality Standards,* SWB–13–1.05.
8. *Ibid.,* SWB–7–1.03, 1.05.
9. K. M. Mackenthun, "Acceptable Water Quality in Mixing Areas," *FWPCA Presentations: ORSANCO Engineering Committee,* May 13–14, 1969, p. A.1.
10. *Ibid.,* pp. A.1–A.2.
11. Illinois Board, *Water Quality Standards,* SWB–8–103.b (emphasis ours).
12. *Ibid.,* SWB–9–1.06.
13. *Ibid.,* SWB–8–103b (emphasis ours).
14. *Ibid.,* SWB–9–103c (emphasis ours).
15. *Ibid.,* SWB–11–1.05.

Chapter 15

1. *United States v. Standard Oil,* 384 U.S. 224, 230 (1966).
2. *United States v. Republic Steel Corp.,* 362 U.S. 482 (1960); *Wyandotte Transportation Co. v. United States,* 389 U.S. 191, 203–204, fn. 15 (1967).
3. Hearings before the Subcommittee on the Environment, Committee on Commerce, U.S. Senate, 92nd Congress, 1st Session, February 19, 1971, pp. 102–3.
4. *Ibid.,* p. 99.
5. 33 C.F.R. Part 209. 131, *Federal Register,* Vol. 35, No. 253, December 31, 1970, pp. 2005–9.
6. Hearings before the Subcommittee on the Environment [1971], p. 140.

Chapter 16

1. U.S. Congress, House of Representatives, *Congressional Record,* February 23, 1966, p. 3667.
2. Casey Bukro, "Hickel Tells Plan to Finance U.S. Fight on Lake Pollution," *Chicago Tribune,* September 20, 1969.
3. FWPCA, *Cost of Clean Water* [1968], Summary Report, p. 3.
4. FWQA, *Economics of Clean Water* [1970], Summary Report, p. 14.
5. U.S. Congress, *Congressional Record,* July 8, 1970, p. 6377.
6. Hearings before the Subcommittee on Air and Water Pollution, Committee on Public Works, U.S. Senate, 91st Congress, 2nd Session, April 20, 1970, p. 10.
7. U.S. Congress, House of Representatives, *Congressional Record,* February 23, 1960, p. 3257.
8. Gladwin Hill, "Localities Told to Clean Water," *New York Times,* November 8, 1969.

Chapter 17

1. The study which the U.S. General Accounting Office (GAO) conducted led to three reports to the Subcommittee (as well as a number of others to Congress as a whole) entitled: *Administration of the construction grant program for abating, controlling, and preventing water pollution; Operation and maintenance of municipal waste treatment plants;* and *Personnel, staffing, and administration of the Federal Water Pollution Control Administration.* All were made public in November 1969.

2. FWPCA, *Cost of Clean Water* [1968], Vol. II, pp. 75–76.

3. GAO, *Federal Grants for Facilities which benefit Industrial Users,* p. xx.

4. FWQA, *Economics of Clean Water* [1970], Vol. I, pp. 124, 126, 134.

5. FWPCA, *The Cost of Clean Water and Its Economic Impact,* 1969, Vol. III "Sewerage Charges," p. 30.

6. *Federal Register,* Vol. 35, No. 128, July 2, 1970, p. 10757, C.F.R. 601.34(c).

7. *Ibid.,* C.F.R. 601.34(a).

8. *Ibid.,* C.F.R. 601.35.

9. *Manpower,* May 1970, p. 18, citing 1969 survey by the National Society for Professional Engineers.

10. GAO, *Operation and maintenance of municipal waste treatment plants,* p. 20.

11. FWPCA, *Manpower Needs at Water Pollution Control Facilities in New England,* June 1969, p. 45.

12. Columbus Smith, "City Tops Industry in River Pollution," *Dayton Daily News,* November 2, 1969.

13. C.F.R. 601.34(b).

14. GAO, *Administration of the construction grant program,* pp. 19–20.

15. C.F.R. 601.32, 601.33.

Chapter 18

1. "Trains Loaded for Nerve Gas Shipment," by United Press International, *New York Times,* August 8, 1970. Hearings were before the Subcommittee on Oceanography of the Committee on Merchant Marine and Fisheries, House of Representatives, 91st Congress, 2nd Session, August 3, 4, 6 and 7.

2. *Ibid.*

3. Hearings before the Subcommittee on International Organizations and Movement of the Committee on

Foreign Affairs, House of Representatives, 91st Congress, 1st Session, May 14, 1969, pp. 124–130.

4. *Ibid.,* p. 35.
5. *Ibid.,* pp. 73–75.
6. *Ibid.,* p. 86.
7. *Ibid.,* p. 82.
8. *Proceedings* [Grand Calumet Conference], December 1968, Vol. I, p. 56.
9. First Report by the Subcommittee on Conservation and Natural Resources, Committee on Government Operations, House of Representatives, 91st Congress, 1st Session, Report No. 91–75, p. 23.
10. Fifteenth Report by the Subcommittee on Natural Resources and Power, Committee on Government Operations, House of Representatives, 89th Congress, 1st Session, Report No. 555, p. 4.
11. Thirty-first Report by the Committee on Government Operations, House of Representatives, 89th Congress, 2nd Session, Report No. 1644, p. 8.
12. First Report by the Subcommittee on Conservation and Natural Resources, House Report No. 91–75, p. 23.
13. *Ibid.*
14. *Proceedings* [Grand Calumet Conference], December 1968, Vol. I, p. 41.

Chapter 19

1. Hearings before a Special Subcommittee on Air and Water Pollution, Committee on Public Works, U.S. Senate, 89th Congress, 1st Session, Part 1, May 20, 1965, pp. 92–112.
2. *Ibid.,* p. 99.
3. FWQA, *Municipal Waste Facilities in the United States,* p. 14.
4. David G. Stephan, "Water renovation—some advanced treatment processes," *Civil Engineering,* September 1965, p. 46.
5. J. W. Gerlich, "Better than the Pilot Model," *American City,* October 1967, p. 94.
6. Dr. John R. Sheaffer, at Hearings before the Subcommittee on Conservation and Natural Resources, Committee on Government Operations, House of Representatives, 91st Congress, 1st Session, December 15, 1969, p. 217. (Hereafter cited as "1969 House Hearings.")
7. *Ibid.,* pp. 211, 213, 217.
8. R. R. Parizek, L. T. Kardos et al., *Waste Water Renovation and Conservation,* The Pennsylvania State Uni-

versity Studies No. 23, 1967, p. 37, reprinted in Hearings before the Subcommittee on Air and Water Pollution of the Committee on Public Works, U.S. Senate, 91st Congress, 2nd Session, May 26, 1970, Part 3, p. 1045.

9. Sheaffer, 1969 House Hearings, p. 214.
10. Muskegon County Board and Department of Public Works, *Engineering Feasibility Demonstration Study for Muskegon County, Michigan Water Treatment System,* September 1970, report prepared in fulfillment of FWQA Grant No. 11010.
11. U.S. Department of Interior news release, "Muskegon County Michigan Gets $2 Million in FWQA Grants to Renovate Wastewater and Recover Marginal Land," September 18, 1970.
12. U.S. Army Corps of Engineers, Office, Chief of Engineers, "A Pilot Wastewater Management Program," March 1971.

Chapter 20

1. See testimony of Professor Henry Jacoby, commenting on a national effluent charge bill introduced by Senator William Proxmire (D.-Wisc.), in Hearings before the Subcommittee on Air and Water Pollution of the Committee on Public Works, U.S. Senate, 91st Congress, 2nd Session, April 29, 1970, Part 2, pp. 488–495.
2. An incentive scheme like this one was first proposed by J. H. Dales in his *Pollution, Property, and Prices* (Toronto, 1968). Additional work has been done on this idea, towards making it into a workable abatement plan, by Harvard University economists Henry Jacoby and Grant Schaumburg.
3. E. W. Kenworthy, "Panel to Study Change in Pollution Control Bill," *New York Times,* March 1, 1971.
4. James M. Naughton, "Nixon Reassures Industry, Bars Pollution Scapegoat," *New York Times,* February 11, 1971.

Index

Index